AF753314

Physics Research and Technology

nova science publishers

www.novapublishers.com

Physics Research and Technology

Electromagnetic Waves: Advances in Applications and Research
Manuel B. Hutchinson (Editor)
2022. ISBN: 979-8-88697-254-2 (Softcover)
2022. ISBN: 979-8-88697-265-8 (eBook)

Radiation and Matter
Marian Apostol, PhD
2022. ISBN: 979-8-88697-246-7 (Hardcover)
2022. ISBN: 979-8-88697-267-2 (eBook)

Second Harmonic Generation: Pathways of Nonlinear Photonics
Abdel-Baset M. A. Ibrahim, PhD
Pankaj Kumar Choudhury, PhD
2022. ISBN: 978-1-68507-888-1 (Hardcover)
2022. ISBN: 979-8-88697-002-9 (eBook)

Advances in Sustainable Materials and Technology
Abhishek Kanoungo, PhD
Sandeep Singh, PhD
Er. Shristi Kanoungo, M.E
Ajay Goyal, PhD
2022. ISBN: 978-1-68507-967-3 (Hardcover)
2022. ISBN: 979-8-88697-070-8 (eBook)

Quantum Field Theory and Applications
Natale Palerma (Editor)
2022. ISBN: 978-1-68507-930-7 (Hardcover)
2022. ISBN: 978-1-68507-957-4 (eBook)

More information about this series can be found at
https://novapublishers.com/product-category/series/physics-research-and-technology/

Parveen Saini
Editor

An Introduction to Permittivity

nova
science publishers

www.novapublishers.com

Copyright © 2022 by Nova Science Publishers, Inc.
https://doi.org/10.52305/NVBF3091

All rights reserved. No part of this book may be reproduced, stored in a retrieval system or transmitted in any form or by any means: electronic, electrostatic, magnetic, tape, mechanical photocopying, recording or otherwise without the written permission of the Publisher.

We have partnered with Copyright Clearance Center to make it easy for you to obtain permissions to reuse content from this publication. Simply navigate to this publication's page on Nova's website and locate the "Get Permission" button below the title description. This button is linked directly to the title's permission page on copyright.com. Alternatively, you can visit copyright.com and search by title, ISBN, or ISSN.

For further questions about using the service on copyright.com, please contact:
Copyright Clearance Center
Phone: +1-(978) 750-8400 Fax: +1-(978) 750-4470 E-mail: info@copyright.com.

NOTICE TO THE READER

The Publisher has taken reasonable care in the preparation of this book, but makes no expressed or implied warranty of any kind and assumes no responsibility for any errors or omissions. No liability is assumed for incidental or consequential damages in connection with or arising out of information contained in this book. The Publisher shall not be liable for any special, consequential, or exemplary damages resulting, in whole or in part, from the readers' use of, or reliance upon, this material. Any parts of this book based on government reports are so indicated and copyright is claimed for those parts to the extent applicable to compilations of such works.

Independent verification should be sought for any data, advice or recommendations contained in this book. In addition, no responsibility is assumed by the Publisher for any injury and/or damage to persons or property arising from any methods, products, instructions, ideas or otherwise contained in this publication.

This publication is designed to provide accurate and authoritative information with regard to the subject matter covered herein. It is sold with the clear understanding that the Publisher is not engaged in rendering legal or any other professional services. If legal or any other expert assistance is required, the services of a competent person should be sought. FROM A DECLARATION OF PARTICIPANTS JOINTLY ADOPTED BY A COMMITTEE OF THE AMERICAN BAR ASSOCIATION AND A COMMITTEE OF PUBLISHERS.

Additional color graphics may be available in the e-book version of this book.

Library of Congress Cataloging-in-Publication Data

ISBN: 979-8-88697-042-5

Published by Nova Science Publishers, Inc. † New York

Contents

Preface

Permittivity is a very important quantity often used to quantify the behavior of a system under a given electric field distribution in space. The dielectric polarization and associated losses are two main associated effects that are connected through the complex permittivity via its real and imaginary parts respectively. Therefore, the quantitative measure of permittivity is essential for predicting the response of a system, e.g., charge storage characteristics of capacitor/supercapacitor, flux through a charge carrying closed surface, exciton binding energy in soar cells/LEDs, properties of liquid crystals and LCDs, relaxation of dielectrics, microwave absorption effectiveness of shielding materials, response of antenna systems, governing phase velocity of electromagnetic radiation through given medium, designing of metamaterials etc. Permittivity has special relevance in energy storage, energy harvesting, energy conversion, energy transmission, energy dissipation, and even stealth technology; therefore, its thorough understanding is required especially in the context of practical applications. This prompted me to compile an interdisciplinary research-oriented dedicated book covering the elementary concepts and recent advancements in the area, to expose current challenges, highlight future perspectives and stimulate thought processes for the evolution of novel materials & technologies. This edited book is the first of its kind that provides a single-source solution to specifically address the fundamentals and applications of permittivity through a comprehensive literature account. Accordingly, the book is organized into 9 dedicated chapters.

Chapter 1 begins with the definition & significance of permittivity, its real and imaginary parts, absolute and relative values, etc. Subsequently, the various models for permittivity & governing equations; relevance of permittivity in energy storage, energy harvesting, energy conversion, energy transmission, dissipation and stealth; and literature account of the same are discussed. Finally, at the end, a summary, challenges and prospects are outlined. Chapter 2 gives a brief introduction to textile materials, theoretical studies regarding the dielectric properties of textile materials and a comprehensive overview of the experimental studies on this subject. A critical assessment of the effect of different factors like frequency, moisture content, temperature, etc. on the permittivity of such materials has been done and an outline of the application of dielectric properties in textiles has also been provided. Chapter 3 discusses ceramic materials for microwave firing applications considering that microwave radiation can be used as an alternative or as auxiliary heating technology for porcelain and stoneware firing. It specifically highlights that the microwave radiation absorption by some materials leads to local heat generation, which depends, among other factors, on the material's dielectric constant and dielectric losses. Dielectric properties control how the microwave radiation interacts with the matter, not only through the absorbed power but also through the penetration depth. Knowledge of these properties, and how they depend on temperature during the different firing stages, is thus of major relevance for a better understanding and improving microwave processing, and crucial for accurate microwave heating simulations, which are discussed in the chapter. In Chapter 4, a study on dielectrics is made. Specifically, the electric dipole in homogeneous and inhomogeneous electric fields is studied as well as polar and non-polar molecules. The four types of polarization of dielectric materials, those being electronic, atomic, bipolar and interfacial polarization, are also discussed and analyzed. This is followed by a review of dielectric permittivity and its dependence on frequency, temperature, pressure and humidity. A capacitor with parallel plates as well as RC circuits in series and in parallel is highlighted. Finally, the equations of the relaxation model of Debye, Cole-Cole, Cole-Davidson and Havriliak–Negami are presented in the context of permittivity. Chapter 5 outlines the Bismuth-based materials as promising low-firing temperature ceramics, namely the bismuth niobate, BiNbO4, which has been a subject of several studies, aiming the improvement of its structural, morphological and dielectric properties. The, (Bi1-xCrx)NbO4 system prepared using the sol-gel method was analysed and the detailed dielectric characterization in the microwave region was made at room temperature, using the small perturbation theory, with a resonant cavity operating in TE1,0,11 mode, at the frequency of 5 GHz. The dielectric properties, such as dielectric constant (ε') and dielectric losses (ε''), were described as a function of composition and treatment temperature. Chapter 6 explains the dielectric and electro-optical properties of FLC in combination with gold nanoparticles and liquids (solvent of gold nanoparticles), particularly water and ethanol. In the first section, the introduction of each type such as FLC, water, and ethanol is discussed. In the later sections, some results of the modification in the basic properties of FLC by employing the gold nanoparticles, water, and ethanol are presented. Chapter 7 points out that with the fast development of power electronics, dielectric materials with high energy-storage density, low loss, and good temperature stability are eagerly desired for the potential application in advanced pulsed capacitors. It serves to give an overall summary of the state-of-the-art progress on electric energy-storage performance in these materials. Moreover, some general future prospects are also provided from the existed theoretical and experimental results in order to propel their application in practice. Chapter 8 focuses on the basic aspects of permittivity, relative permittivity or dielectric constant, and the role of dielectrics in electronic devices (Field Effect Transistors (FETs) and Single Electron Transistors (SETs)) and energy storage devices (capacitor, supercapacitor and battery) as well as solar cells. The high-k dielectrics come in handy during the process of device miniaturization and enhance certain performance parameters of the MOSFETs and SETs, which are discussed in detail in this chapter. Besides, the role of dielectrics in a capacitor and the choice of solvent for supercapacitors and batteries have been discussed in reference to the dielectric constant. Chapter 9 enlightens the fundamentals of EMI shielding, measurement techniques, governing equations and potential material candidates with special reference to polymer-based hybrids and composites and the correlation between their dielectric properties and EMI shielding performance.

The above chapters cover all perspectives of permittivity, its relevance and practical applications through various materials including multiphase systems that have been complemented by a comprehensive literature account covering past & present research, key developments and state-of-the-art in the area. Extra care has been taken to simplify the matter and to facilitate the understanding via the specifically designed and selected graphics, i.e., illustrations, schematics and figures, tabulated information, case studies and many more literature examples. Due attention has also been given at the end of specific chapters to address the current challenges and future aspects, to stimulate the grey matter and ignite thought processes. I feel that a broader picture will evolve after reading all the chapters that will eventually help to improve the existing understanding of the subject and stimulate further innovation and interdisciplinary research. I hope this compilation will also act as a smooth entry point for new researchers to become acquainted with the field, and facilitate advanced readers to enrich their existing understanding. The key challenges and opportunities have also been thrown open for related academicians, material scientists, nanotechnologists and industry personals, to dig deeper into the subject and come out with cheaper, viable and efficient versions of the existing materials and technologies with desirable permittivity response.

I cannot end here without expressing my regards and gratitude to all those who have made any direct or indirect contributions to this book. I am greatly indebted to renowned authors from diverse backgrounds, who kindly agreed to extend their expertise for the respective chapters. Their valuable contributions provided an excellent overview of these multiphase systems in terms of academic importance, technological relevance and potential applications. I am extremely grateful to Nadya S. Columbus, President, Nova Science Publishers, Inc., for giving me this opportunity and for her tremendous faith and support at each stage. I would also like to thank the entire Nova editorial team who helped me to overcome numerous difficulties and extended strong technical support at every stage of publication. I am extremely thankful to all my peers, colleagues, associates, students and friends, who extended all possible help at every stage of the book and made it become a reality eventually. Thanks are also due to CSIR, DST, and CSIR-NPL, India, for providing the financial support, necessary research facilities and conducive research environment. Special thanks to past and present directors of CSIR-NPL for their support and facilitation at each stage.

My special regards and thanks to my parents, sisters and wife who stood by there for me in all shades of life and showed immense faith in my abilities. I am extremely thankful to my critics and all those who pushed me beyond my limits that actually and eventually helped me to improve at each stage. Finally, I would like to thank the almighty GOD who has shown me the path.

Parveen Saini
जय हिन्द
New Delhi, India
February 2022

Chapter 1

Permittivity and Its Significance for Devices and Applications

Parveen Saini*

Conducting Polymers, Graphene Technology and Waste Management Group, Photovoltaic Metrology Section, Advance Materials and Devices Metrology Division, CSIR-National Physical Laboratory, Dr. K. S. Krishnan Marg, New Delhi, India

Abstract

Permittivity is a very important quantity often used to quantify the behavior of a system under given electric field distribution in space. The dielectric polarization and related losses are two main associated effects, which are connected through the complex permittivity via its real and imaginary parts respectively. The quantitative measure of permittivity is essential for predicting response of a system e.g., charge storage characteristics of capacitor/supercapacitor; flux through a charge carrying closed surface, exciton binding energy in soar cells/LEDs, properties of liquid crystals and LCDs, relaxation of dielectrics, microwave absorption effectiveness of shielding materials, response of antenna systems, governing phase velocity of electromagnetic radiation through given medium, designing of metamaterials etc. This chapter begins with the definition & significance of permittivity, its real and imaginary parts, absolute and relative values etc. In the next section, various models for permittivity & governing equations; relevance of permittivity in energy storage, energy harvesting, energy conversion, energy transmission, dissipation & stealth; and literature account of the same is discussed. Finally, at the end, summary, challenges and prospects are outlined.

---

* Corresponding Author's Email: pksaini@nplindia.org.

In: An Introduction to Permittivity
Editor: Parveen Saini
ISBN: 979-8-88697-042-5
© 2022 Nova Science Publishers, Inc.

Keywords: complex permittivity, relative permittivity, dielectric constant, dielectric loss, polarization, Gauss theorem, Coulomb's Law, parallel plate capacitor, Debye relaxation, Cole-Cole plots, Havriliak–Negami relaxation, Cole-Davidson relaxation, impedance analyzer, CR meter, vector network analyzer (VNA), coaxial probe technique, transmission method, free space technique, resonant cavity method, Nicolson-Ross-Weir algorithm, supercapacitor, solar cell, exciton, microwave absorption, metamaterials

1.1. Introduction

In the electromagnetism, the term "Permittivity" is denoted by the Greek letter ε (epsilon), and is defined as the measure of the electric polarizability of a material or medium [1, 2]. Therefore, a given material with high permittivity value is known to display more polarization in response to same externally applied electric field than a material with low permittivity. As the polarized state reflects bound yet separated static charges with associated energy, therefore, in electrostatics, permittivity plays an important role in determining the energy storage or charge holding capacity called as capacitance of a capacitor. As permittivity is a thermodynamic function of state, it depends on the external factors like frequency, magnitude, and direction of the applied field; temperature, pressure, humidity and other environmental conditions [1]. The SI unit for permittivity is farad per meter (F/m). Often, the permittivity is represented as the dimensionless quantity called as relative permittivity (ε_r) which is the ratio of the absolute permittivity (ε) of the given medium and the permittivity of air or more precisely that of vacuum or free space (ε_0). The calculations show that a perfect vacuum has a relative permittivity of exactly 1 whereas at STP, air has a relative permittivity of ≈ 1.0006, which nearly same yet higher than that of vacuum. In general practices, terms relative permittivity and dielectric constant [3] are used interchangeably; however, permittivity is actually a complex quantity and can be expressed as [4, 5]:

$$\varepsilon = [\varepsilon' - j\varepsilon''] \tag{1}$$

where ε' is its real part (relative to vacuum) that represent energy storage whereas the imaginary part (ε'') is a measure of energy losses, as shown in Figure 1.

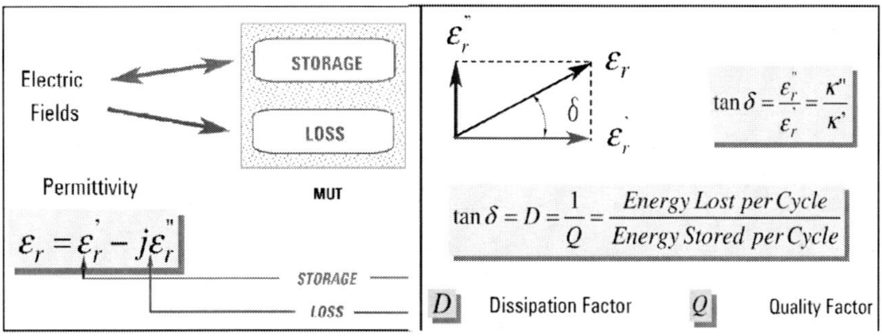

Figure 1. Complex permittivity of a material under test (left panel) and real/imaginary parts of permittivity and their relation with energy storage and losses (right panel).

The relative value of real permittivity w.r.t. vacuum (ε_r) represents the dielectric constant (k) that gives an idea about relative polarization by a dielectric substance or charge storage by a capacitor. In contrast, ε'' (imaginary permittivity) is a measure of dielectric dissipation or losses. The extent of losses can be quantified by calculating the quantity dielectric loss tangent (tan δ) that can be represented as [6]:

$$\tan\delta = \frac{\varepsilon''}{\varepsilon'} \tag{2}$$

The very high lossy materials display high **tan** δ value (which is inverse of quality factor, Q). In practical cases, material for capacitor is expected to have high Q value or very low **tan** δ value. In contrast, a good microwave absorbing material is expected to have very high **tan** δ value that will be discussed in detail in the next sections.

1.2. Classification of Materials

It is worth mentioning here that based on the sign of permittivity and permeability and their magnitudes, the materials can be divided into four categories [7-9]. As seen from the Figure 2, meta-materials may have either negative permittivity or permeability or both.

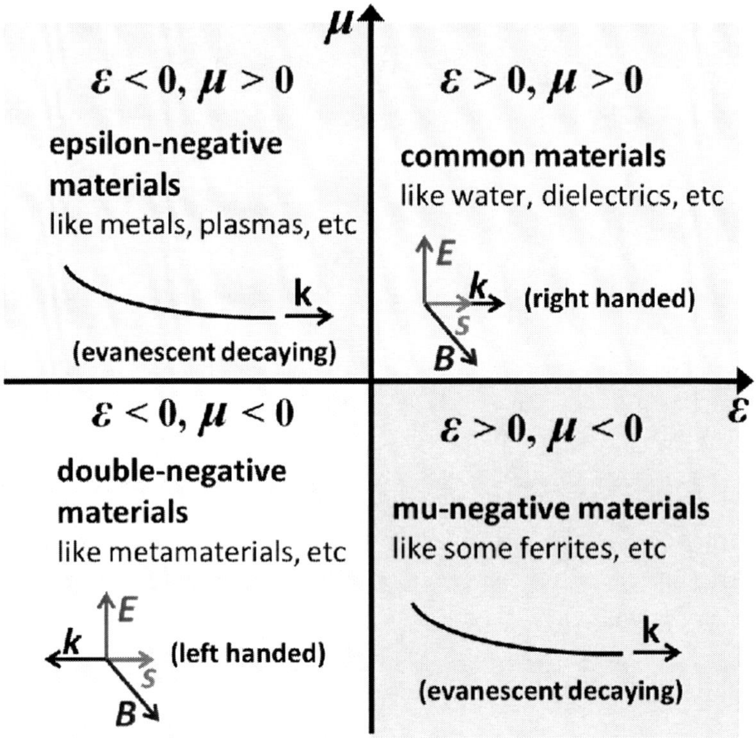

Figure 2. Classification of materials on the basis of values of permittivity and permeability.

Quadrant 1: This region contains materials which are naturally occurring and not fall into category of meta-materials. It is characterized by positive permittivity and permeability values. The conventional dielectric materials used for capacitor and microwave applications fall into this category.

Quadrant 2: This quadrant represents the class of meta-materials called as artificial dielectrics, known to display negative permittivity and positive permeability value. Due it's highly negative dielectric value, these materials have applications in microwave technology, particularly for reducing the size of antennas. Typical examples are metals and plasmas.

Quadrant 3: This region contains the meta-materials possessing both negative permittivity and permeability. These are referred as left handed materials (LHMs) due to backward wave propagation.

Quadrant 4: This region consists of the materials characterized by negative permeability below plasma frequency along with positive permittivity value. This class of meta-material is also called artificial magnetics. These meta-materials have applications in electromagnetics specifically to increase the gain and to reduce the size of antenna.

1.3. Relative Permittivity, Electric Susceptibility, Dielectric Polarization and Electric Displacement

The relative permittivity (ε_r) is the ratio of the absolute permittivity and free space permittivity [2, 4, 10-13]:

$$k = \varepsilon_r = \frac{\varepsilon}{\varepsilon_o} \tag{3}$$

It's a dimensionless quantity used to compare the polarization of various medium under same electric field distribution. Being characteristic of a given material, in electromagnetics and electrostatics, it's also known as dielectric constant (κ). The εr and κ is directly related to electric susceptibility (χ) as:

$$\chi = (k-1) \tag{4}$$

Therefore, if εr or χ values are known, the absolute permittivity can be calculated as:

$$\varepsilon = \varepsilon_r . \varepsilon_o = (1+\chi)\varepsilon_o \tag{5}$$

where εo is permittivity of free space. In electromagnetism, inside a given material and under applied electric field (E), the electric displacement field (**D**) represents the distribution of electric charges with the medium, including charge migration and electric dipole reorientation. In case of linear, *homogeneous, isotropic* materials with *"instantaneous" response* to changes in electric field, the D and E are related through the ε by equation:

$$D = \varepsilon E \tag{6}$$

where the permittivity ε is a scalar. If the medium is anisotropic, the permittivity is a second rank tensor. In a nonlinear medium, the permittivity can depend on the strength of the electric field. As already pointed, permittivity is a function of frequency and can take on real or complex values. In electromagnetism, term susceptibility is defined as the constant of proportionality (which may be a tensor) relating an electric field **E** to the induced dielectric polarization density (**P**) such that:

$$P = \varepsilon_o \chi E \tag{7}$$

where ε_0 is the electric permittivity of free space. The susceptibility of a medium is related to its permittivity of medium (ε) and that of free space (ε_o) by the relation:

$$\chi = \left(\frac{\varepsilon}{\varepsilon_O} - 1 \right) \tag{8}$$

So in the case of a vacuum,

$$\chi = 0 \; and \; \varepsilon = \varepsilon_O$$

The susceptibility is also related to the polarizability of individual particles in the medium by the Clausius-Mossotti relation. The electric displacement (**D**) is related to the polarization density **P** by

$$D = (\varepsilon_o E + P) = \varepsilon_o (1 + \chi) E = \varepsilon E \tag{9}$$

1.3.1. Permittivity of Free Space or Vacuum

The vacuum permittivity ε_0 (also called permittivity of free space or the electric constant) is the ratio **D**/E in free space [10-12].

$$\varepsilon_o = \frac{D}{E} (in\, vacuum) \tag{10}$$

Therefore, the ε_0 can be expressed in terms of speed of electromagnetic wave (c_o) and permeability (μ_0) of free space or vacuum as [14]:

$$\varepsilon_o = \frac{1}{\mu_o . c_o^{\ 2}} \tag{11}$$

Taking $\mu_0 = 4\pi \times 10^{-7}$ H/m is and c_o = 299792458 m/s, the value of ε_0 comes out to be $8.854187817 \times 10^{-12}$ F/m

1.4. Relevance of Permittivity in Determining Performance of Medium and Devices

Permittivity is a used to quantify the behavior of a system under the given electric field distribution in space. Therefore, the permittivity value (sign as well as magnitude) plays governing role for predicting response of a system e.g., charge storage characteristics of capacitor/supercapacitor devices; magnitude flux through a charge carrying closed surface, exciton binding energy in solar cells/LEDs, properties/performance of liquid crystals and LCD devices, relaxation of dielectrics, microwave absorption effectiveness of shielding materials, response of antenna systems, governing phase velocity of electromagnetic radiation through given medium, performance of metamaterials etc., some of which are discussed in next section with suitable examples [4, 10, 12, 15-18].

1.4.1. Phase Velocity in a Medium

The permittivity of a given medium along with the permeability value plays a governing role in the determining the propagation velocity of the incident electromagnetic radiation through the medium. The linear permittivity of a homogeneous material is usually given relative to that of free space or vacuum, in the form of relative permittivity (ε_r). However, in an anisotropic material, the relative permittivity may be a tensor, causing birefringence. The permittivity ε and permeability μ of a given medium together determine the phase velocity v of electromagnetic radiation through that medium as per equation [4, 6, 19]:

$$\varepsilon\mu = \frac{1}{v^2} \quad or \quad v = \frac{1}{\sqrt{\varepsilon\mu}} \tag{12}$$

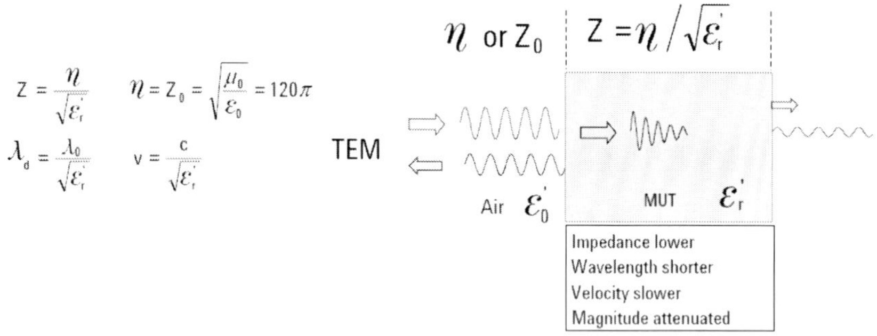

Figure 3. Sample-EM radiation interaction, reflected/transmitted signals & dielectric constant.

Therefore, from above formalism it's evident that the velocity scales inversely as square root of product of product of permittivity and permeability. Consequently, a high permittivity medium like a dielectric slab or ionic region, act as resistance and resulted in reduction of propagation velocity through the medium. The wavelength λ of a signal is inversely proportional to its frequency *f* (λ = *c/f*), such that as the frequency increases, the wavelength decreases. For example, in free space a 10 MHz signal has a wavelength of 30 m, while at 10 GHz it is just 3 cm. Many aspects of wave propagation are dependent on the permittivity and permeability of a material. Let's use the "optical view" of dielectric behaviour. Consider a flat slab of material (MUT) in space, with a TEM wave incident on its surface (Figure 3). There will be incident, reflected and transmitted waves. Since the impedance of the wave in the material Z is different (lower) from the free space impedance η (or Z_0) there will be impedance mismatch and this will create the reflected wave. Part of the energy will penetrate the sample. Once in the slab, the wave velocity *v*, is slower than the speed of light c. The wavelength λ_d is shorter than the wavelength λ_o in free space. Since the material will always have some loss, there will be attenuation or insertion loss.

1.4.2. Gauss Law: Flux and Permittivity

According to gauss law, the electric flux (ϕ) through a closed Gaussian surface carrying the charge (q) for a medium of permittivity (ε) can be expressed as [12]:

$$\phi = \frac{q}{\varepsilon} = \frac{q}{\varepsilon_r.\varepsilon_o} \tag{13}$$

Therefore, a high permittivity or high dielectric constant medium inversely reduces the flux. It is clean that for a given enclosed charge, the flux is lower for medium having higher permittivity value.

1.4.3. Coulomb's Law: Electrostatic Force and Permittivity

According to Coulomb's law, the electrostatic force (F) of attraction or repulsion between two point charges (q1 and q2), separated by a distance (r), which is large compared to their dimensions, in a given medium of permittivity (ε) can be expressed as [20]:

$$F = \frac{q_1 \cdot q_2}{4\pi\varepsilon.r^2} = \frac{q_1 \cdot q_2}{4\pi\varepsilon_r\varepsilon_o.r^2} \tag{14}$$

Therefore, it's evident that force inversely scales with the medium's permittivity so that the force is reduced in presence of high relative permittivity or dielectric constant medium. This is implications in deciding the binding energy between the charges especially in the area of electronics. More precisely, in the optoelectronic devices like solar cells, wherein the binding energy of photo-generated electron-hole pairs (better known n as excitons) inversely scales with the effective dielectric constant of the active layer. In this context, in the case of inorganic semiconductor systems like silicon based solar cells, the high dielectric constant of the system, resulted in very low binding energy, thus free charge carriers (electrons and holes) and, consequently, band type conduction and ultimately good charge carrier collection and high light to electricity conversion efficiency. In contrast, in case of organic semiconductors, the dielectric constant teds to be very low, resulting in relatively higher binding energy for excitons. Consequently, the recombination rates are higher and intimately placed donor-acceptor architecture has to be adopted to effectively break the excitons. It has also been highlighted that development of solar cell materials having high dielectric constant, result in reduction of excitonic binding energy and efficient charge collection. Similar to solar cells, the exploitation of high dielectric constant materials is known to regulate the charge injection and enhance the performance of light emitting diodes (LEDs) and Organic light emitting transistors (OLETs).

1.4.4. Capacitance of Capacitor and Supercapacitors

In a simple case of parallel plate capacitor containing two plates of given area (A) and separated distance (d), the capacitance (C) can be written as [12, 16, 20]:

$$C = \frac{A\varepsilon}{d} = \frac{A\varepsilon_r\varepsilon_o}{d} \tag{15}$$

Therefore, the permittivity of medium between the plates plays governing role and capacitance directly scales with the same, thus can be regulated by medium characteristics. It apparent that a supercapacitor device built using high permittivity or dielectric constant medium gives higher capacitance than the one built using lower permittivity system. For example, barium titanate is considered a good material for capacitor due to good dielectric properties. It is also clear from the above equation that by using a porous conducting system, high surface area (A) and low charge separation (d) can be realized ultimately resulting in huge increase in capacitance in the realized supercapacitor device. Again the exploitation of high dielectric constant electrolytes is known to further push the capacitance and energy density of supercapacitor devices. For example, it is known that use of organic electrolytes (having lower dielectric constant) gives the lower capacitance than aqueous medium electrolytes (relatively higher dielectric constant). The exploitation of ionic liquids based electrolytes (molten salts thus give high dielectric constant) is known to further enhance the capacitance of supercapacitor devices.

1.4.5. Permittivity and Microwave Absorption

It is important to highlight that permittivity also plays governing role in determining he microwave absorption behavior of given EMI shielding material. Especially under the transmission line theory and plane wave shielding phenomenon, the shielding effectiveness of sample is strongly dependent upon dielectric properties. In order to understand the same, let's assume a uniform plane wave characteristic by E and H that vary within a plane only with x direction as showed in Figure 4. The Maxwell's curl equations give [6, 19]:

$$\frac{dE}{dx} = -j\omega\mu H \text{ and } \frac{dH}{dx} = -(\sigma + j\omega\varepsilon)E \tag{16}$$

where μ is the permeability of the material and $\mu = \mu_o\mu_r$. μ_o and μ_r denote the permeability of air (*or free space*) and shield material respectively, σ is the conductivity of material in *S/m*. where ε is the permittivity of the material and $\varepsilon = \varepsilon_o\varepsilon_r$. ε_o and ε_r represent the permittivity of air (or free space) and shield material respectively, $\omega = 2\pi f$. $\omega\,(f)$ is angular frequency (linear frequency) in Hz.

All homogenous materials are characterized by a quantity known as the intrinsic impedance:

$$\eta = \sqrt{\frac{j\omega\mu}{\sigma + j\omega\varepsilon}} \tag{17}$$

When an electromagnetic wave propagates through the material, the wave impedance approaches the intrinsic impedance of the material. For dielectric material, the conductivity is extremely small ($\sigma \ll \omega\varepsilon$) and the intrinsic impedance of Eq. (17) becomes:

$$\eta = \sqrt{\frac{\mu}{\varepsilon}} \tag{18}$$

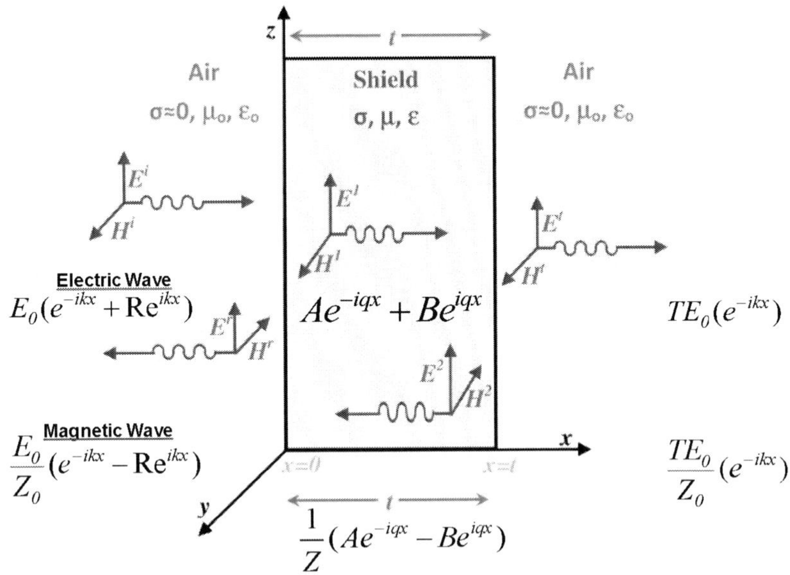

Figure 4. Propagation of electromagnetic waves and its interaction with the material.

For a conductor used below optical frequencies defined by $\sigma \gg \omega\varepsilon$, the intrinsic impedance of Eq. (17) can be written as:

$$\eta = \sqrt{\frac{j\omega\mu}{\sigma}} = (1+j)\sqrt{\frac{\pi\mu f}{\sigma}} \tag{19}$$

It is customary to define propagation constant (γ) in the media such that:

$$\gamma = (\alpha + j\beta) = \sqrt{j\omega\mu(\sigma + j\omega\varepsilon)} \tag{20}$$

where α is attenuation constant and β is phase constant. A good conductor is a medium for which $\sigma / \omega\varepsilon \gg 1$. Under this condition the Eq. (20) becomes:

$$\gamma = \sqrt{j\omega\mu} = (1+j)\sqrt{\pi\mu f \sigma} \tag{21}$$

Therefore, we can write $\alpha = \beta = 1/\delta = \sqrt{\pi\mu f\sigma}$, where quantity δ represents skin depth which is defined as the distance required by the wave to be attenuated to $1/e$ or 37% of its original strength. For a dielectric plane sheet $\sigma/\omega\varepsilon \ll 1$ and Eq. (20) becomes:

$$\gamma = \sqrt{-\omega^2\mu\varepsilon} = j\omega\sqrt{\mu\varepsilon} \tag{22}$$

The above expression can be serrated into reflection, absorption and multiple-reflection (re-reflection) components as [5, 6, 19, 21–23]:

$$SE_R(dB) = -20\log_{10}\left(\frac{|1+\eta|^2}{4|\eta|}\right) \tag{23}$$

$$SE_A(dB) = -20\alpha t\log_{10} e = -8.68\alpha t \tag{24}$$

$$SE_M = -20\log_{10}\left(1 - \left\{\frac{|1-\eta|^2}{|1+\eta|^2}\right\}.\exp(-2\gamma t)\right) \tag{25}$$

$$\alpha = \frac{1}{\delta} = \frac{2\pi}{\lambda_o}\sqrt{\frac{\varepsilon_r\left(\sqrt{1+\tan\delta} \mp 1\right)}{2}} \tag{26}$$

$$\eta = \sqrt{\frac{\varepsilon_r\left(\sqrt{1+\tan\delta} \pm 1\right)}{2}} + i\sqrt{\frac{\varepsilon_r\left(\sqrt{1+\tan\delta} \mp 1\right)}{2}} \tag{27}$$

$$\gamma = \frac{2\pi}{\lambda_o}\left[\sqrt{\frac{\varepsilon_r\left(\sqrt{1+\tan\delta} \mp 1\right)}{2}} + i\sqrt{\frac{\varepsilon_r\left(\sqrt{1+\tan\delta} \pm 1\right)}{2}}\right] \tag{28}$$

Here, η is the complex refractive index, γ is propagation constant, α is attenuation constant and t is thickness of the shielding material. The parameter skin depth ($\delta = 1/\alpha$) is defined as the depth of penetration at which the incident EM radiation is reduced to 33% of its original strength. Now from the knowledge $\varepsilon_r = \varepsilon'$, $\tan\delta = \frac{\varepsilon''}{\varepsilon'}$ and $\sigma_T = (\sigma_{ac} + \sigma_{dc})\omega\varepsilon_o\varepsilon''$ we get $\tan\delta = \frac{\sigma_T}{\omega\varepsilon_o\varepsilon'}$ where σ_{ac} and σ_{dc} are frequency ($f = \omega/2\pi$) dependent (ac) and independent (dc) components of total microwave conductivity (σ_T) whereas ε' (μ') and ε'' (μ'') represent real and imaginary parts of complex permittivity (permeability) respectively.

$$\lambda_o = \frac{2\pi c}{\omega} = \frac{2\pi}{\omega\sqrt{(\varepsilon\mu)}} = \frac{2\pi}{\omega\sqrt{(\varepsilon'\mu').(\varepsilon_o\mu_o)}} \tag{29}$$

The above equations revealed that electrical conductivity as well as dielectric and magnetic properties are important for absorption loss. Thus permittivity has a governing role in determining the shielding due to reflection, absorption and multiple reflection phenomenon.

1.4.6. Return Loss (R_L)

Another important parameter that correlates the absorption characteristics with the relative permeability and permittivity for a given frequency and absorber thickness is return loss (R_L) that can be expressed as [6]:

$$Z_{in} = Z_o\left(\frac{\mu_r}{\varepsilon_r}\right)^{1/2}\tanh\left[j\left(\frac{2\pi fd}{c}\right).\left(\frac{\mu_r}{\varepsilon_r}\right)^{1/2}\right] \tag{30}$$

$$R_L = 20\log\left|\frac{(Z_{in} - Z_o)}{(Z_{in} + Z_o)}\right| \tag{31}$$

where *f*, is the frequency of the EM-wave, *d*, the thickness of the absorber, *c*, the velocity of light, Z_0, the impedance of free space and Z_{in}, the input impedance of the absorber. It is important to note that permittivity of the same again plays a governing role in deciding the return loss and impedance matching conditions.

1.4.7. Permittivity and Electronic Devices

Similar to previous cases, the permittivity plays a crucial role in designing and performance governance of electronic devices like dielectric resonator antenna (DRA), dielectric wireless receivers and dielectric waveguides. For example, the dimension of a DRA inversely scales with the square root of the dielectric constant of the resonator material. Thus, by choosing a high value of dielectric constant, the size of the DRA can be significantly reduced [17].

1.4.8. Dielectric Metamaterials

Low-loss electromagnetic responses covering all four quadrants of possible permittivities and permeabilities has been demonstrated using completely transparent and high-refractive-index dielectric building blocks [7, 18, 24].

1.5. Permittivity of Complex Systems

1.5.1. Permittivity of Mixed Phase Systems

In case single phase material, the permittivity is presented by earlier mentioned processes and mechanisms. However, in case of multi-phase systems including admixtures, blends, composites, alloys and hybrids, the permittivity is governed by the nature of materials, their electrical properties, respective fractions and interfacial interactions between them. In such systems having homogenous filler distribution and non-interacting phases, the permittivity follows the rules of mixtures and can be expressed in terms of summation of individual permittivity and fractions of phases as [10, 25]:

$$\varepsilon_{eff} = \sum_{i=1}^{n}(\varepsilon_i . f_i) \tag{32}$$

where is the effective permittivity of the system, εi is the permittivity and f_i is the fraction of the ith phase. In more refined form, for a binary system consisting of a filled inclusion of permittivity and filled volume fraction v_f homogenously dispersed inside a matrix, the effective permittivity can be expressed by Maxwell Garnett equation [10, 26]:

$$\varepsilon_{eff} = \varepsilon_m \left[1 + \frac{3v_f(\varepsilon_f - \varepsilon_m)}{\{\varepsilon_f + 2\varepsilon_m - v_f(\varepsilon_f - \varepsilon_m)\}} \right] \tag{33}$$

In general terms, the Maxwell Garnett EMA is expected to be valid at low volume fractions, since it is assumed that the domains are spatially separated and electrostatic interaction between the chosen inclusions and all other neighboring inclusions is neglected.

1.5.2. Interfacial or Space Charge Polarization

Electronic, atomic, and orientation polarization occur when charges are locally bound in atoms, molecules, or structures of solids or liquids. Charge carriers also exist that can migrate over a distance through the material when a low frequency electric field is applied. Interfacial or space charge polarization occurs when the motion of these migrating charges is impeded. The charges can become trapped within the interfaces of a material. Motion may also be impeded when charges cannot be freely discharged or replaced at the electrodes. The field distortion caused by the accumulation of these charges increases the overall capacitance of a material which appears as an increase in er'. Mixtures of materials (e.g., in admixtures, alloys, blends or composites) with electrically conducting regions that are not in contact with each other (separated by non-conducting regions e.g., conjugated polymers, conducting fillers loaded polymer/ceramic composites etc.) exhibit the Maxwell-Wagner effect at low frequencies [5, 16, 27]. If the charge layers are thin and much smaller than the particle dimensions, the charge responds independently of the charge on nearby particles. At low frequencies the charges have time to accumulate at the borders of the conducting regions causing ε_r' to increase. At higher frequencies the charges do not have time to accumulate and polarization does not occur since the charge displacement is small compared to the dimensions of the conducting region. As the frequency increases, ε_r' decreases and the losses exhibit the same 1/f slope as normal ionic conductivity.

1.6. Permittivity and Classification of Materials

As already discussed earlier, the permittivity (and permeability) of the material plays governing role in determining their properties, response to external stimuli and interaction with electromagnetic waves. Therefore, various materials, especially the non-magnetic dielectric substances, can be classified (Table 1) according to their complex permittivity ε, more precisely upon comparison of its real ε' and imaginary ε'' components in terms of their ratio $\varepsilon''/\varepsilon'$ or loss tangent ($\tan\delta$) [28–30].

Table 1. Classification of materials based on permittivity

$\varepsilon_r''/\varepsilon_r'$	**Current conduction**	**Field propagation**
0	No conduction	Perfect dielectric: lossless medium
$\ll 1$	Low-conductivity material poor conductor	Low-loss medium: good dielectric
≈ 1	Lossy conducting material	Lossy propagation medium
$\gg 1$	High-conductivity material good conductor	High-loss medium: poor dielectric
∞	Perfect conductor	High conduction loses

The conductivity, σ, for many systems is related to loss par (ε'') as $\sigma = \omega\varepsilon''$ such that a *perfect conductor* has infinite conductivity, $\sigma = \infty$, while a *perfect dielectric* is a material that has no conductivity at all, $\sigma = 0$. Therefore, a perfect dielectric poses only real permittivity component and zero imaginary component, thus behaved as a *lossless media*. Such materials are considered as ideal materials for electrodes gap filling in capacitors. Generally, when $\sigma/\omega\varepsilon' \ll 1$ we consider the material to be a *low-loss dielectric* (although not exactly lossless), whereas $\sigma/\omega\varepsilon' \gg 1$ is associated with a *good conductor*; such materials with non-negligible conductivity yield a large amount of loss that inhibit the propagation of electromagnetic waves, thus are also said to be *lossy media*. Such materials have applications in the microwave absorbing materials. Those materials that do not fall under either limit are considered to be general media. In the case of a lossy medium, i.e., when the conduction current is not negligible, the total current density flowing is:

$$J_{total} = j\omega\varepsilon E = j\omega E(\varepsilon' - j\varepsilon'') = j\omega E(\varepsilon' - j\sigma/\omega) = (j\omega\varepsilon' E + \sigma E) \quad (34)$$

In this formalism, the complex permittivity is defined as [6, 19]:

$$\varepsilon = (\varepsilon' - j\sigma/\omega) = \varepsilon'\left(1 - \frac{j\sigma}{\omega\varepsilon'}\right) \quad (35)$$

1.7. Complex Permittivity and Dielectric Losses

In analogy to the equation, $P = \varepsilon_o \chi E$, the polarization is a function of frequency for given frequency dependent electric filed, and in such a regime, the frequency dependence of the susceptibility leads to frequency dependence of the permittivity. It is important to point that the frequency dependence unveils that material's polarization does not change instantaneously (or in phase) when an electric field is applied and is marked by a phase difference. Therefore, the permittivity is often treated as a complex function of the (angular) frequency ω of the applied field:

$$D_o e^{-j\omega t} = \varepsilon(\omega).E_o e^{-j\omega t} \quad (36)$$

Where D_0 and E_0 are the amplitudes of the displacement and electric fields, respectively, and j is the imaginary unit, $j^2 = -1$. The response of a medium to static electric fields is described by the low-frequency limit of permittivity, also called the static permittivity ε_s (also ε_{DC}):

$$\varepsilon_{static} = \varepsilon_{dc} = \lim_{\omega \to 0} \varepsilon(\omega) \quad (37)$$

At the high-frequency limit (meaning optical frequencies), the complex permittivity is commonly referred to as ε_∞ (or sometimes ε_{opt}). At the plasma frequency and below, dielectrics behave as ideal metals, with electron gas behavior. The static permittivity is a good approximation for alternating fields of low frequencies, and as the frequency increases a measurable phase difference δ emerges between **D** and **E**. The frequency at which the phase shift becomes noticeable depends on temperature and the details of the medium. For moderate field strength (E_0), **D** and **E** remain proportional, and

$$\varepsilon = \frac{D_o}{E_o} = |\varepsilon| e^{-j\delta} \tag{38}$$

Since the response of materials to alternating fields is characterized by a complex permittivity, it is natural to separate its real and imaginary parts, which is done by convention in the following way:

$$\varepsilon(\omega) = \varepsilon' - j\varepsilon'' = \left|\frac{D_o}{E_o}\right| (\cos\delta - j\sin\delta) \tag{39}$$

Where, ε' is the real part of the permittivity; ε'' is the imaginary part of the permittivity and δ is the loss angle. The choice of sign for time-dependence, $e^{-i\omega t}$, dictates the sign convention for the imaginary part of permittivity. The signs used here correspond to those commonly used in physics, whereas for the engineering convention one should reverse all imaginary quantities.

The Figure 5 shows the dispersion of real and imaginary permittivity as function of frequency. Due to such frequency dependence these quantities may also be called as transient or ac permittivity values. We can see that at very low frequencies the real permittivity is highest and this value when extrapolated to zero frequency is known as static permittivity, optical permittivity or dc permittivity.

As the frequency increases, real permittivity falls with involvement of different types of polarization and all associated losses e.g., ionic polarization at very low frequencies, dipolar polarization near microwave frequencies, atomic polarization in high end of microwave to infrared frequencies, and finally electronic polarization at high frequencies in visible and UV range. In general, the absorption of electromagnetic energy by dielectrics is involves different mechanisms that influence the scaling of permittivity as a function of frequency.

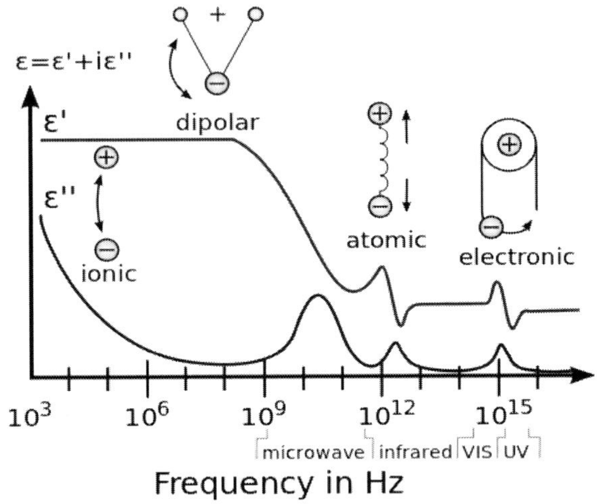

Figure 5. A dielectric permittivity spectrum over a wide range of frequencies. ε' and ε'' denote the real and the imaginary part of the permittivity, respectively. Various processes are labeled on the image: ionic and dipolar relaxation, and atomic and electronic resonances at higher energies [4].

1.7.1. Relaxation Effects

These effects are associated with behavior of permanent and induced molecular dipoles due to transient fields. At low frequencies, the field changes slow enough to allow dipoles to reach equilibrium before the field has measurably changed. However, at higher frequencies, dipole orientations cannot follow the applied field due to the viscosity of the medium, resulting in energy dissipation and related absorption phenomenon. The mechanism of dipoles relaxing is called dielectric relaxation and for ideal dipoles is described by classic Debye relaxation.

1.7.2. Resonance Effects

The resonance process occurs due to the rotations or vibrations of atoms, ions, or electrons. These processes are observed in the neighborhood of their characteristic absorption frequencies.

As already pointed, the Dielectric relaxation is the momentary delay (or lag) in the dielectric constant of a material. This is usually caused by the delay in molecular polarization with respect to a changing electric field in a dielectric medium (e.g., inside capacitors or between two large conducting surfaces). Dielectric relaxation in changing electric fields could be considered analogous to hysteresis in changing magnetic fields (e.g., in inductor or transformer cores). Relaxation in general is a delay or lag in the response of a linear system, and therefore dielectric relaxation is measured relative to the expected linear steady state (equilibrium) dielectric values. The time lag between electrical field and polarization implies an irreversible degradation of Gibbs free energy. In physics, dielectric relaxation refers to the relaxation response of a dielectric medium to an external, oscillating electric field. This relaxation is often described in terms of permittivity as a function of frequency, which can, for ideal systems, be described by the Debye equation. On the other hand, the distortion related to ionic and electronic polarization shows behavior of the resonance or oscillator type. The character of the distortion process depends on the structure, composition, and surroundings of the sample. According to EM theory, the dielectric loss may be contributed by the processes like natural resonance and relaxation effects.

1.7.3. Natural Resonance: Electronic and Atomic Polarization

Electronic polarization occurs in neutral atoms when an electric field displaces the nucleus with respect to the electrons that surround it. Atomic polarization occurs when adjacent positive and negative ions "stretch" under an applied electric field. For many dry solids, these are the dominant polarization mechanisms at microwave frequencies, although the actual resonance occurs at a much higher frequency. In the infrared and visible light regions the inertia of the orbiting electrons must be taken into account. Atoms can be modelled as oscillators with a damping effect similar to a mechanical spring and mass system (Figure 7). The amplitude of the oscillations will be small for any frequency other than the resonant frequency. Far below resonance, the electronic and atomic mechanisms contribute only a small constant amount to ε' and are almost lossless. The resonant frequency is identified by a resonant response in ε' and a peak of maximum absorption in ε''. Above the resonance, the contribution from these mechanisms disappears.

1.7.4. Orientation/Dipolar Polarization

A molecule is formed when atoms combine to share one or more of their electrons. This rearrangement of electrons may cause an imbalance in charge distribution creating a permanent dipole moment. These moments are oriented in a random manner in the absence of an electric field so that no polarization exists. The electric field *E* will exercise torque *T* on the electric dipole, and the dipole will rotate to align with the electric field causing orientation polarization to occur (Figure 5). If the field changes the direction, the torque will also change. The friction accompanying the orientation of the dipole will contribute to the dielectric losses. The dipole rotation causes a variation in both ε_r' and ε_r'' at the relaxation frequency which usually occurs in the microwave region. As mentioned, water is an example of a substance that exhibits a strong orientation polarization.

Dipole orientation and ionic conduction interact strongly at microwave frequencies. Water molecules, for example, are permanent dipoles, which rotate to follow an alternating electric field. These mechanisms are quite lossy – which explains why food heats in a microwave oven. Atomic and electronic mechanisms are relatively weak, and usually constant over the microwave region. Each dielectric mechanism has a characteristic "cut-off frequency." As frequency increases, the slow mechanisms drop out in turn, leaving the faster ones to contribute to ε'. The loss factor (ε'') will correspondingly peak at each critical frequency. The magnitude and "cut-off frequency" of each mechanism is unique for different materials. Water has a strong dipolar effect at low frequencies – but its dielectric constant rolls off dramatically around 22 GHz. Teflon, on the other hand, has no dipolar mechanisms and its permittivity is remarkably constant well into the millimeter-wave region.

1.7.5. Debye Relaxation

Debye relaxation is the dielectric relaxation response of an ideal, non-interacting population of dipoles to an alternating external electric field. It is usually expressed in the complex permittivity ε of a medium as a function of the field's angular frequency (ω) as [4, 6, 13, 31]:

$$\varepsilon(\omega) = \left\{ \varepsilon_\infty + \frac{(\varepsilon_s - \varepsilon_\infty)}{(1 + j\omega\tau)} \right\} = \varepsilon_r^* = (\varepsilon' + j\varepsilon'') \tag{40}$$

where ε_∞ is the permittivity at the high frequency limit, $\Delta\varepsilon = \varepsilon_s - \varepsilon_\infty$ where ε_s is the static, low frequency permittivity, and τ is the characteristic relaxation time of the medium. Separating into the real part ε' and the imaginary part ε'' of the complex dielectric permittivity yields:

$$\varepsilon' = \left\{ \varepsilon_\infty + \frac{(\varepsilon_s - \varepsilon_\infty)}{(1 + \omega^2\tau^2)} \right\} \tag{41}$$

$$\varepsilon'' = \left\{ \frac{(\varepsilon_s - \varepsilon_\infty)\omega\tau}{(1 + \omega^2\tau^2)} \right\} \tag{42}$$

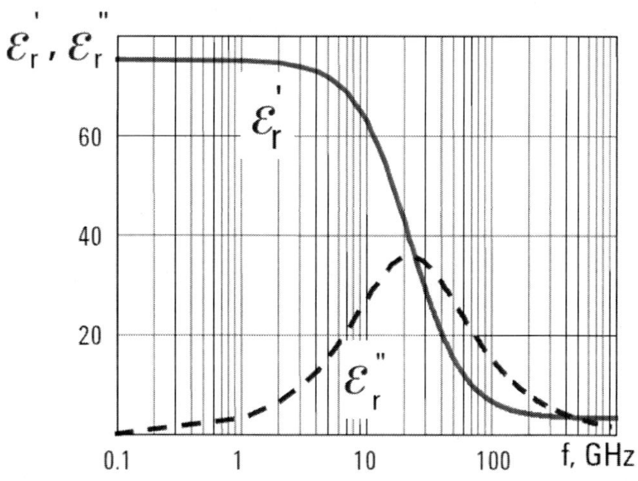

Figure 6. Debye plot for water [4].

Materials that exhibit a single relaxation time constant can be modelled by the Debye relation, which appears as a characteristic response in permittivity as a function of frequency (Figure 6). $\varepsilon'$is constant above and below the relaxation with the transition occurring near the relaxation frequency (22 GHz). Additionally, ε'' is small above and below relaxation and peaks in the transition region at the relaxation frequency.

According to equation (41) and (42), the following relationship between ε' and ε'' is derived [6]:

$$\left[\varepsilon' - \frac{(\varepsilon_s + \varepsilon_\infty)}{2}\right]^2 + [\varepsilon'']^2 = \left[\frac{(\varepsilon_s + \varepsilon_\infty)}{2}\right]^2 \tag{43}$$

Thus, the plot of ε' versus ε'' would be a single semicircle, generally denoted as the Cole–Cole semicircle. Each semicircle corresponds to one Debye relaxation process. Nevertheless, the dielectric loss is also represented by the loss tangent [13, 31]:

$$\tan\delta = \frac{\varepsilon''}{\varepsilon'} = \left\{\frac{(\varepsilon_s - \varepsilon_\infty)\omega\tau}{(\varepsilon_s + \varepsilon_\infty\omega^2\tau^2)}\right\} \tag{44}$$

This relaxation model was introduced by and named after the physicist Peter Debye (1913). It is characteristic for dynamic polarization with only one relaxation time. There are various variants of the Debye equation, which are discussed in the following text.

1.7.6. Cole–Cole Equation

This equation is used when the dielectric loss peak shows symmetric broadening. The Cole–Cole equation is a relaxation model that is often used to describe dielectric relaxation in polymers and can be expressed by the equation [32, 33]:

$$\varepsilon(\omega) = \left\{\varepsilon_\infty + \frac{(\varepsilon_s - \varepsilon_\infty)}{(1 + (j\omega\tau)^{1-\alpha})}\right\} \tag{45}$$

where $\varepsilon(\omega)$ is the complex dielectric constant, ε_s and ε_∞ are the "static" and "infinite frequency" dielectric constants, ω is the angular frequency and τ is a time constant. The exponent parameter α, which takes a value between 0 and 1, allows the description of different spectral shapes. When $\alpha = 0$, the

Cole-Cole model reduces to the Debye model. When $\alpha > 0$, the relaxation is *stretched*. That is, it extends over a wider range on a logarithmic ω scale than Debye relaxation. The separation of the complex dielectric constant was reported in the original paper by Cole and Cole as follows [13, 32, 32]:

$$\varepsilon' = \left\{ \varepsilon_\infty + (\varepsilon_s - \varepsilon_\infty) . \frac{[1 + (\omega\tau)^{1-\alpha} . \sin(\alpha\pi / 2)]}{[1 + 2(\omega\tau)^{1-\alpha} . \sin(\alpha\pi / 2) + (\omega\tau)^{2(1-\alpha)}]} \right\} \tag{46}$$

$$\varepsilon'' = \left\{ (\varepsilon_s - \varepsilon_\infty) . \frac{[(\omega\tau)^{1-\alpha} . \cos(\alpha\pi / 2)]}{[1 + 2(\omega\tau)^{1-\alpha} . \sin(\alpha\pi / 2) + (\omega\tau)^{2(1-\alpha)}]} \right\} \tag{47}$$

These equations reduce to the Debye expression when $\alpha = 1$. Cole–Cole relaxation constitutes a special case of Havriliak–Negami relaxation when the symmetry parameter $\beta = 1$, that is, when the relaxation peaks are symmetric. The complex permittivity may also be shown on a Cole-Cole diagram by plotting the imaginary part (ε''’) on the vertical axis and the real part (ε') on the horizontal axis with frequency as the independent parameter (Figure 7).

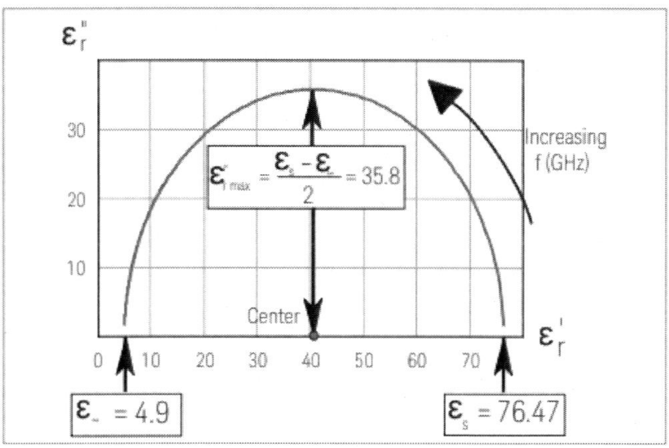

Figure 7. Cole-Cole plot for water [4].

A Cole-Cole diagram is, to some extent, similar to the Smith chart. A material that has a single relaxation frequency as exhibited by the Debye relation will appear as a semicircle with its center lying on the horizontal ε_r'' = 0 axis and the peak of the loss factor occurring at $1/\tau$. A material with multiple relaxation frequencies will be a semicircle (symmetric distribution) or an arc (nonsymmetrical distribution) with its center lying below the horizontal ε_r''= 0 axis. The curve in Figure 10 is a half circle with its center on the x-axis and its radius. The maximum imaginary part of the dielectric constant ε'_r max will be equal to the radius. The frequency moves counter clockwise on the curve.

Another special case of Havriliak–Negami relaxation [34] where $\beta < 1$ and $\alpha = 1$ is known as Cole–Davidson relaxation. For an abridged and updated review of anomalous dielectric relaxation in disordered systems, see Kalmykov.

1.7.7. Cole–Davidson Equation

In 1950 Davidson and Cole gave another expression for $\varepsilon^*(\omega)$ given as [13, 35]:

$$\varepsilon(\omega) = \varepsilon_\infty + \frac{(\varepsilon_s - \varepsilon_\infty)}{(1 + j\omega\tau)^\beta} \tag{48}$$

This equation is used when the dielectric loss peak shows asymmetric broadening. This expression reduces to the Debye equation for $\beta = 1$. Since

$$1 + j\omega\tau_o = e^{i\phi}\sqrt{1 + \omega^2\tau^2} = e^{i\phi} / \cos\phi \tag{49}$$

where $\varphi = \text{arctg}\,\omega\,\tau_0$, separation of the real and imaginary parts is easy, leading to the following expressions for ε' and ε'':

$$\varepsilon'(\omega) = \varepsilon_\infty + (\varepsilon_s - \varepsilon_\infty)\cos\phi^\beta \cos\beta\phi \tag{50}$$

$$\varepsilon'(\omega) = (\varepsilon_s - \varepsilon_\infty)\cos\phi^\beta \sin\beta\phi \tag{51}$$

1.7.8. Havriliak–Negami Relaxation

This equation considers both symmetric and asymmetric broadening. The Havriliak–Negami relaxation [29, 34] n is an empirical modification of the Debye relaxation model in electromagnetism. Unlike the Debye model, the Havriliak–Negami relaxation accounts for the asymmetry and broadness of the dielectric dispersion curve. The model was first used to describe the dielectric relaxation of some polymers, by adding two exponential parameters to the Debye equation:

$$\varepsilon(\omega) = \left\{ \varepsilon_\infty + \frac{(\varepsilon_s - \varepsilon_\infty)}{[1 + (j\omega\tau)^{1-\alpha}]^\beta} \right\} \tag{52}$$

where ε_∞ is the permittivity at the high frequency limit, ε_s is the static, low frequency permittivity, and τ is the characteristic relaxation time of the medium. The exponents α and β describe the asymmetry and broadness of the corresponding spectra. Depending on application, the Fourier transform of the stretched exponential function can be a viable alternative that has one parameter less.

For β=1 the Havriliak–Negami equation reduces to the Cole–Cole equation, for α =1 it reduces to the Cole–Davidson equation.

The storage part $\varepsilon'(\omega)$ and the loss part $\varepsilon''(\omega)$ of the permittivity can be calculated as

$$\varepsilon'(\omega) = \left\{ \varepsilon_\infty + (\varepsilon_s - \varepsilon_\infty).[1 + (2\omega\tau)^\alpha.\cos(\alpha\pi/2) + (\omega\tau)^{2\alpha}]^{-\beta/2}.\cos(\beta\phi) \right\} \tag{53}$$

and

$$\varepsilon''(\omega) = \left\{ (\varepsilon_s - \varepsilon_\infty).[1 + (2\omega\tau)^\alpha.\cos(\alpha\pi/2) + (\omega\tau)^{2\alpha}]^{-\beta/2}.\sin(\beta\phi) \right\} \tag{54}$$

with

$$\phi = \arctan\left\{ \frac{(\omega\tau)^{\alpha}.\sin(\alpha\pi/2)}{1+(\omega\tau)^{\alpha}.\cos(\alpha\pi/2)} \right\} \tag{55}$$

The maximum of the loss part lies at

$$\omega_{\max} = \left\{ \frac{\sin\left(\dfrac{\pi\alpha}{2(\beta+1)}\right)}{\sin\left(\dfrac{\pi\alpha\beta}{2(\beta+1)}\right)} \right\}^{1/\alpha} .\tau^{-1} \tag{56}$$

1.8. Standards for Measurement of Permittivity

There are various international standards for measurement of permittivity of various materials in terms of dielectric constant, and dissipation factors.

1.8.1. ASTM D2149-13(2021) [36]

Standard Test Method for Permittivity (Dielectric Constant) and Dissipation Factor of Solid Dielectrics at Frequencies to 10 MHz and Temperatures to 500°C, more specifically, 50 Hz to 10 MHz over a range of temperatures from −80 to 500°C. Two procedures are included as follows:

- Procedure A—Using Micrometer Electrode.
- Procedure B—Using Precision Capacitor.

1.8.2. ASTM D150-98 [37]

Standard Test Methods for AC Loss Characteristics and Permittivity (Dielectric Constant) of Solid Electrical Insulation. Particularly, these test methods cover the determination of relative permittivity, dissipation factor, loss index, power factor, phase angle, and loss angle of specimens of solid electrical insulating materials when the standards used are lumped impedances. The frequency range that can be covered extends from less than 1 Hz to several hundred megahertz.

1.8.3. ASTM D2520-21 [38]

Standard Test Methods for Complex Permittivity (Dielectric Constant) of Solid Electrical Insulating Materials at Microwave Frequencies and Temperatures to 1650 °C; Particularly, these test methods cover the determination of relative (Note 1) complex permittivity (dielectric constant and dissipation factor) of nonmagnetic solid dielectric materials. Although these test methods are used over the microwave frequency spectrum from around 0.5 to 50.0 GHz, each octave increase usually requires a different generator and a smaller test waveguide or resonant cavity.

Test Method A is for specimens precisely formed to the inside dimension of a waveguide.

Test Method B is for specimens of specified geometry that occupy a very small portion of the space inside a resonant cavity.

Test Method C uses a resonant cavity with fewer restrictions on specimen size, geometry, and placement than Test Methods A and B.

According to type of sample, intended application, frequency & temperature range, technique & instrument selected for dielectric measurements, appropriate standard and test method has to be identified and followed. At infrared and optical frequencies, a common technique is ellipsometry. Dual polarization interferometry is also used to measure the complex refractive index for very thin films at optical frequencies. The dielectric properties at intermediate and high frequencies is measured using instruments called network analyzer (NA).

1.9. Measurement of Permittivity

The relative permittivity of a material can be measured by a variety of static electrical techniques. The complex permittivity is evaluated over a wide range of frequencies by using dielectric spectroscopy technique, covering nearly 21 orders of magnitude frequency range i.e., from 10^{-6} to 10^{15} hertz [4]. Various microwave measurement techniques are outlined in Chen et al. [28]. Typical errors for the Hakki-Coleman method employing a puck of material between conducting planes are about 0.3%.

- Low-frequency time domain measurements (10^{-6} to 10^3 Hz)
- Low-frequency frequency domain measurements (10^{-5} to 10^6 Hz)
- Reflective coaxial methods (10^6 to 10^{10} Hz)
- Transmission coaxial method (10^8 to 10^{11} Hz)
- Quasi-optical methods (10^9 to 10^{10} Hz)
- Terahertz time-domain spectroscopy (10^{11} to 10^{13} Hz)
- Fourier-transform methods (10^{11} to 10^{15} Hz)

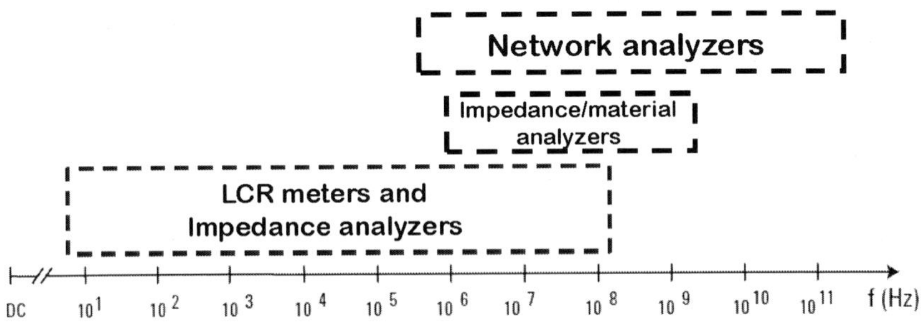

Figure 8. Frequency range and applicability of different types of analyzers.

Impedance analyzers and LCR meters (Figure 8) are used to measure the material properties at lower frequencies. The material is stimulated with an AC source and the actual voltage across the material is monitored. Material test parameters are derived by knowing the dimensions of the material and by measuring its capacitance and dissipation factor. In contrast, the network analyzers are used to measure the properties intermediate to high frequency range i.e., sub MHz range to several hundred GHz. It is important to note that before the dielectric properties of a material can be measured with network analyzer, impedance analyzer, or LCR meter, a measurement fixture (or sample holder) is required to apply the electromagnetic fields in a predictable way and to allow connection to the measurement instrument. The type of fixture required will depend on the chosen measurement technique and the physical properties of the material (solid, liquid, powder, gas).

1.9.1. Parallel Plate Capacitor and LCR Meter Method

The parallel plate capacitor method involves sandwiching a thin sheet of material between two electrodes to form a capacitor. A typical measurement system using the parallel plate method consists of an LCR meter or impedance analyzer and a dielectric test fixture, which operates up to 30 MHz.

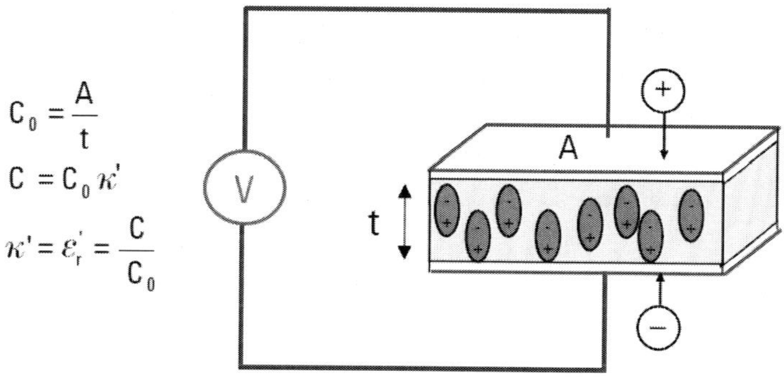

Figure 9. Parallel plate capacitor configuration for measurement of dielectric constant under application of DC voltage.

where C and C_0 are capacitance with and without dielectric, $k' = \varepsilon_r'$ is the real dielectric constant or permittivity, and A and t are the area of the capacitor plates and the distance between them (Figure 9). The dielectric material increases the storage capacity of the capacitor by neutralizing charges at the electrodes, which ordinarily would contribute to the external field. The capacitance of the dielectric material is related to the dielectric constant as indicated in the above equations [4].

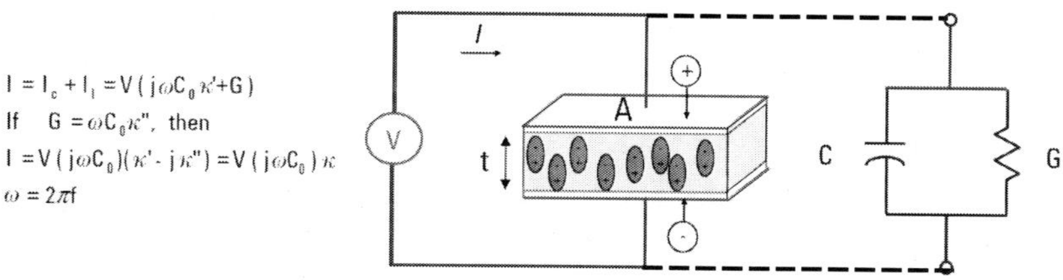

Figure 10. Parallel plate capacitor configuration for measurement of dielectric constant under application of AC voltage.

If an AC sinusoidal voltage source V is placed across the same capacitor (Figure 10), the resulting current will be made up of a charging current I_c and a loss current I_l that is related to the dielectric constant. The losses in the material can be represented as a conductance (G) in parallel with a capacitor (C). The complex dielectric constant k consists of a real part k' which represents the storage and an imaginary part k" which represents the loss. The following notations are used for the complex dielectric constant interchangeably [4]:

$$k = k^* = \varepsilon_r = \varepsilon_r^* \tag{57}$$

From the point of view of electromagnetic theory, the definition of electric displacement (electric flux density) D_f is:

$$D_f = \varepsilon E \tag{58}$$

where $\varepsilon = \varepsilon^* = \varepsilon_r . \varepsilon_o$ is the absolute permittivity (or permittivity), ε_r is the relative permittivity, $\varepsilon_o \approx \frac{1}{36\pi} x10^{-9}$ *F/m* is the free space permittivity and *E* is the electric field. Permittivity describes the interaction of a material with an electric field *E* and is a complex quantity.

$$k = \frac{\varepsilon}{\varepsilon_o} = \varepsilon_r = \varepsilon'_r - j\varepsilon''_r \tag{59}$$

Dielectric constant (k) is equivalent to relative permittivity (ε_r) or the absolute permittivity (ε) relative to the permittivity of free space (ε_o). The real part of permittivity (ε_r') is a measure of how much energy from an external electric field is stored in a material. The imaginary part of permittivity (ε_r'') is called the loss factor and is a measure of how dissipative or lossy a material is to an external electric field. The imaginary part of permittivity (ε_r'') is always greater than zero and is usually much smaller than ($\varepsilon_{r'}$). The loss factor includes the effects of both dielectric loss and conductivity. When complex permittivity is drawn as a simple vector diagram (as already shown in Figure 1), the real and imaginary components are 90° out of phase. The vector sum forms an angle δ with the real axis (ε_r'). The relative "lossiness" of a material is the ratio of the energy lost to the energy stored. The loss tangent or tanδ is defined as the ratio of the imaginary part of the dielectric constant to the real part. *D* denotes dissipation factor and *Q* is quality factor. The loss tangent tanδ is called tan delta, tangent loss or dissipation factor. Sometimes the term "quality factor or *Q*-factor" is used with respect to an electronic microwave material, which is the reciprocal of the loss tangent. For very low loss materials, since $\tan\delta \approx ta\delta$, the loss tangent can be expressed in angle units, milliradians or microradians. The parallel plate method is quite accurate and best in the low frequency region, for thin and flat sheets.

It is to be noted that at intermediate (several MHz) and very high frequencies, the losses become very high, and electrical methods (LCR meter or impedance analyzer) may not be much useful. In such situation, transmission methods using network analyzers (NA) are more useful. The vector NA (VNA) which measures the magnitude and associated phase, carry information about attenuation constant as well as dielectric/magnetic properties.

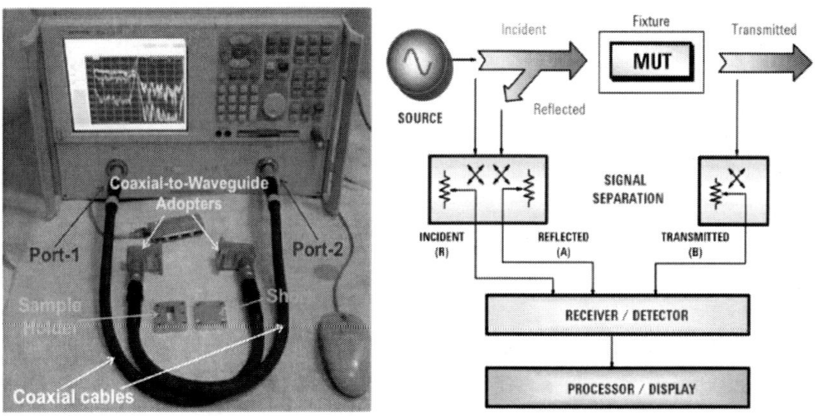

Figure 11. A two port VNA (left) and its internal block diagram (right).

The incident and transmitted waves in a two port VNA (Figure 11) can be mathematically represented by complex scattering parameters (or S-parameters) i.e., S_{11} (or S_{22}) and S_{12} (or S_{21}), which in-turn can be conveniently correlated with reflectance (R) and transmittance (T) i.e., $T = |E_T/E_I|^2 = |S_{12}|^2 = |S_{21}|^2$, $R = |E_R/E_I|^2 = |S_{11}|^2 = |S_{22}|^2$, that ultimately allows the measurement of dielectric properties using suitable algorithm. There are many techniques developed for measuring these S-parameters like Transmission/Reflection method, Open ended coaxial probe technique, Free space technique, Resonant cavity method and Parallel plate technique [4, 6].

1.9.2. Coaxial Probe Technique

The open-ended coaxial probe is a cut off section of transmission line. The material is measured by immersing the probe into a liquid or touching it to the flat face of a solid (or powder) material.

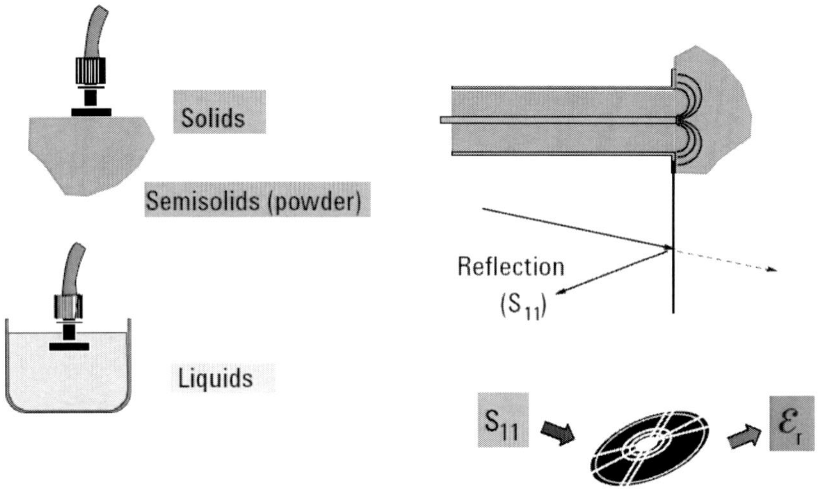

Figure 12. Coaxial probe method for measurement of relative permittivity .

The fields at the probe end "fringe" into the material and change as they come into contact with the MUT (Figure 12). The reflected signal (S_{11}) can be measured and related to ε_r* via a suitable algorithm. This method enables broadband measurements, convenient and non-destructive. It is best suited for lossy samples like liquids or semi-solids.

1.9.3. Transmission-Reflection Technique

Transmission line methods involve placing the material inside a portion of an enclosed transmission line. The line is usually a section of rectangular waveguide or coaxial airline (Figure 13). The testing involves the measurement of both the reflected and transmitted waves in terms of corresponding scattering parameters (S-parameters e.g., S_{11} or S_{22} and S_{21} or S_{12}).

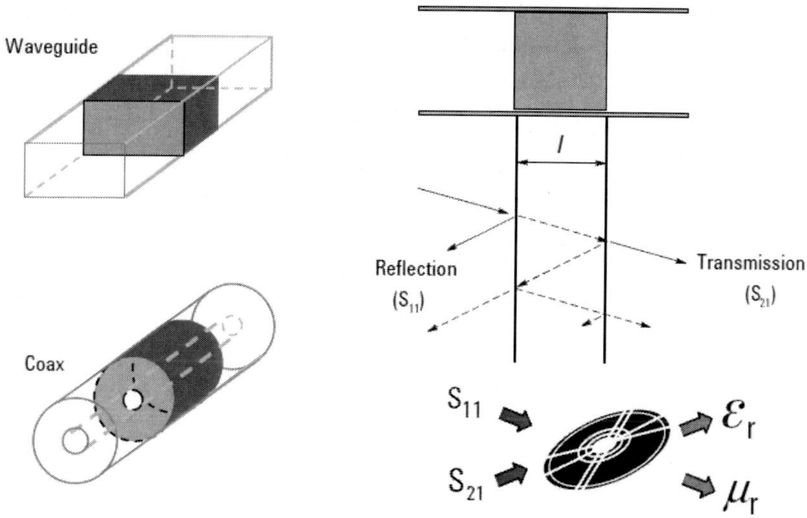

Figure 13. Transmission-reflection method for measurement of relative permittivity & permeability .

The ε_r* and μ_r* are computed from the measurement of the reflected signal (S_{11}) and transmitted signal (S_{21}) via using various available algorithms. Coaxial transmission lines cover a broad frequency range, but a toroid-shaped sample is more difficult to manufacture. Waveguide fixtures extend to the mm-wave frequencies and the samples are simpler to machine, but their frequency coverage is banded. This method allows broadband measurement and best suited for lossy to low loss samples.

1.9.4. Free Space Method

Free-space methods use antennas to focus microwave energy at or through a slab of material without the need for a test fixture.

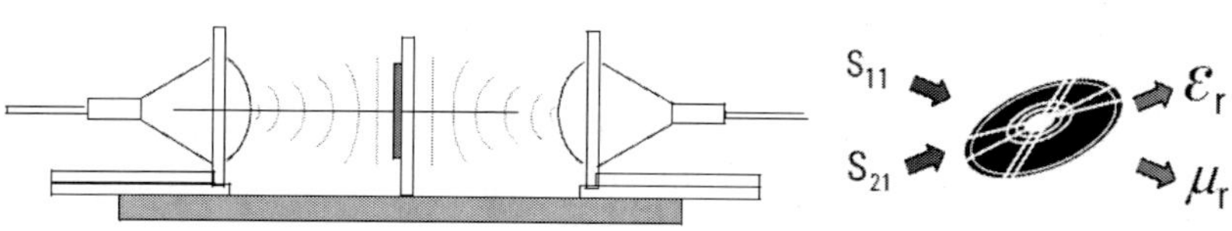

Figure 14. Free space method for measurement of relative permittivity & permeability.

Again the reflection and transmission scattering parameters are measured and the permittivity is calculated through suitable algorithm. This method is non-contacting and can be applied to materials to be tested under high temperatures and hostile environments. It's a non-contacting technique and best for high temperature measurements on large and flat samples.

1.9.5. Resonant Cavity Method

Resonant cavities are high *Q* structures that resonate at certain frequencies. A piece of sample material affects the center frequency (*f*) and quality factor (*Q*) of the cavity. From these parameters, the complex permittivity (ε_r) or permeability (μ_r) of the material can be calculated at a single frequency.

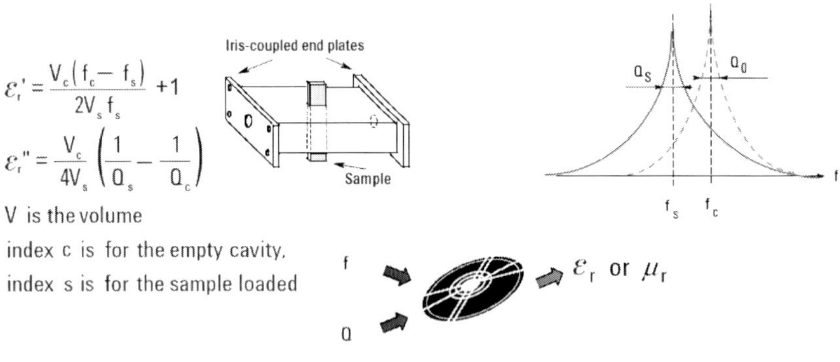

Figure 15. Resonant cavity technique for measurement of relative permittivity& permeability .

There are many different types of cavities and methods. Here, the most widely used cavity perturbation method, as described in ASTM 252010, is considered. This method uses a rectangular waveguide with iris-coupled end plates, operating in TE10n mode (Figure 15). For a dielectric measurement the sample should be placed in a maximum electric field and for a magnetic measurement, in a maximum magnetic field. The cavity perturbation method requires a very small sample such that the fields in the cavity are only slightly disturbed to shift the measured resonant frequency and cavity *Q*. This assumption allows simplifying the theory to use the equations above to calculate the dielectric properties of the material. This technique is very accurate and best suited for low loss small samples.

Among these techniques Transmission/Reflection method is the most popular as it simultaneously measures of all four S-parameters and gives complex permittivity as well as magnetic permeability by using suitable algorithms or models developed for obtaining the permittivity and permeability from the recorded S-parameters. Table 2 gives an overview of some of the conversion techniques, S-parameters & their evaluation capability for output attributes i.e., complex permittivity and permeability.

Table 2. Conversion techniques, S-parameters & output attributes

Conversion technique	**Input S-parameters**	**Output parameters**
Nicolson-Ross-Weir	S_{11}, S_{21}, S_{12} and S_{22}	Both ε* and μ*
NIST iterative	S_{11}, S_{21}, S_{12} and S_{22}	Only ε*
New non-iterative	S_{11}, S_{21}, S_{12} and S_{22}	Only ε*
Short circuit line	S_{11} or S_{22}	Only ε*

Each of the above conversion technique has different advantages and limitations. The selection of the technique depends on several factors such as the measured S-parameters, sample length, desired output properties, speed of conversion and accuracies in the converted results. Among above-mentioned procedures, Nicholson-Ross-Weir (NRW) technique [27, 39, 40] is the most widely used regressive/iterative analysis as it provides direct calculation of both the permittivity (ε^*) and permeability (μ^*) from the input S-parameters. It is the most commonly used technique for performing such conversions where the measurement of reflection (Γ) and transmission (T) coefficient requires all four (S_{11}, S_{21}, S_{12}, S_{22}) or a pair (S_{11}, S_{21}) of S-parameters of the material under test to be measured. The procedure proposed by NRW method is deduced from the following set of equations:

$$S_{11} = \frac{\Gamma\left(1-T^2\right)}{\left(1-\Gamma^2 T^2\right)} \text{ and } S_{21} = \frac{T\left(1-\Gamma^2\right)}{\left(1-\Gamma^2 T^2\right)} \tag{60}$$

Once these S-parameters are extracted from the network analyzer, simultaneous solving of equation set (61) gives the reflection coefficient as:

$$\Gamma = X \pm \sqrt{X^2 - 1} \tag{61}$$

The condition [$|\Gamma| < 1$] is used for finding the correct root of the quadratic equation so that parameter X can be expressed as:

$$X = \frac{S_{11}^2 - S_{21}^2 + 1}{2S_{11}} \tag{62}$$

Therefore, the transmission coefficient can be written as:

$$T = \frac{S_{11} + S_{21} - \Gamma}{1 - (S_{11} - S_{21})\Gamma} \tag{63}$$

The permeability is then given by:

$$\mu^* = \frac{1+\Gamma}{\Lambda(1-\Gamma)\sqrt{\frac{1}{\lambda_o^2} - \frac{1}{\lambda_c^2}}} \tag{64}$$

where λ_o and λ_c are free space and cutoff wavelength respectively and Λ is given by [4, 6]:

$$\frac{1}{\Lambda^2} = \left(\frac{\varepsilon^* \mu^*}{\lambda_o^2} - \frac{1}{\lambda_c^2}\right) = -\left[\frac{1}{2\pi L}\ln\left(\frac{1}{T}\right)\right]^2 \tag{65}$$

Therefore, the permittivity can be written as:

$$\varepsilon^* = \frac{\lambda_o^2}{\mu^*}\left(\frac{1}{\lambda_c^2} - \left[\frac{1}{2\pi L}\ln\left(\frac{1}{T}\right)\right]^2\right) \tag{66}$$

Equations (9.44) & (9.45) have infinite number of roots since the imaginary part of the term $\ln(1/T)$ is equal to $i(\theta + 2\pi n)$ where $n = 0, \pm 1, \pm 2\ldots$, i.e., the integral multiples of ratio L/λ_g, where L is sample length and λ_g, is wavelength inside the sample. This brings phase ambiguity and the correct value of 'n' can be determined either by analysis of group delay or by phase unwrapping method [6].

Conclusion: Challenges and Prospects

In this chapter permittivity and its significance has been explained using related terminologies, examples, models and mathematical formulas/expressions. It has been highlighted that permittivity plays crucial role in designing a system or deciding its response to extremal stimuli e.g., charge storage characteristics of capacitor/supercapacitor; exciton binding energy in solar cells, charge injection in organic LEDs, rheological characteristics of electrorheological fluids, properties of liquid crystals & display devices, microwave absorption effectiveness of shielding materials, and designing of metamaterials. It is realized that there is lot to be done in terms of realizing desired dielectric response materials by use of nanoscience and nanotechnology driven modification, new materials synthesis, designing of nanoscale & quantum confined systems, formation of hybrids & nanocomposite materials, and formation of desired morphology/textured thin films, and integration of these materials/systems with existing technologies. The areas like nanocomposite based dielectrics for technological applications; high dielectric solid state electrolytes for batteries/supercapacitors; high dielectric active layers for solar cells; thin dielectric films for LEDs and transparent polymer based dielectric for metamaterials, are gaining importance and lot can be done in the direction.

Acknowledgment

Author would like to thank Director CSIR-NPL for all the support for this work and providing the necessary infrastructure.

References

[1] Landau, L. D., Lifšic, E. M., Pitaevskij, L. P. & Landau, L. D. *Electrodynamics of continuous media*. (Elsevier Butterworth-Heinemann, 2009).

[2] Permittivity Definition, In *Wikipedia*. https://en.wikipedia.org/wiki/Permittivity.

[3] Braslavsky, S. E. Glossary of terms used in photochemistry, 3rd edition (IUPAC Recommendations 2006). *Pure Appl. Chem.* 79, 293–465 (2007).

[4] Agilent Basics of Measuring the Dielectric Properties of Materials: Application Note.

[5] Saini, P. & Aror, M. Microwave Absorption and EMI Shielding Behavior of Nanocomposites Based on Intrinsically Conducting Polymers, Graphene and Carbon Nanotubes. in *New Polymers for Special Applications* (ed. De Souza Gomes, A.) (InTech, 2012). doi:10.5772/48779.
[6] Saini, P. Intrinsically Conducting Polymer-Based Blends and Composites for Electromagnetic Interference Shielding: Theoretical and Experimental Aspects. in *Fundamentals of Conjugated Polymer Blends, Copolymers and Composites* (ed. Saini, P.) 449–518 (John Wiley & Sons, Inc., 2015). doi:10.1002/9781119137160.ch9.
[7] Metamaterial, In *Wikipedia*. https://en.wikipedia.org/wiki/Metamaterial.
[8] Tian, J., Yang, Y., Qiu, M., Laurell, F., Pasiskevicius, V., & Jang, H. All-dielectric KTiOPO_4 metasurfaces based on multipolar resonances in the terahertz region. *Opt. Express* 25, 24068 (2017).
[9] Chen, Z., Guo, B., Yang, Y. & Cheng, C. Metamaterials-based enhanced energy harvesting: A review. *Phys. B Condens. Matter* 438, 1–8 (2014).
[10] Kremer, F., & Schönhals, A. (Eds.). *Broadband dielectric spectroscopy*. (Springer, 2003).
[11] Kittel, C., McEuen, P., & John Wiley & Sons. *Introduction to solid state physics*. (John Wiley & Sons, 2019).
[12] Verma, H. C. *Concepts of Physics*. (Bharti Bhavan).
[13] *Dielectric,* In *Wikipedia*. https://en.wikipedia.org/wiki/Dielectric#Debye_relaxation.
[14] *Vacuum, electric permittivity;* NIST refeecne on costants, units and uncertainity: https://physics.nist.gov/cgi-bin/cuu/Value?ep0.
[15] Wang, F., Liu, S. & Zhang, C. The dielectric constant of materials effect the property of the OLED. *Microelectron. J.* 38, 259–261 (2007).
[16] Saini, P. (Ed.). *Fundamentals of Conjugated Polymer Blends, Copolymers and Composites: Synthesis, Properties and Applications*. (John Wiley & Sons, Inc., 2015). doi:10.1002/9781119137160.
[17] *Dielecric Resonator Antenna,* https://en.wikipedia.org/wiki/Dielectric_resonator_antenna.
[18] Jahani, S. & Jacob, Z. All-dielectric metamaterials. *Nat. Nanotechnol.* 11, 23–36 (2016).
[19] Schulz, R. B., Plantz, V. C. & Brush, D. R. Shielding theory and practice. *IEEE Trans. Electromagn. Compat.* 30, 187–201 (1988).
[20] Duff, A. W. *Text-Book of Physics*. (Forgotten Books, 2019).
[21] Colaneri, N. F. & Schacklette, L. W. EMI shielding measurements of conductive polymer blends. *IEEE Trans. Instrum. Meas.* 41, 291–297 (1992).
[22] Shacklette, L. W. & Colaneri, N. F. EMI shielding measurements of conductive polymer blends. in *[1991] Conference Record. IEEE Instrumentation and Measurement Technology Conference* 72–78 (IEEE, 1991). doi:10.1109/IMTC.1991.161543.
[23] Shacklette, L. W., Colaneri, N. F., Kulkarni, V. G. & Wessling, B. EMI shielding of intinsically conductive polymers. *J. Vinyl Addit. Technol.* 14, 118–122 (1992).

[24] *Dielectric Metamaterials,* https://www.researchgate.net/publication/289521230_All-dielectric_metamaterials.
[25] Sihvola, A. H. & Kong, J. A. Effective permittivity of dielectric mixtures. *IEEE Trans. Geosci. Remote Sens.* 26, 420–429 (1988).
[26] Wagner, K. W. Erklärung der dielektrischen Nachwirkungsvorgänge auf Grund Maxwellscher Vorstellungen. *Arch. Für Elektrotechnik* 2, 371–387 (1914).
[27] Saini, P., Choudhary, V., Singh, B. P., Mathur, R. B. & Dhawan, S. K. Polyaniline–MWCNT nanocomposites for microwave absorption and EMI shielding. *Mater. Chem. Phys.* 113, 919–926 (2009).
[28] Chen, L. (Ed.). *Microwave electronics: measurement and materials characterisation*. (John Wiley, 2004).
[29] Kao, K.-C. *Dielectric phenomena in solids: with emphasis on physical concepts of electronic processes*. (Academic Press, 2004).
[30] Paul, C. R. *Electromagnetics for engineers: with applications to digital systems and electromagnetic interference*. (John Wiley & Sons, 2004).
[31] Debye, P. (1913), *Ver. Deut. Phys. Gesell.* 15, 777; reprinted 1954 in collected papers of Peter J. W. Debye. Interscience, New York.
[32] Cole, K. S. & Cole, R. H. Dispersion and Absorption in Dielectrics I. Alternating Current Characteristics. *J. Chem. Phys.* 9, 341–351 (1941).
[33] Cole, K. S. & Cole, R. H. Dispersion and Absorption in Dielectrics II. Direct Current Characteristics. *J. Chem. Phys.* 10, 98–105 (1942).
[34] Havriliak, S. & Negami, S. A complex plane representation of dielectric and mechanical relaxation processes in some polymers. *Polymer* 8, 161–210 (1967).
[35] Iglesias, T. P., Vilão, G. & Reis, J. C. R. An approach to the interpretation of Cole–Davidson and Cole–Cole dielectric functions. *J. Appl. Phys.* 122, 074102 (2017).
[36] ASTM D 2149, https://www.astm.org/d2149-13r21.html.
[37] ASTM D 150, https://www.astm.org/d0150-98.html.
[38] ASTM D 2520, https://www.astm.org/d2520-21.html.
[39] Nicolson, A. M. & Ross, G. F. Measurement of the Intrinsic Properties of Materials by Time-Domain Techniques. *IEEE Trans. Instrum. Meas.* 19, 377–382 (1970).
[40] Weir, W. B. Automatic measurement of complex dielectric constant and permeability at microwave frequencies. *Proc. IEEE* 62, 33–36 (1974).

Chapter 2

Advances in Dielectric Properties of Textile Materials

Srijan Das and Prabir K. Mukherjee*
Government College of Engineering and Textile Technology, Serampore, Hooghly, India

Abstract

The dielectric properties of textile materials have gained importance with respect to their application in varied fields of textile testing, electromagnetic shielding, textile antennas, and supercapacitors. In this chapter, firstly, a brief introduction has been given about textile materials. Theoretical studies regarding the dielectric properties of textile materials have been reviewed, and a comprehensive overview of the experimental studies on this subject has been provided. A critical assessment of the effect of different factors like frequency, moisture content, and temperature on the permittivity of such materials has been done, and an outline of the applications of dielectric properties in textiles has also been given.

Keywords: dielectric permittivity, textile materials, fiber, capacitance

1. Introduction

The dielectric properties of textiles have long been investigated. The first incentive for looking into such properties of textiles came from their use as insulating materials [1]. Since then, research on such properties of textiles has

* Corresponding Author's Email: pkmuk1966@gmail.com.

In: An Introduction to Permittivity
Editor: Parveen Saini
ISBN: 979-8-88697-042-5
© 2022 Nova Science Publishers, Inc.

bloomed. Textile materials are generally composed of fibers. Fibers, in turn, are composed of long-chain polymers. Fibers are the basic material for developing various textile products; hence a brief introduction to textile fibers would provide a better insight into the characteristics of textile materials.

Textile fibers have structures with a length to diameter ratio of at least 1000 [2]. Coss sectional shape of the fibers varies from being circular, kidney-shaped, triangular, etc., to specially designed cross-sectional shapes. The earliest fibers known to man were the natural fibers. Such fibers are available from plants like cotton, jute, hemp, flax, etc., and animals such as wool, silk, etc. The building blocks of fibers are polymer macromolecules. For example, cotton fiber obtained from cotton seeds is composed chiefly of α-cellulose. A molecule of cellulose consists of many glucose units. Figure 1 is a schematic representation of cotton fiber. Fibers like jute, hemp, and flax are multicellular, lignocellulosic fibers. Such fibers are obtained from the basts of plants. The fibers are mostly composed of cellulose, lignin, and hemicellulose. Silk and wool are natural protein fibers. The long protein molecules are made of several different amino acids joined together by amide linkage. There is a large number of possibilities for the arrangement of amino acids in the protein molecules. Such arrangements and the presence of different amino acids in the protein molecule itself lead to different properties of fibers.

Figure 1. Schematic representation of cotton fiber (Reproduced from Kajiwara and Ohta [3]).

Figure 2. Micrograph of Kevlar® 29 (Reproduced from Ohta and Kajiwara [4]).

Fibers have not remained confined to only those obtained from nature. Natural products like cellulose can be solubilized in suitable solvents and extruded through a spinneret to form regenerated fibers. Examples of some of such fibers are viscose, cellulose acetate, etc. The first commercially successful class of synthetic fibers were the polyamides, followed closely by polyesters, polyacrylics, and polyurethanes. Figure 2 displays the micrograph of Kevlar® 29 fiber.

However, fibers are not directly used to produce knitted or woven fabrics. Fibers are arranged in a parallel manner and are cohered by twisting axially or by other means to form yarns which are then used for producing fabrics. Fabrics can again be manufactured following different methods. Woven fabrics are formed by the interlacement of yarns following different patterns. Knitted fabrics are formed by interlooping of yarns, and nonwovens are formed by bonding fibrous webs employing different techniques. Figure 3 is a schematic representation of woven fabric, knitted fabric, and nonwoven.

Textiles are flexible, porous, and lightweight yet have appreciable strength. This allows the application of such materials in diverse fields. Knowledge of dielectric properties is essential regarding the application of textiles in E-textiles, composites, sensors, etc. Moreover, the dependence of the dielectric properties of textiles on various factors has been exploited for various textile testing related applications. Dielectric properties also give an idea about the structure of the textile material. Hence, research regarding the dielectric properties of textiles has been extensively carried out.

In the following sections, the theoretical approaches regarding dielectric properties of textiles, characterization techniques, the dependence of such property on different factors, relevant applications, and fabrication techniques have been discussed.

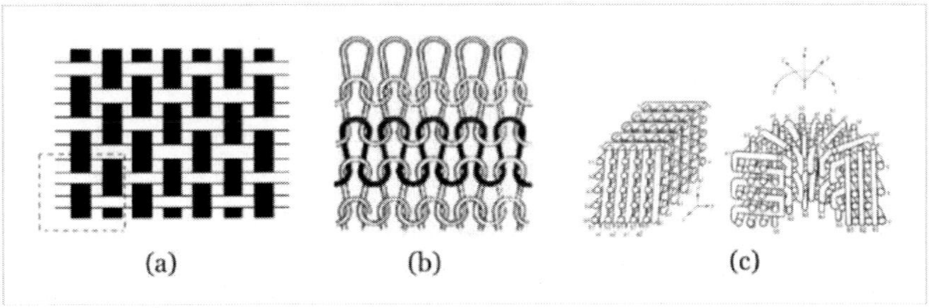

Figure 3. Schematic representation of a) woven fabric, b) knitted fabric, c) nonwoven fabric (Reproduced from Hong et al. [2]).

2. Theory

A dielectric is a non-conducting material that, when placed between the two plates of a capacitor, increases the capacitance. The relative permittivity (ε) of a material is thus defined in terms of capacitance (C) of a condenser with the dielectric material placed in between two parallel plates of area, A and separated by a distance, d, as per the following equation:

$$C = \frac{\varepsilon \varepsilon_0 A}{d} \tag{1}$$

where ε_0 (8.854×10^{-12} F/m) is the permittivity of free space, which is a universal constant.

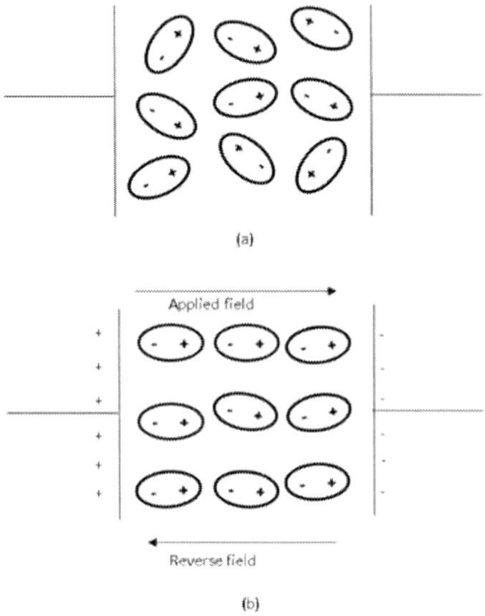

Figure 4. State of dipoles a) before application of electrical field and b) after application of electrical field.

When a dielectric material is subjected to an electric field, polarization occurs. This may be due to the presence of permanent dipoles or because of induced dipoles. This polarization gives rise to an electric field, opposite in direction to the applied electric field, as shown in Figure 4, thereby reducing the potential difference between the charged conductor plates [1]. When a sinusoidal voltage is applied, the total current flowing through the dielectric is found to be a vector sum of a charging current which leads the applied voltage by $90^{0,}$ and a loss current which is in phase with the applied voltage [5]. Relative permittivity is then represented by:

$$\varepsilon = \varepsilon_r - i\varepsilon_m \quad (2)$$

where, ε_r is the real part of permittivity and ε_m is the imaginary part of permittivity.

The dielectric constant of some textile fibers at 65% RH is given in Table 1.

Table 1. Dielectric constants of some common textile fibers (Adapted from Hong et al. [2])

Fiber	Dielectric Constant
Cotton	18
Wool	5.5
Viscose	8.4
Acetate	4.0
Acrylic staple	3.5
Nylon staple	3.7
Polyester staple	2.3

Mukherjee [6] theoretically studied the dielectric properties of textile materials using Landau theory and Flory-Huggins theory. This study showed that relative humidity affects both relative permittivity and loss tangent of the textile materials. The equations can explain the variation of permittivity of textile materials with frequency and temperature also. The causes of such variations are discussed in the section. The derived equations for real and imaginary permittivity are:

$$\varepsilon_r(\omega, T, x) = \varepsilon_1 + \frac{1}{u(1+\frac{\omega^2 v^2}{u^2})} \quad (3)$$

$$\varepsilon_m(\omega, T, x) = \varepsilon_2 + \frac{wv}{u^2(1+\frac{\omega^2 v^2}{u^2})} \quad (4)$$

where,

$$u = z - \frac{C_1^2}{A_1^2 + \frac{w^2}{\Gamma_1^2}} \quad (5)$$

$$v = \frac{1}{\Gamma_2} + \frac{C_1^2}{\Gamma_1[A_1^2 + \frac{w^2}{\Gamma_1^2}]} \quad (6)$$

$$z = \frac{1}{2\chi_0} - \frac{1}{2}\gamma\psi_0 \quad (7)$$

where,

$$A = a(T - x - T^*) \quad (8)$$

$$A_1 = A - B\psi_0 + \frac{3}{2}C\psi_0^2 \quad (9)$$

$$B_1 = \frac{1}{2\chi_0} - \frac{1}{2}\gamma\psi_0 \quad (10)$$

$$C_1 = \gamma P_0 \quad (11)$$

Here, x is the concentration of moisture content, T^* is the supercooled temperature, T is the temperature, γ is a coupling constant assumed to be positive, χ_0 is the dielectric susceptibility, ψ_0 and P_0 are the equilibrium values of the crystal order parameter and induced macroscopic polarization in the fibers, Γ_1 and Γ_2 are the kinetic coefficients, ε_1 and ε_2 are both the sum of electronic and ionic permittivities, ε_r and ε_m are the real and imaginary part of the dielectric permittivity and a, B, and C are positive Landau constants.

The complex dielectric permittivity can thus be expressed as

$$\varepsilon(\omega, T, x) = \varepsilon_r(\omega, T, x) - i\varepsilon_m(\omega, T, x) \quad (12)$$

The loss angle, a measure of dielectric losses, can be expressed as

$$tan\delta = \frac{\varepsilon_m}{\varepsilon_r} \quad (13)$$

Further, the dielectric permittivity of the water-polyester mixture was calculated based on the Flory-Huggins theory. The permittivity of the water polyester mixture can be expressed according to the following equations:

$$\varepsilon_r^p(\omega, T, \varphi) = \frac{1}{u_1(1+\frac{\omega^2 v_1{}^2}{u_1{}^2})} \tag{14}$$

$$\varepsilon_m^p(\omega, T, \varphi) = \frac{\omega v_1}{u_1^2(1+\frac{\omega^2 v_1{}^2}{u_1{}^2})} \tag{15}$$

where,

$$u_1 = z - \frac{c_1^2}{A_1^2+\frac{w^2}{\Gamma_1^2}} \tag{16}$$

$$v_1 = \frac{1}{\Gamma_2} + \frac{c_1^2}{\Gamma_1[A_1^2+\frac{w^2}{\Gamma_1^2}]} \tag{17}$$

$$z_1 = \frac{1}{2\chi_0} - \frac{1}{2}\gamma\varphi\psi_0 \tag{18}$$

where,

$$a_1 = a(T-(1-T)T^*) - B\psi_0 + \frac{3}{2}C\psi_0^2 \tag{19}$$

$$b_1 = \frac{1}{2\chi_0} - \frac{1}{2}\gamma\varphi\psi_0 \tag{20}$$

$$c_1 = \gamma\varphi P_0 \tag{21}$$

Here, φ and $(1-\varphi)$ are the volume fractions of water and polyester, respectively.

The relaxation time was also calculated and can be expressed as follows [7]

$$\tau_q = \frac{\Gamma}{D(q^2+\xi^{-2})} \tag{22}$$

where Γ is the kinetic coefficient and $\boldsymbol{\xi}$ is the correlation length which can be expressed as:

$$\xi = \sqrt{\frac{D}{a(T-x-T^*)}} \tag{23}$$

3. Modeling and Simulation

The models of dielectric properties of textiles have been aimed at calculating the dielectric permittivity of individual fibers at different conditions [8], calculating the dielectric constant of fibers from the permittivity of air fiber mixture [1]. Alekseeva [9] also developed a theoretical method to determine the capacitance of fibers in different configurations. It has been of great interest to estimate the permittivity of yarns and filaments from the permittivity of fabrics. Bal proposed three mathematical models for predicting the dielectric properties of fabrics [10]. Fabrics were considered to be two-phase mixtures of air and fiber. The first model fabric assumed the fabric to be a uniform homogeneous polymeric film that had the same value of dielectric constant, density, and volume as the fiber. The second model considered the fabric to be a mesh of cylindrical yarns with two sets of yarns lying perpendicular to each other without interlacement. The third model was based on Pierce's model of woven fabric that assumed yarns to be incompressible and flexible and have a circular cross-section. The models were verified with experimental data obtained from the dielectric characterization of high density polyethylene fabrics. The variability between the results for the woven fabrics was found to be least in the case of the third model. The authors concluded that the second model could be used for Printed Circuit Boards that used non-crimped yarns, and the third model could be used to predict the permittivity of monofilaments from the dielectric constant of fabric samples. A computer simulation of a 3D model transmission stripline was carried out by Leśnikowski to obtain the permittivity of yarns from the permittivity of the fabric [11]. The procedure is based on the method proposed by Barry to estimate the permittivity of a flat fabric from s parameters of a stripline. According to the author, the accuracy of the method depends chiefly on how well the geometric structure of the textile material is adopted.

4. Measurement Techniques

Apart from the conventional methods, new methodologies have also been used for the characterization of electromagnetic properties of textiles. Such are well documented and reported by Loss [12] and Leśnikowski [13]. The characterization methods are discussed briefly in the following sub-sections.

4.1. Characterization Methods

4.1.1. Parallel Plate Method

Measurement of permittivity by parallel plate capacitor method essentially involves placing the textile substrate between the plates of a condenser and then measuring the impedance [1]. The capacitance of the structure is measured by using an RLC meter [12, 13]. However, the measurement of the dielectric constant is limited to a maximum frequency of 1 GHz using this method [13]. Figure 5 is a schematic representation of the parallel plate method of evaluation of permittivity.

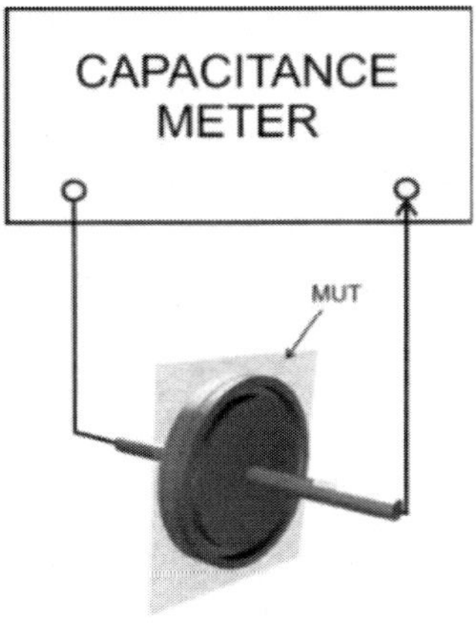

Figure 5. Schematic representation of the parallel plate method of evaluation of permittivity (Reproduced from Leśnikowski [13]).

4.1.2. Free Space Method

The free-space method requires two horn antennas, a sample holder, and Vector Network Analyzer. The dielectric constant is estimated using the S parameters (transmission and reflectance coefficients). Though free-space methods can be used under different environmental conditions, a comparatively larger area of the sample is required for low-frequency measurement, and exact system calibration is required for getting accurate results [12]. Figure 6 is a schematic representation of the free space method of evaluation of permittivity.

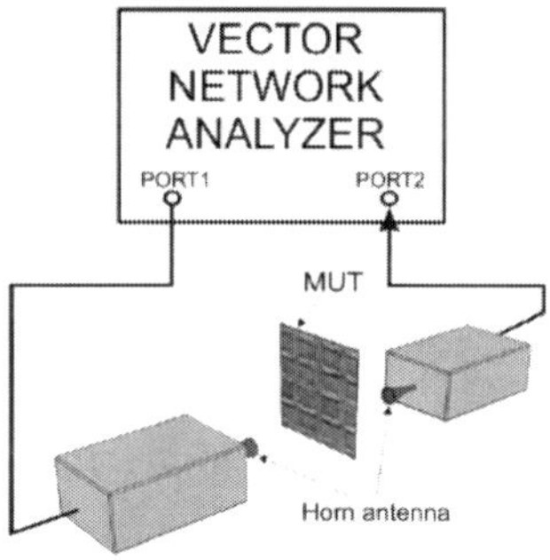

Figure 6. Schematic representation of the free space method of evaluation of permittivity (Reproduced from Leśnikowski [13]).

4.1.3. Planar Transmission Line

Measurements using planar transmission lines are also based on estimating the dielectric constant via S parameters using the Vector Network Analyzer. Though the testing method is rapid, the sample preparation is complex and can result in erroneous and non-reproducible results [12]. Transmission lines using waveguides also have the disadvantage of having a cutoff frequency that decides the minimum value of frequency where dielectric permittivity of the material can be obtained [13].

4.1.4. Cavity Perturbation Method

The resonance frequency varies when a dielectric material is put in the cavity, and the quantitative value of such change in resonance frequency can be attributed to the material's dielectric properties, thereby enabling us to measure the material's permittivity.

4.1.5. Microstrip Patch Sensor

The microstrip patch sensor consists of a patch antenna covered with the dielectric material under test. This technique is based on the resonance-perturbation method, where the permittivity value is calculated knowing the shift in frequency caused by the introduction of the material [12].

4.1.6. Microstrip Ring Resonator

On account of not having open ends, the microstrip ring resonator method decreases the radiation loss, thereby increasing the quality factor, making it more accurate as compared to the straight and circular resonators. The dielectric property of the material under test can be characterized by attaching the conductive parts, made of copper foil, to the textile dielectric probe under test [12, 13].

4.1.7. Microstrip Resonator Patch

This is mainly based on the method proposed by Sankarligam and Bhasker [14]. They designed a patch antenna based on the dielectric values of the materials estimated from the literature review. The final dielectric permittivity value could be calculated by measuring the shift in the resonant frequency of the manufactured antenna. However, manufacturing techniques could introduce errors [12].

4.2. Examples of Permittivity Measurements for Textiles

Textile materials are difficult to deal with because of their structure [15]. Hence standard methods for measuring dielectric properties could not be directly used in the case of textiles. This has called for modifications of condensers for measuring the permittivity of different arrangements of textile materials under different conditions. Also, as pointed out by Bal [15], textile materials are usually three-phase mixtures of fiber, moisture, and air. Permittivity has been found to vary with both the moisture content and the volume fraction of fibers in the mixture. Hence measuring the true permittivity of fibers has been a challenge. One technique that overcame such a problem was the immersion technique used by Errera and Sack [16]. The sample condenser to be used for determining the dielectric constant was filled with a mixture of liquids. The dielectric constant of the mixture was varied until the addition or removal of fibers did not change the capacitance of the system. This would indicate that the permittivity of the fibers and that of the mixed liquid is the same, thereby giving the value of permittivity of the fibers. This approach, though, successfully got rid of the problem of evaluating the dielectric properties of the fibers from the dielectric properties of the air-fiber mixture, which proved to be tedious and difficult to execute. Balls used a resonance circuit for measuring the capacitance of cotton [17]. The condenser used consisted of cotton fibers packed between plates of duralumin. The dielectric constants of cotton in longitudinal and transverse directions were measured. Cotton requires a long time to attain moisture equilibrium. Hence, provision for air circulation through the perforations of the condenser was made to reduce the error raised due to the aforementioned cause. The live plate of the condenser was suspended in between two cotton layers, and the external plates were grounded. The entire condenser could be compressed using a vice, and the maximum packing thus obtained was 50% by volume. Measurements were done at a frequency of 1775 kHz. The dielectric constant could be computed by changing the inter-electrode distance. Shaw and Windle developed a modified cavity resonator for measuring the dielectric properties of polymers in different forms at microwave frequencies [18]. The cavity resonator was used to calculate the dielectric constants of nylon, wool, and cellophane. Hearle used the capacitance bridge method for the measurement of capacitance and power factor of several fibers [19]. The test condensers were made from two aluminum cores. The inner cone had a layer of yarn wound around it with tension, and the outer cone was pressed on it. Nylon, acetate rayon, cotton, wool, viscose rayon, orlon, and others were used for testing. Variations of permittivity of fibers with the frequency, packing density, and moisture content were recorded and plotted. The effect of variation of each factor on the permittivity of the fibers was discussed in detail. Hearle also showed that dielectric permittivity and resistance are correlated, and he addressed the problem of obtaining the dielectric permittivity of fibers only from air–fiber mixtures. Hearle extended his work in a later paper by repeating the experiments at frequencies between 100 kHz and 10 MHz [20]. Kirkwood et al. measured the permittivity and dissipation factor of different varieties of cotton using a Schering-type bridge [21]. Measurements were done at a frequency of 200 kc/sec and at 45%, 55%, and 65% relative humidity. Ishida et al. also used the mutual inductance bridge for obtaining the dielectric constant and loss factor of cellulose fibers [22]. Dielectric properties of viscose, cuprammonium rayon, and cotton slivers were measured. The range of frequency used was from 500 Hz to 3MHz, and the temperature range was -60^{0}C to +20^{0}C. Kumar and Smith used a modified perturbation method to study the dielectric properties of textile materials at 9.8 GHz [23]. At first, the yarn was placed perpendicular to the field in the midplane of the TE_{011} mode of a cylindrical transmission type cavity, and in the second part, the material was taken in the form of an annular ring and placed in the midplane of the cavity. The complex permittivity was calculated from the shift in resonant frequency and change in Q-factor. The principle of operation was provided, and modified perturbation equations for the complex permittivity were calculated. Dielectric permittivity of wet polyester fibers was also measured at other frequencies [24-26]. Kumar also showed that in the case of a polyester-water mixture, Weiner's theory of mixtures could be applied to the experimental observations [24, 26]. Chand used a bridge method and an LCR meter for the dielectric characterization of fibers [27]. Dielectric parameters of polyacrylonitrile and polyester fibers were measured. Cerovic et al. used an L.C.R meter to study the variation of permittivity and dissipation factor of cotton, wool, polyester, hemp, jute, cotton-polyester blend, and flax in the temperature range between -50^0C and +50^0C at a test frequency of 1 MHz [28]. They also measured the variation of permittivity with relative humidity keeping the temperature constant, and the value of permittivity at different frequencies keeping temperature and relative humidity constant. Ray Chaudhuri used LC resonance technique for measuring the dielectric properties of fabric samples [15]. He concluded that degumming of silk and scouring of cotton changed their respective dielectric constants. The dielectric constant of finished cotton was also found to be different, its value lying between that of raw and scoured cotton. A matrix-pencil two-line method was proposed by Declercq et al. for electromagnetic characterization of textile materials [29]. The proposed method got rid of the perturbations in the calculated effective permittivity and loss tangent. Such perturbations are usually caused by imperfect de-embedding and inhomogeneities of the textile microstrip line structure. Measurements were done for two fabrics of aramid and copper blends but of different blend proportions; one nonwoven of polypropylene and the other of fleece. Copper sheet and Flectron® fabric were used as the conductive part to manufacture the transmission lines. Kapilevich et al. presented an angular-invariant approach to the free space method [30]. Kaplevich et al. measured the parameters by varying the angle of the material under test. The authors reported the angular-invariant limit for an incidence angle to be between 25 to 30 degrees.

5. Factors Affecting the Dielectric Property

The dielectric properties differ from fiber to fiber for their different chemical structures and also for the impurities present. Change in frequency for measurement, temperature, packing fraction, and moisture content change the value of permittivity measured. Hence the variation of permittivity with such factors has been studied by various researchers. Such properties of textile materials have also been exploited for several applications. An interesting fact to note is the interdependence of the factors resulting in complex interactions that call for intensive research in areas of demanding applications.

5.1. Frequency

Frequency is the number of times the current changes direction per unit of time. In general, as the frequency used for measurement increases, the permittivity of textile materials decreases. Under the action of an electric field, the dipoles take a specific amount of time to reorient themselves along the direction of the field. This time is known as relaxation time. At lower frequencies, the dipoles of such materials can reverse their direction when the direction of the field is reversed, leading to a higher value of permittivity. As the frequency is gradually increased, at a certain point, the intervals required for reversal of the field become the same as that of the relaxation time of the dipoles of the material concerned. At higher frequencies, the dipoles are unable to react to the changes leading them to not follow the changes at all. Thus at higher frequencies, there is a further decrease in the value of permittivity. These changes in the value of permittivity do not occur in discreet steps. Consequently, there is a range of relaxation times to be found in textile materials [15]. Figure 7 shows the effect of frequency on dielectric permittivity for different textile fibers. Figure 8 shows the effect of frequency on the loss factor for cotton fiber. Figure 7 and Figure 8 are the theoretical and experimental results [6, 20]. Figures 7 and 8 show that dielectric permittivity and loss factor decrease with an increase in frequency. According to Hearle, at frequencies less than 5 kHz, permittivity values increase with the decreasing frequency, which can be due to interfacial polarization [1]. For woven jute fabrics, Ivanovska reported permittivity values to decrease appreciably with increasing frequency in the range of 30 Hz to 3 kHz [31]. After 3 KHz change in permittivity value was not appreciable. Ishida et al. observed a single relaxation peak for cotton in the graph of loss factor against the log of frequency which became narrower with the increase in temperature [22]. For polyester woven fabrics, Cerovic noted negligible change with an increase in frequency which could be because of the low polarity of polyester molecules [32].

At frequencies equal to the relaxation time of the polymer sections, the value of permittivity decreases, and the power factor shows a peak. Hence peaks in the loss factor curves indicate the presence of relaxation times of different sections or groups of the polymer. Cerovic concluded the relaxation in the samples of cotton, cotton-polyester, and polyester fabrics to be secondary relaxations based on the temperature of the maximum value of the loss factor and its corresponding frequency [32]. Relaxations occurring due to side group motions are associated with γ relaxations, while relaxations occurring at comparatively higher temperatures are due to main-chain motions, and are associated with β relaxations.

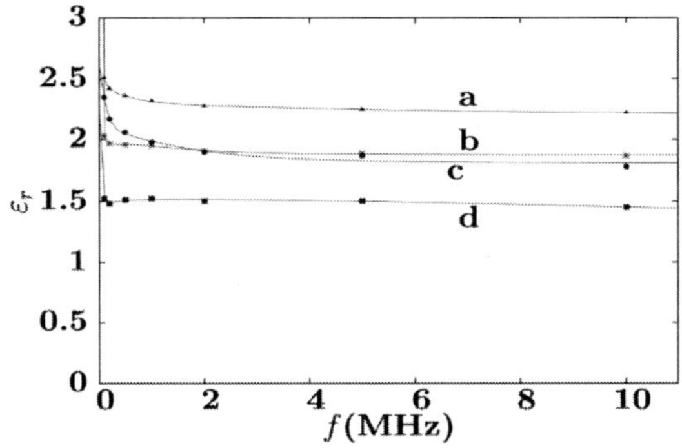

Figure 7. Effect of frequency on dielectric permittivity for a: Nylon; b: Viscose; c: Cotton; d: Wool (Reproduced from Mukherjee [6]).

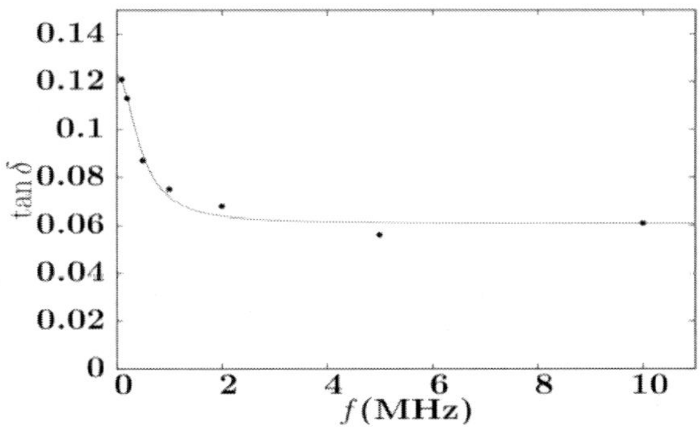

Figure 8. Effect of frequency on loss factor for cotton (Reproduced from Mukherjee [6]).

5.2. Moisture

Textile materials, when left exposed to the atmosphere, absorb moisture. This is mainly because of the hydrophilic groups present in different fibers. The quantity of moisture absorbed by textile materials hence depends on the nature of the fiber, time of exposure, the previous history of the material, and relative humidity. The two terms commonly used to denote the amount of moisture present in a material are moisture regain (M_r) and moisture content (M_c) which can be described below [33]

$$M_r = \frac{weight\ of\ water}{oven\ dry\ weight\ of\ material} \times 100\% \tag{24}$$

$$M_c = \frac{weight\ of\ water}{weight\ of\ water + oven\ dry\ weight\ of\ material} \times 100\% \tag{25}$$

The dielectric constant of water is around 78 when measured at 2.45 GHz and 25°C, and the permittivity of water is essentially more stable and greater than the permittivity of most of the fibers [34]. The addition of water to textile materials is therefore equivalent to the addition of more polar groups into the fiber leading to an increase in permittivity. Also, the addition of water may lead to the generation of polar groups and ions in the fiber itself due to the breakage of bonds, causing a further increase in permittivity. Again, fibers swell due to the absorption of water. This reduces the porosity of the material, thereby increasing permittivity. In addition to that, absorption of moisture leads to the evolution of heat, termed as the heat of wetting. Thus, the absorption of moisture also results in a decrease in temperature, especially in the case of fibers with high moisture regain, thereby affecting the permittivity of the material.

King studied the dielectric constant of keratin for keratin-water mixtures [35]. It was concluded that the increase in permittivity for keratin water systems could not be attributed to the permittivity of the added water only. According to King, bound water also acts as a plasticizer, breaking the crosslinks or the hydrogen bonds between the protein chains, which allows the polar groups of the polymer chains to rotate more freely, thereby increasing the value of permittivity of the system further. Windle and Shaw studied the variation of permittivity of wool between 0 to 28% of the weight of dry wool [36]. Complex permittivity was found to increase with the increase in moisture. They analyzed the result in light of the multilayer absorption theory as proposed by Cassie. Hearle studied and plotted the variation of dielectric constant and loss factor with moisture content for cotton, wool, viscose, nylon, acetate rayon, and orlon. According to Hearle, the non-absorbing fibers like Saran and Dacron showed no variation in permittivity with increasing moisture, while vinyon and fiberglass still showed appreciable change at a low frequency which Hearle attributed to surface effects [19]. Algie measured the dielectric constant of single wool fibers at 1529 Hz and 35^0C [37]. Based on the reported results, he proposed that increase in permittivity to 11% RH is mainly due to the polarization of polypeptide chains that have been relieved of the cross-links and hydrogen bonds. It is followed immediately by a bond reformation process in the presence of water. According to the author, there is an increase in the number of water molecules available in the region of 11% to 80% RH, and beyond 80% RH, there is a formation of an ice-like state where the water molecules provide a conduction network. Cerovic measured the value of dielectric permittivity and loss factor of cotton, jute, polyester, hemp, flax, cotton-polyester blend, and wool at relative humidities of 40%, 60%, 80% at a constant temperature [28]. Cerovic also measured the permittivity at 0%, 35%, 65% RH at 21^0C [32]. The decrease in permittivity values at higher frequencies was more pronounced in the case of higher relative humidity values. Bal proposed a method for measuring the dielectric properties of fibers containing moisture [38]. The role of moisture content was studied using theoretical models and also verified using experimental results. Hertleer et al. also plotted permittivity versus moisture regain for 100% cotton, blend of flax and cotton, blend of cotton and nylon, polyester, and polyurethane [39]. It was concluded that the relative humidity had a greater effect in the case of the natural fiber-based materials, whereas the variation in relative permittivity due to the change in relative humidity for the polyester and polyurethane fabrics was less pronounced. This was because the moisture regains of both polyester and polyurethane are less, and hence the amount of moisture taken up by them with an increase in relative humidity is also less compared to the natural fibers. A quadratic relation was also found between relative humidity and permittivity. Thus fibers with lower moisture regain values were found to be suitable as substrates for textile antennas.

5.3. Temperature

For solid dielectrics, permittivity generally increases with an increase in temperature. This is because, with an increase in temperature, the dipoles become less restrained and hence can more freely align to the direction of the field, increasing permittivity. There is a decrease in dipole-dipole interaction with an increase in temperature, which makes rotation and polarization of the side groups and other flexible portions easier. Figure 9 shows the temperature dependence of permittivity for cotton, and Figure 10 shows the temperature dependence of loss factor for cotton.

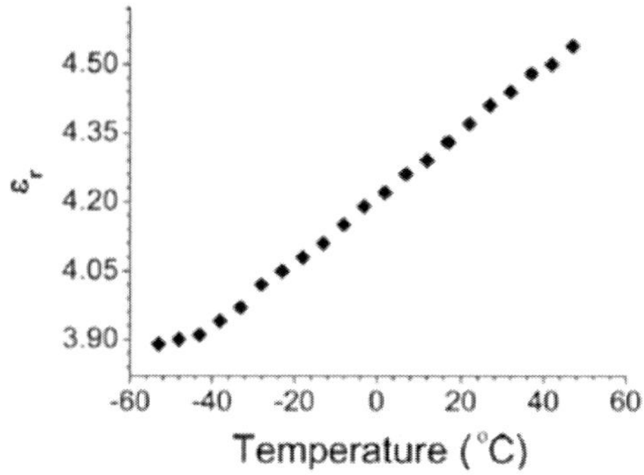

Figure 9. Temperature dependence of permittivity for cotton (Reproduced from Cerovic et al. [28] with the permission of AIP Publishing).

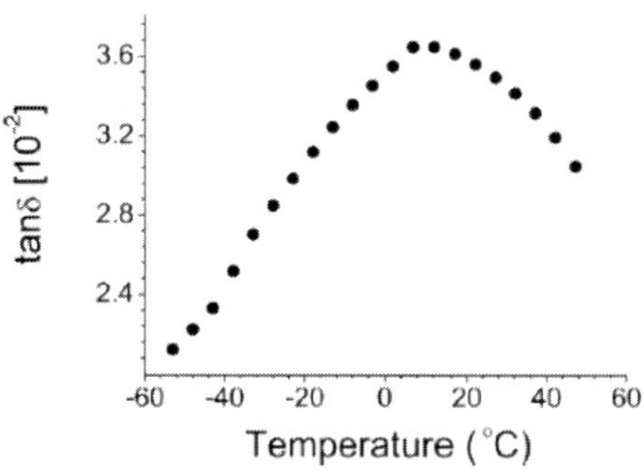

Figure 10. Temperature dependence of loss factor for cotton (Reproduced from Cerovic et al. [28] with the permission of AIP Publishing).

The values of the dielectric constant remain almost constant in the temperature range. Bora reported the permittivity values to remain nearly constant at temperatures 338 K to 490 K for ramie, 360 K to 450 K for jute, and 343 K to 470 K for cotton in the air [40]. Beyond the temperature up to which the values remain constant, degradation of the fibers starts leading to the formation of voids in the polymeric structure, which contribute to the interfacial polarization of the fiber and the gases evolved during decomposition. The dielectric constant reached maximum value at 533 K for ramie, 540 K for jute, and 544 K for cotton.

Chand [27] and Cerovic [28] found relaxation peaks corresponding to β relaxation in the case of measurement of loss factor with increasing temperature. Cerovic reported that maxima peaks of loss factor for hemp, flax, and jute were shifted to lower temperatures compared to that of cotton. According to the authors, such a shift may be because cotton is denser as compared to the others [28].

5.4. Packing Fraction

Textile materials like yarns, fabrics, nonwovens and fiber bundles are porous. Neglecting the amount of moisture present, they are mixtures of fiber and air. Therefore, as the volume fraction of the textile material in the fiber-air mixture increases, the value of permittivity is also found to increase. Theoretical deductions by Boyd [41] and experimental observations by Hearle [20] and Bal [42] relate to the aforementioned statement.

The anisotropy of the fiber structure, natural variations, and variations introduced during several processing steps make it impossible to correctly predict the properties. However, the measured capacitance is a function of the mass of the material present inside or the volume fraction of the textile material present. This makes it possible to predict the properties of the mixture considering the volume fraction of each element. A mathematical approach was taken by Bal to relate volume fraction with effective permittivity [40]. Mixing laws that mainly used fractional volume as a parameter to relate the dielectric permittivity of fibers and that of a two-phase mixture of fiber and air were tested. They concluded that considering only volume fraction may be enough for predicting the dielectric properties. However, as pointed out by the author, the models were applicable in the case of fibers that have low moisture regain.

6. Fabrication

There have been several studies that have concentrated on the technical specifications of the yarn used and the fabric produced. Such studies help develop fabrics with optimum dielectric properties based on the technical properties of the yarn and fabric.

6.1. Effect of Structural Parameters

Warp and weft yarns are two sets of yarns that, when interlaced, form the woven fabrics. According to studies by Mustata et al., the degree of yarn crimping, especially in the weft direction, the surface mass of the fabric, and the technological density in the direction of the warp and the weft, had an appreciable effect on the dielectric permittivity values [43]. For thin weaves with surface mass up to 350 g/m^2, the value of dielectric permittivity was found to be below four, whereas, for the weaves with surface mass higher than 350 g/m^2, the increase in permittivity values was more pronounced. Figure 11 shows the surface mass dependence of permittivity for natural cellulosic fibers.

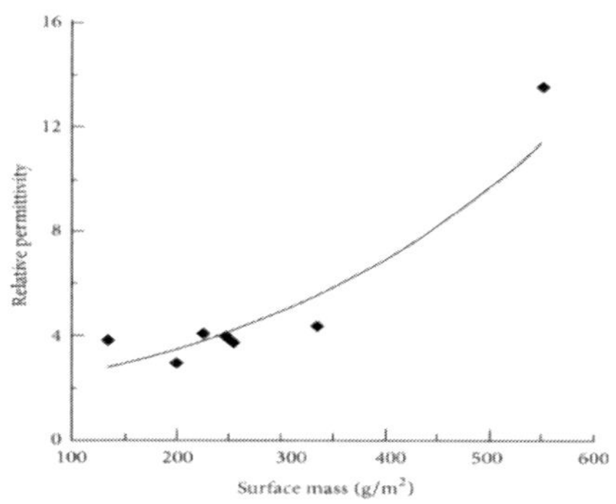

Figure 11. Surface mass dependence of permittivity (Reproduced from Mustata et al. [43]).

Studies by Jiong et al. showed that density and warp yarn count have no significant effect on the permittivity and the dielectric loss tangent [44]. Weft yarn density and count were found to have a nonlinear relationship with the dielectric constants. Fabric with more interweaving points and a compact structure had a larger value of dielectric permittivity and a higher dielectric loss, which can be attributed to higher packing fraction in more compact weaves. Cerovic also noted that samples consisting of yarns made from short fibers had a lower value of dielectric permittivity as compared to samples made of monofilaments and multifilaments [45]. Loss related the dielectric permittivity of textiles with the structural parameters like surface roughness and surface and bulk porosity [46].

Lv found a linear relationship between the real part of permittivity and the bulk density of the fiber aggregation between 100 kHz to 20 MHz [47].

6.2. Effect of Alkali Treatments

Chemical treatments have also been found to affect the value of dielectric permittivity of textile materials. Ivanovska et al. showed the effect of alkali treatment on the dielectric properties of woven jute fabric [31]. Treatments of jute with alkali increased the value of permittivity, which occurred due to structural changes and a decrease in the content of hemicellulose. With decreasing content of hemicellulose, more polar groups are available for water absorption, and hence with increased content of moisture, the permittivity value increases.

6.3. Effect of Blending of Yarns

In the case of blends of polypropylene and viscose nonwovens, the permittivity value increased with an increase in the percentage of viscose fibers [48]. This can be attributed to the higher moisture regain of viscose fibers when compared to polypropylene. Similar results were obtained for composites made of lignocellulosic fibers by Markiewicz et al. and George et al. and for cotton polypropylene composites by Tomlal et al. [48].

6.4. Incorporation of Intrinsically Conducting Polymers

Intrinsically conducting polymers like polypyrrole, polyaniline, polythiophene, etc., are polymers that have conjugated double bonds. This enables them to conduct current. The doping of such polymers further increases their conductivity. Such polymers enable the textile structure to remain flexible yet impart appreciable conductivity to textile materials. Disadvantages include poor mechanical strength and brittleness [49]. The incorporation of intrinsically conducting polymers increases the dielectric constant of textile materials. Liu et al. prepared cotton/polypyrrole composites using ferric chloride as an oxidant and *p*-toluenesulfonic acid as a dopant. The real and imaginary values of permittivity were found to be highest at 0.6 mol/L concentration of pyrrole, and composites prepared using 0.4 mol/L and 1.2

mol/L concentration showed higher loss tangent values compared to the others [50]. For cotton/polypyrrole composites, the dielectric constant was found to be highest while using *p*-toluenesulfonic acid [51]. Studies by Cetiner showed polypyrrole-coated cotton fabrics with ammonium persulphate as oxidant and dodecylbenzenesulfonic acid sodium salt as a dopant to show appreciable dielectric properties [52]. The dielectric constant increased with the increase in coating thickness. Reaction time, temperature, and aniline concentration during polymerization were found to greatly influence the dielectric properties of polyaniline/nylon composites [53]. Polyaniline coated cellulose fiber/polyvinyl alcohol composites were reported to have a high dielectric constant. Also, it was possible to vary the permittivity and loss tangent by varying the size of fibers used [54].

However, the stability of the deposited conductive layer on textile structures is found to be low, thereby leading to poor abrasion resistance and wash fastness. Approaches to obtain better fastness properties include treating the substrate like cotton using NaOH/urea solution, using TiO_2 nanoparticles [55], and plasma treatment [56].

6.5. Incorporation of Carbon

Carbon materials like carbon nanotubes, carbon black, graphene, carbon fibers, and nanofibers have gained interest in forming composites with textile materials for multifarious applications.

In the case of carbon fiber filled low density polyethylene composites, it was reported that the dielectric constant increased with an increase in carbon fiber contents gradually to a volume fraction of 0.2, after which the change in the dielectric constant was abrupt [57]. The rise is because of the additional polarization of carbon fiber/low density polyethylene interfaces. Similarly, for carbon fiber/ epoxy composites dielectric constant increased with an increase in carbon fiber content. This could be attributed to an increase in orientation polarization and the presence of interfacial polarization [58]

The presence of carbon nanotubes has also been found to increase the dielectric permittivity of textile materials. The change in permittivity is dependent on the aspect ratio of carbon nanotubes, length, doping, and content [59]. Acrylic resin and activated carbon composites also show increased values of dielectric permittivity [60]. Carbon fabrics coated with polyaniline showed better thermal stability with an increase in the value of real permittivity with an increase in polyaniline content [61]. However, the imaginary permittivity was found to decrease with the increase in polyaniline loading. Polyaniline/carbon fabric composites with 5% weight of polyaniline content were found to show a higher dielectric loss as compared to carbon fibers. Multi-wall carbon nanotube/polypyrrole nanocomposites also showed stable and high dielectric constant of about 44 and low dielectric loss [62]. Ling-LiXu et al. prepared Graphene/cotton composite fabrics employing the dipping and dropping process, which showed appreciable capacitance as well as mechanical flexibility [63]. Wang et al. produced short carbon fiber, and aramid fiber blended nonwoven and reported that nonwovens with finer carbon fibers show better dielectric properties [64]. NiAl-layered double hydroxide nanoflakes on cotton derived carbon microfiber produced using the hydrothermal method also showed excellent capacitative performance [65].

7. Applications

7.1. Textile Testing Related Applications

Applications related to textile testing include unevenness and moisture testing. Testing of the amount of moisture present in textiles and also the moisture management properties of textiles can be done employing testing methodologies based on capacitance. Studies by Bailey and Philips [66], Spencer and Smith [67], Peeters [68], Côté et al. [69], and Lv [70] were aimed at using capacitance methods successfully for predicting moisture content. Bailey and Phelps used a Burton-Pitt apparatus to measure the permittivity of cotton, wool, rayon, linen, and silk at different moisture contents [66]. However, to properly execute the method, humidity, temperature, history of the material, and also the distribution of the material between the electrodes would need to be precisely controlled. The method invented by Spencer-Smith involved measuring the capacitance of a fabric covering a specific area and measuring its weight. The measured value would then indicate the increase in capacitance value for a particular weight, and from a calibrated graph, the moisture present could be obtained. The method was verified against values obtained from fibers like linen, cotton, etc., at different stages of processing. The method designed by Peeters required the measurement of capacitance at low and high frequencies. The output signals were to be plotted against the regains. After approximation by a set of equations, finally, the quantity of moisture could be obtained. Dielectric properties have also been used to detect moisture transport [71], and breakage of yarns in high-speed winding [72].

David et al. used dielectric response to study the effect of aging on Kevlar® (poly
p-phenylene terephthalamide, PPTA) and PBI (poly benzimidazole) fibers. [73] Major changes could be detected in the dielectric response of the aged samples. They concluded that dielectric spectroscopy could be instrumental in detecting structural changes in polymeric materials.

7.2. Dielectric Heating and Welding

Dielectric properties find applications in dielectric heating and dielectric welding of textiles [74]. Dielectric heating generates heat uniformly throughout the material, negating the ill effects of standard processes of heating. Dielectric heating can be done at microwave and radio frequencies [15]. Dielectric welding employs high frequency alternating frequency at around 27 MHz to generate heat for welding. Polymers like poly(vinyl chloride), poly(ethylene-co-vinyl acetate), and thermoplastic polyurethane can undergo dielectric welding. Nylon can also undergo dielectric welding provided it is preheated to 50^0C [74].

7.3. Textile Based Antennas

Dielectric properties find application in building textile-based antennas. Conductive textiles, as well as the regularly used textiles, have been used for the same. The low dielectric constant of textiles leads to a reduction in surface wave losses. Lowering the dielectric constant increases spatial waves, which in turn also increases the impedance bandwidth of the antenna. The resonance frequency of the antenna also increases [34].

7.4. Electromagnetic Interference Shielding

Electromagnetic waves that are extensively used in the wireless communications sector disturb the usual functioning of electronic devices and even can cause potential harm to human life [75, 76]. Electromagnetic shielding is used to suppress such electromagnetic energy from leaking out of the area that is under the protection of the electromagnetic shield and also to stop electromagnetic waves from an external source from reaching the protected area [2]. Electrically conductive woven or knitted fabrics can be used in applications where such properties are required. The permittivity of a material determines how such electromagnetic waves interact with it. Hence, it is important to manufacture materials with a specific complex permittivity over a suitable frequency range for effective electromagnetic interference shielding [77]. Metal wrapped yarns, metal filled yarns, and plied yarns have been used to produce electromagnetic interference shielding structures [77]. However, such products are comparatively heavy in weight, costly and metals are inherently susceptible to corrosion [76]. Similarly metal coated fabrics used for such purposes also suffer from poor flexibility, wash fastness, and abrasion resistance [75]. Carbon nanotubes are able to form conductive networks in fabrics and have excellent electrical properties and a high aspect ratio. However, their high cost and difficulties in manufacturing limit their applications [77]. Conducting polymers can alternatively be used to produce effective electromagnetic interference shielding structures by impregnating textiles with such polymers. Such conductive polymers have been the most promising materials yet because of their lightweight nature, flexibility, tunable shielding response, and durability [76].

In order to further increase the EMI shielding effectiveness of fabrics, Ti_3SiC_2 [78], $BaTiO_3$, and Fe_3O_4 nanoparticles [76] may be added as fillers. Hoghoghifard et al. studied the effect of aging and washing on the surface resistivity of polyaniline coated polyester fabrics [79]. The influence of efficient doping and redoping procedure on the dielectric properties of such fabrics was also studied. They were found to change the surface resistivity and permittivity and thereby affect the shielding effectiveness of the developed fabrics.

7.5. Sensing

Capacitative strain sensors consist of two textile based electrode layers separated by a dielectric. The capacitance of the sensor is dependent on the geometrical parameters like the area of electrodes and the distance between the two electrodes [80]. Therefore, the capacitance changes by changing the two parameters during stretching. Textile sensors should therefore retain mechanical properties even on large deformation and should withstand cyclic loading and unloading. The sensitivity of the sensor is measured using guage factor, which denotes the change in capacitance during straining as per the equation

$$GF = \frac{\Delta C/C_0}{\varepsilon_s} \tag{26}$$

Here GF represents the guage factor, C_0 is the capacitance at the unstretched state, and ΔC is the difference between C_0 and capacitance at the stretched state. ε_s is the applied strain ratio.

Choi et al. developed a twistable and stretchable strain sensor consisting of a core dielectric rubber layer sandwiched between functionalized buckled carbon nanotube electrodes. The role of the core rubber was to inhibit irreversible changes during mechanical deformations [81]. Cooper et al. fabricated a strain sensor consisting of hollow elastomeric fibers with liquid metals injected into the core of the fibers [82]. The resulting intertwined helical fibers could act as strain, torsion, and touch sensors. Atalay et al. developed a strain sensor consisting of conductive knit fabric as electrode and silicone elastomer as dielectric [83]. Liu et al. developed a strain sensor that consisted of sheath-core conducting fibers. The fibers were produced by wrapping carbon nanotube sheets that were oriented in the fiber direction on stretched rubber fibers [84].

Textile based capacitors have also been used as pressure sensors. Holleczek et al. developed a textile pressure sensor based on the principle of variable capacitance [85]. The electrodes were prepared using conductive textiles coated with silver. Meyer et al. also developed a pressure sensor to measure pressure distribution in the human body for sitting posture classification [86].

Komazaki and Uemura developed a humidity-sensitive Polydimethyle siloxane/calcium chloride microcomposite which consisted of deliquescent calcium chloride encapsulated within microcavities of polydimethyl siloxane. The sensor was developed by screen printing of the microcomposite on textile substrate [87]. Corchia et al. also proposed a chipless humidity sensor based on the variation of permittivity of textiles with changes in moisture [88].

Bahoumina et al. proposed an inkjet-printed capacitative chemical gas sensor based on a carbon composite film that targets ethanol vapour-like volatile organic compounds. When the target species are absorbed, the electric behavior of the material is modified according to it [89].

7.6. Energy Storage

As compared to fuel cells and batteries, capacitors have high power densities but lower energy density which implies that capacitors usually store less energy per unit mass or volume as compared to conventional batteries but discharge faster. Supercapacitors are inherently normal capacitors but with a much larger plate area and less distance between electrodes, thus giving rise to higher capacitance and energy [90]. Unlike conventional energy generators, those to be used in wearable electronics should be flexible, biocompatible, and lightweight. Hence, textile-based supercapacitors have come into prominence. Zhou et al. developed a flexible supercapacitor by using reduced graphene carbonized cotton fabric which exhibited high capacitance and cyclic stability [91]. Sundriyal and Bhattacharya developed a supercapacitor using MnO_2–$NiCo_2O_4$ as a positive electrode, reduced graphene oxide as a negative electrode, and LiCl/Polyvinyl alcohol gel as a solid-state electrolyte over the bamboo fabric [92]. Cheng et al. developed a supercapacitor employing the spray coating method to coat poly (3,4-ethylenedioxythiophene):poly(styrenesulfonate) on dyed polyester fabric [93]. Such supercapacitors, along with thermoelectric or triboelectric or piezoelectric energy generators, can be used as wearable power sources [94].

Conclusion and Future Scope

Textiles are diverse materials capable of application in many challenging areas. They usually have low dielectric constants, and the value of the constant depends on frequency, moisture content, temperature, volume fraction, and other factors. Though many new techniques for measurement of the dielectric property of textiles have come up, accurate characterization of dielectric properties for such porous materials is still difficult and hence calls for new techniques to be applied. With the advent of new applications of such properties for EMI shielding, development of sensors, energy harvesting devices, etc., further investigation of the dielectric properties of textiles, theoretically as well as employing experimental techniques, is essential.

References

[1] Morton, William E., and John W. S. Hearle. 1993. *Physical Properties of Textile Fibers*. Manchester: Textile Inst.

[2] Xiao, Hong, Meiwu Shi, and Jianying Chen. 2020. "Electromagnetic Function Textiles." In *Electromagnetic Materials and Devices*, edited by Man-Gui Han, Intechopen. https://www.intechopen.com/books/electromagnetic-materials-and-devices/electromagnetic-function-textiles.

[3] Kajiwara, K., and Y. Ohta. 2009. "Synthetic Textile Fibers: Structure, Characteristics and Identification." In *Identification of Textile Fibers*, edited by Max M. Houck, 68–87. Cambridge: Woodhead Publishing Limited in association with The Textile Institute.

[4] Ohta, Y., and K. Kajiwara. 2009. "High Performance Fibers: Structure, Characteristics and Identification." In *Identification of Textile Fibers*, edited by Max M. Houck, 88-110. Cambridge: Woodhead Publishing Limited in association with The Textile Institute.

[5] Bal, K., and V. K Kothari. 2012. "Modeling and Simulation of Dielectric Permittivity and Electromagnetic Shielding Efficiency of Fibrous Material" in *Modeling and Simulation in Fibrous Materials* edited by Asis Paatnaik, 183-215. New York: Nova Science Publishers, Inc.

[6] Mukherjee, P. K. 2019. "Dielectric properties of textile materials: a theoretical study." *The Journal of Textile Institute* 110: 211-214.

[7] Mukherjee, Prabir K., and Srijan Das. 2020. "Improved Analysis of the Dielectric Properties of Textile Materials." *The Journal of The Textile Institute*, December, 1–6. https://doi.org/10.1080/00405000.2020.1859781.

[8] Mack, C. 1955. "The Field Due to an Infinite Dielectric Cylinder between Two Parallel Conducting Planes." *British Journal of Applied Physics* 6: 59–62.

[9] Alekseeva, L. V. 2007. "Theoretical Aspects of Predicting the Electrostatic Properties of Textile Materials." *Fibre Chemistry* 39:225–26.

[10] Bal, K., and V. K. Kothari. 2009. "Study of Dielectric Behaviour of Woven Fabric Based on Two Phase Models." *Journal of Electrostatics* 67: 751–58.

[11] Leśnikowski, J. 2018. "Prediction of Electric Permittivity of Threads in Woven Fabric." *Autex Research Journal* 18: 80–85.

[12] Loss, Caroline. 2017. "Study of the Electromagnetic Properties of Textiles: Development of Textile Antennas for Energy Harvesting" PhD diss., Universidade Da Beira Interior.

[13] Leśnikowski, J. 2012. "Dielectric permittivity measurement methods of textile substrate of textile transmission lines." *Przegląd Elektrotechniczny 88: 148–151*

[14] Sankaralingam, S., and G. Bhaskar. 2010. "Determination of Dielectric Constant of Fabric Materials and Their Use as Substrates for Design and Development of Antennas for Wearable Applications," *IEEE Transactions on Instrum. Meas* 59:3122–3130.

[15] Bal, K. and V. K. Kothari. 2009. "Measurement of dielectric properties of Textile materials and their applications." *Indian Journal of Fibre and Textile Research* 34:191–199.

[16] Errera, Jacques, and Henri S. Sack. S. 1943. "Dielectric Properties of Animal Fibers." *Industrial & Engineering Chemistry* 35:712–16.

[17] Balls, W. Lawrence. 1946. "Dielectric Properties of Raw Cotton." *Nature* 158: 9–11.

[18] Shaw, T. M., and J. J. Windle. 1950. "Microwave Techniques for the Measurement of the Dielectric Constant of Fibers and Films of High Polymers." *Journal of Applied Physics* 21: 956–61.

[19] Hearle, J. W. S. 1954. "Capacity, Dielectric Constant, and Power Factor of Fiber Assemblies." *Textile Research Journal* 24:307–21.

[20] Hearle, J. W. S. 1956. "The Dielectric Properties of Fiber Assemblies." *Textile Research Journal* 26:108–11.

[21] Kirkwood, C. E., N. S. J. Kendrick, and H. M. Brown. 1954. "Measurement of dielectric constant and dissipation factor of raw cottons." *Textile Research Journal* 24:841–847.

[22] Ishida, Yoichi, Masatsugu Yōshino, Motowo Takayanagi, and Fujio Irie. 1959. "Dielectric Studies on Cellulose Fibers." *Journal of Applied Polymer Science* 1: 227–35.

[23] Kumar, A., and D. G. Smith. 1977. "Microwave Properties of Yarns and Textiles Using a Resonant Microwave Cavity." *IEEE Transactions on Instrumentation and Measurement* 26:95–98.

[24] Kumar, A. 1978. "The Dielectric Properties of Wet Polyester Fibre at 9.8 GHz." *The Journal of The Textile Institute* 69:116–17.

[25] Kumar, A. 1982. "The Determination of the Moisture Content of Wet Polyester-Fibre Fabric at 10 GHz." *The Journal of The Textile Institute* 73:287–89.

[26] Kumar, A. 1980. "The Complex Permittivity and Attenuation of Wet Polyester Fibre at 2.45 GHz." *The Journal of The Textile Institute* 71: 210–12.

[27] Chand, N. 1979. "Investigations on the Structure of Polymers and Fibers through Dielectric Relaxations" PhD diss., Indian Institute of Technology, Delhi.

[28] Cerovic, Dragana D., Jablan R. Dojcilovic, Koviljka A. Asanovic, and Tatjana A. Mihajlidi. 2009. "Dielectric Investigation of Some Woven Fabrics." *Journal of Applied Physics* 106: 084101.

[29] Declercq, F., H. Rogie, and C. Hertleer. 2008. "Permittivity and Loss Tangent Characterization for Garment Antennas Based on a New Matirx-Pencil Two-Line Method." *IEEE Transactions Antennas Propag* 56:2548–2554.

[30] Kapilevich, B., B. Litvak, M. Anisimov, D. Hardon, and Y. Pinhasi. 2012. "Complex Permittivity Measurements of Textiles and Leather in a Free Space: An Angular-Invariant Approach," *International Journal of Microwave Science and Technolology:* 1–7

[31] Ivanovska, A., D. Cerovic, S. Maletic, I. Jankovic Castvan, K. Asanovic, and M. Kostic. 2019. "Influence of the Alkali Treatment on the Sorption and Dielectric Properties of Woven Jute Fabric." *Cellulose* 26: 5133–46.

[32] Cerovic, D. D., Koviljka, A. Asanovic, Slavica B. Maletic, and, Jablan R. Dojcilovic. 2013. "Comparative Study of the Electrical and Structural Properties of Woven Fabrics." *Composites Part B: Engineering* 49: 65–70.

[33] Booth, J. E. *Principles of Textile Testing: an Introduction to Physical Methods of Testing Textile Fibres, Yarns and Fabrics*. London: Butterworths, 1986.
[34] Salvado, Rita., Caroline Loss, Ricardo Gonçalves, and Pedro Pinho. 2012. "Textile Materials for the Design of Wearable Antennas: A Survey." *Sensors* 12: 15841–57.
[35] King, G. 1947. "Electric Polarisation in Keratin-Water and Keratin-Methyl Alcohol Systems." *Transactions of the Faraday Society* 43: 601.
[36] Windle, J. J., and T. M. Shaw. 1956. "Dielectric Properties of Wool-Water Systems. II. 26 000 Megacycles." *The Journal of Chemical Physics* 25: 435–39.
[37] Algie, J. E. 1964. "Dielectric Constant and Conductance Changes in Wool Fibers Produced by Step Changes in the Relative Humidity." *Textile Research Journal* 34: 477–86.
[38] Bal, K., and V. K. Kothari. 2014. "Dielectric Behaviour of Polyamide Monofilament Fibers Containing Moisture as Measured in Woven Form." *Fibers and Polymers* 15: 1745–51.
[39] Hertleer, Carla, Annelien Van Laere, Hendrik Rogier, and Lieva Van Langenhove. 2009. "Influence of Relative Humidity on Textile Antenna Performance." *Textile Research Journal* 80: 177–83.
[40] Bora, M. N., G. C. Baruah, and C. I. Talukdar. 1993. "Studies on the Dielectric Properties of Some Natural (Plant) and Synthetic Fibres in Audio Frequency Range and Their DC Conductivity at Elevated Temperature." *Thermochimica Acta* 218: 435–43.
[41] Boyd, G. N. 1949. "26—An Electronic Instrument For Measuring Weight Variations In Slivers, Rovings and Yarns." *Journal of the Textile Institute Transactions* 40: T407-T423.
[42] Bal, K., and V. K. Kothari. 2010. "Permittivity of Woven Fabrics: A Comparison of Dielectric Formulas for Air-Fiber Mixture." *IEEE Transactions on Dielectrics and Electrical Insulation* 17: 881–89.
[43] Mustata, Florin St. C., and A. Mustata. 2014. "Dielectric Behaviour of Some Woven Fabrics on the Basis of Natural Cellulosic Fibers." *Advances in Materials Science and Engineering* 2014: 1–8.
[44] Jiyong, Hu, Jiang Hongyan, Tu Huating, Bai Huan, Hong Hong, Zhang Yaya, and Yang Xudong. 2017. "Influence of Woven Fabric Specification and Yarn Constitutions on the Dielectric Properties at Ultrahigh Frequency." *Materials Research Express* 4: 116308.
[45] Cerovic, Dragana D., Jablan R. Dojcilovic., and Slavica B. Maletic. 2012. "Dielectric Response of Fibrous Polyethyleneterephtalate." *European Polymer Journal* 48: 850–56.
[46] Loss, Caroline, Ricardo Gonçalves, Pedro Pinho, and Rita Salvado. 2018. "Influence of Some Structural Parameters on the Dielectric Behavior of Materials for Textile Antennas." *Textile Research Journal* 89: 1131–43.
[47] Lv, Han Ming, and Chong Qi Ma. 2013. "Experimental Study on the Dielectric Spectrum of Cotton Fiber Aggregation." *Advanced Materials Research* 821-822:1475–78.
[48] Asanovic, Koviljka A., Dragana D. Cerovic, Mirjana M. Kostic, Slavica B. Maletic, and Ana D. Kramar. 2018. "Multipurpose Nonwoven Viscose/Polypropylene

Fabrics: Effect of Fabric Characteristics on Sorption and Dielectric Properties." *Journal of Polymer Science Part B: Polymer Physics* 56: 947–57.

[49] Maiti, Syamal, Dipayan Das, and Kushal Sen. 2017. "Flexible Non-Metallic Electro-Conductive Textiles." *Textile Progress* 49 (1): 1–52. https://doi.org/10.1080/00405167.2017.1278875.

[50] Liu, Yuanjun, Yuanchen Liu, and Xiaoming Zhao. 2016. "The Influence of Pyrrole Concentration on the Dielectric Properties of Polypyrrole Composite Material." *The Journal of The Textile Institute*, 1–4. https://doi.org/10.1080/00405000.2016.1238129.

[51] Liu, Yuanjun, Yuanchen Liu, and Xiaoming Zhao. 2016. "The Influence of Dopant on the Dielectric Properties of Flexible Polypyrrole Composites." *The Journal of The Textile Institute*, 1–5. https://doi.org/10.1080/00405000.2016.1240145.

[52] Cetiner, Suat. 2014. "Dielectric and Morphological Studies of Nanostructured Polypyrrole-Coated Cotton Fabrics." *Textile Research Journal* 84, no(14): 1463–75. https://doi.org/10.1177/0040517514523180.

[53] Liu, Yuanjun, Jiarui Sun, and Xiaoming Zhao. 2020. "The Study on the Dielectric Properties and Conductivity of Polyaniline/Nylon Composites." *The Journal of The Textile Institute* 111(11):1697–1704. https://doi.org/10.1080/00405000.2020.1743478.

[54] Anju, V. P., and Sunil K. Narayanankutty. 2016. "Polyaniline Coated Cellulose Fiber/Polyvinyl Alcohol Composites with High Dielectric Permittivity and Low Percolation Threshold." *AIP Advances* 6(1): 015109. https://doi.org/10.1063/1.4940664.

[55] Radoičić, Marija, Gordana Ćirić-Marjanović, Dejan Miličević, Edin Suljovrujić, Milica Milošević, Jadranka Kuljanin Jakovljević, and Zoran Šaponjić. 2020. "Fine-Tuning of Conductive and Dielectric Properties of Polypyrrole/TiO_2 Nanocomposite-Coated Polyamide Fabric." *Composite Interfaces*, 2020, 1–14. https://doi.org/10.1080/09276440.2020.1805219.

[56] Deogaonkar, Smita C. 2020. "Dielectric Barrier Discharge Plasma Induced Surface Modification of Polyester/Cotton Blend Fabrics to Improve Polypyrrole Coating Adhesion and Conductivity." *The Journal of The Textile Institute* 111(10):1530–37. https://doi.org/10.1080/00405000.2019.1710905.

[57] Dang, Zhimin, Yang Shen, Lizhen Fan, Ning Cai, Cewen Nan, and Shujin Zhao. 2003. "Dielectric Properties of Carbon Fiber Filled Low-Density Polyethylene." *Journal of Applied Physics* 93(9): 5543–45. https://doi.org/10.1063/1.1562740.

[58] Elimat, Z. M., M. S. Hamideen, K. I. Schulte, H. Wittich, A. de la Vega, M. Wichmann, and S. Buschhorn. 2010. "Dielectric Properties of Epoxy/Short Carbon Fiber Composites." *Journal of Materials Science* 45(19): 5196–5203. https://doi.org/10.1007/s10853-010-4557-6.

[59] Khan, Waseem, Rahul Sharma, and Parveen Saini. 2016. "Carbon Nanotube-Based Polymer Composites: Synthesis, Properties and Applications." *Carbon Nanotubes - Current Progress of their Polymer Composites*. https://doi.org/10.5772/62497.

[60] Khan, Sharief ud, Manju Arora, M. A. Wahab, and Parveen Saini. 2014. "Permittivity and Electromagnetic Interference Shielding Investigations of

Activated Charcoal Loaded Acrylic Coating Compositions." *Journal of Polymers* 2014 :1–7. https://doi.org/10.1155/2014/193058.

[61] Qiu, Bin, Jiang Guo, Yiran Wang, Xin Wei, Qiang Wang, Dezhi Sun, Mojammel A. Khan, et al. 2015. "Dielectric Properties and Magnetoresistance Behavior of Polyaniline Coated Carbon Fabrics." *Journal of Materials Chemistry C* 3(16): 3989–98. https://doi.org/10.1039/c4tc02578d.

[62] Yang, Cheng, Yuanhua Lin, and C. W. Nan. 2009. "Modified Carbon Nanotube Composites with High Dielectric Constant, Low Dielectric Loss and Large Energy Density." *Carbon* 47(4): 1096–1101. https://doi.org/10.1016/j.carbon.2008.12.037.

[63] Xu, Ling-Li, Mei-Xia Guo, Si Liu, and Shao-Wei Bian. 2015. "Graphene/Cotton Composite Fabrics as Flexible Electrode Materials for Electrochemical Capacitors." *RSC Advances* 5(32) : 25244–49. https://doi.org/10.1039/c4ra16063k.

[64] Wang, Hao, Jin Long, Yi Wang, Yun Liang, Jian Hu, and Haiyang Jiang. 2018. "The Influence of Carbon Fiber Diameter and Content on the Dielectric Properties of Wet-Laid Nonwoven Fabric." *Textile Research Journal* 89(13) 2542–52. https://doi.org/10.1177/0040517518797348.

[65] Xu, Jing, Xiaochen Li, Xuejin Li, Shuo Li, Lianming Zhao, Dandan Wang, Wei Xing, and Zifeng Yan. 2019. "Free-Standing Cotton-Derived Carbon Microfiber@Nickel-Aluminum Layered Double Hydroxides Composite and Its Excellent Capacitive Performance." *Journal of Alloys and Compounds* 787: 27–35. https://doi.org/10.1016/j.jallcom.2019.01.270.

[66] Bailey, Barbara, and Thel L. Phelps. 1939. "The Capacitance of Textile Materials in Relation to Moisture Content a." *Textile Research* 9: 101–13.

[67] Spencer-Smith, J. L. 1935. "27—An Electrical Method For Measuring The Moisture Contents Of Fabrics." *Journal of the Textile Institute Transactions* 26: T336-T340.

[68] Peeters, M. 1969. "Independent Evaluation of the Dry Mass and the Moisture Content of Running Textile Yarns with a Capacity-Type Meter." *Textile Research Journal* 39: 1081–88.

[69] Côté, Benoît, Christine Forget, Normand Thérien, and Arthur D. Broadbent. 1991. "Measuring the Water Content of a Textile Fabric Using a Radio Frequency Sensor." *Textile Research Journal* 61: 724–28.

[70] Lv, Han Ming, Li Ma, Chong Qi Ma, and Hong Jun Fu. 2014. "Estimation of the Moisture Regain of Cotton Fiber Using the Dielectric Spectrum." *Textile Research Journal* 84: 2056–64.

[71] Ito, Hiraku, and, Yoichiro Muraoka. 1993. "Water Transport Along Textile Fibers as Measured by an Electrical Capacitance Technique." *Textile Research Journal* 63: 414–20.

[72] Slater, K., Frebis Hoffmeyer, and D. A. Hood. 1982. "14—Capacitive Detection Of Partial Breaks In High-Speed Winding." *The Journal of The Textile Institute* 73: 130–34.

[73] David, E., C. Arrieta, P. Dolez, T. Vu-Khanh, H. Couderc and M. F. Fréchette. 2010. "Dielectric properties of high performance fibers." 2010 Paper presented at *Annual Report Conference on Electrical Insulation and Dielectic Phenomena*, West Lafayette, IN, October 17-20.

[74] Jones, Ian, and George Stylios. 2013. *Joining Textiles: Principles and Applications*. Oxford: Woodhead Publishing Limited in association with the Textile Institute.

[75] Gahlout, Pragati, and Veena Choudhary. 2019. "Microwave Shielding Behaviour of Polypyrrole Impregnated Fabrics." *Composites Part B: Engineering* 175 (October): 107093. https://doi.org/10.1016/j.compositesb.2019.107093.

[76] Saini, Parveen, Veena Choudhary, N. Vijayan, and R. K. Kotnala. 2012. "Improved Electromagnetic Interference Shielding Response of Poly(Aniline)-Coated Fabrics Containing Dielectric and Magnetic Nanoparticles." *The Journal of Physical Chemistry C* 116 (24): 13403–12. https://doi.org/10.1021/jp302131w.

[77] Rubežienė, Vitalija, and Sandra Varnaitė-Žuravliova. 2020. "EMI Shielding Textile Materials." In *Materials for Potential EMI Shielding Applications*, 357–78. Elsevier. https://doi.org/10.1016/B978-0-12-817590-3.00022-1

[78] Shi, Suilin, Lingzhen Zhang, and Junshou Li. 2009. "Complex Permittivity and Electromagnetic Interference Shielding Properties of Ti_3SiC_2/Polyaniline Composites." *Journal of Materials Science* 44(4): 945–48. https://doi.org/10.1007/s10853-008-3207-8.

[79] Hoghoghifard, Sedigheh, Hossein Mokhtari, and Sajjad Dehghani. 2016. "Improving EMI Shielding Effectiveness and Dielectric Properties of Polyaniline-Coated Polyester Fabric by Effective Doping and Redoping Procedures." *Journal of Industrial Textiles* 47: 587–601.

[80] Seyedin, Shayan, Peng Zhang, Maryam Naebe, Si Qin, Jun Chen, Xungai Wang, and Joselito M. Razal. 2019. "Textile Strain Sensors: a Review of the Fabrication Technologies, Performance Evaluation and Applications." *Materials Horizons* 6(2): 219–49. https://doi.org/10.1039/c8mh01062e.

[81] Choi, Changsoon, Jae Myeong Lee, Shi Hyeong Kim, Seon Jeong Kim, Jiangtao Di, and Ray H. Baughman. 2016. "Twistable and Stretchable Sandwich Structured Fiber for Wearable Sensors and Supercapacitors." *Nano Letters* 16(12): 7677–84. https://doi.org/10.1021/acs.nanolett.6b03739.

[82] Cooper, Christopher B., Kuralamudhan Arutselvan, Ying Liu, Daniel Armstrong, Yiliang Lin, Mohammad Rashed Khan, Jan Genzer, and Michael D. Dickey. 2017. "Stretchable Capacitive Sensors of Torsion, Strain, and Touch Using Double Helix Liquid Metal Fibers." *Advanced Functional Materials* 27(20): 1605630. https://doi.org/10.1002/adfm.201605630.

[83] Atalay, Asli, Vanessa Sanchez, Ozgur Atalay, Daniel M. Vogt, Florian Haufe, Robert J. Wood, and Conor J. Walsh. 2017. "Batch Fabrication of Customizable Silicone-Textile Composite Capacitive Strain Sensors for Human Motion Tracking." *Advanced Materials Technologies* 2(9): 1700136. https://doi.org/10.1002/admt.201700136.

[84] Liu, Z. F., S. Fang, F. A. Moura, J. N. Ding, N. Jiang, J. Di, M. Zhang, et al. (2015). "Hierarchically Buckled Sheath-Core Fibers for Superelastic Electronics, Sensors, and Muscles." *Science* 349(6246): 400–404. https://doi.org/10.1126/science.aaa7952.

[85] Holleczek, Thomas, Alex Ru, Holger Harms, and Gerard Tro. 2010. "Textile Pressure Sensors for Sports Applications." Paper presented at SENSORS, 2010 IEEE, Kona, HI, November 1–4.

[86] Meyer, Jan, Bert Arnrich, Johannes Schumm, and Gerhard Troster. 2010. "Design and Modeling of a Textile Pressure Sensor for Sitting Posture Classification." *IEEE Sensors Journal* 10: 1391–98.

[87] Komazaki, Yusuke, and Sei Uemura. 2019. "Stretchable, Printable, and Tunable PDMS-$CaCl_2$ Microcomposite for Capacitive Humidity Sensors on Textiles." *Sensors and Actuators B: Chemical* 297: 126711. https://doi.org/10.1016/j.snb.2019.126711.

[88] Corchia, Laura, Giuseppina Monti, Egidio De Benedetto, and Luciano Tarricone. 2019. "A Chipless Humidity Sensor for Wearable Applications." *2019 IEEE International Conference on RFID Technology and Applications (RFID-TA).* https://doi.org/10.1109/rfid-ta.2019.8892048.

[89] Bahoumina, Prince, Hamida Hallil, Jean-Luc Lachaud, Aymen Abdelghani, Kamel Frigui, Stephane Bila, Dominique Baillargeat, et al. 2017. "Chemical Gas Sensor Based on a Flexible Capacitive Microwave Transducer Associated with a Sensitive Carbon Composite Polymer Film." *Proceedings* 1(4): 439. https://doi.org/10.3390/proceedings1040439.

[90] Abu-Thabit N. Y., A. S. H Makhlouf. 2016. "Smart Textile Supercapacitors Coated with Conducting Polymers for Energy Storage Applications." In: Hosseini M., Makhlouf A. (eds) *Industrial Applications for Intelligent Polymers and Coatings.* Springer, Cham. https://doi.org/10.1007/978-3-319-26893-4_21.

[91] Zhou, Qianlong, Xingke Ye, Zhongquan Wan, and Chunyang Jia. 2015. "A Three-Dimensional Flexible Supercapacitor with Enhanced Performance Based on Lightweight, Conductive Graphene-Cotton Fabric Electrode." *Journal of Power Sources* 296: 186–96. https://doi.org/10.1016/j.jpowsour.2015.07.012.

[92] Sundriyal, Poonam, and Shantanu Bhattacharya. 2020. "Textile-Based Supercapacitors for Flexible and Wearable Electronic Applications." *Scientific Reports* 10(1). https://doi.org/10.1038/s41598-020-70182-z.

[93] Cheng, Qian, Chen Meng, Yong Qian, Jinxin He, and Xia Dong. 2020. "Energy Capacity Enhancement of All-Organic Fabric Supercapacitors by Organic Dyes: Old Method for New Application." *Progress in Organic Coatings* 138: 105439. https://doi.org/10.1016/j.porgcoat.2019.105439.

[94] Lund, Anja, Karin Rundqvist, Erik Nilsson, Liyang Yu, Bengt Hagström, and Christian Müller 2018. "Energy Harvesting Textiles for a Rainy Day: Woven Piezoelectrics Based on Melt-Spun PVDF Microfibres with a Conducting Core." *npj Flexible Electronics* 2(1) https://doi.org/10.1038/s41528-018-0022-4.

Chapter 3

Permittivity Measurements of Ceramic Materials for Using and Firing in Microwave Furnaces

T. Santos[1,2,*], S. Devesa[3], V. A. F. Costa[2] and L. C. Costa[1]

[1]I3N and Physics Department, University of Aveiro, Aveiro, Portugal
[2]Centre for Mechanical Technology and Automation, Department of Mechanical Engineering, University of Aveiro, Aveiro, Portugal
[3]CFisUC, Physics Department, University of Coimbra, Coimbra, Portugal

Abstract

Microwave firing of ceramic materials, such as porcelain and stoneware, is a promising alternative to conventional firing, usually achieved by natural gas combustion. Microwave heating technology has a considerably lower environmental impact, allowing ceramic materials to be fired in shorter times, at lower temperatures, and therefor enabling higher production rates and lower energy consumptions. Nowadays, microwave radiation can be used as an alternative or as an auxiliary heating technology for porcelain and stoneware firing. This is especially relevant as the greener electricity production from renewable energy sources is considerably increasing.

Microwave radiation absorption by some materials leads to local heat generation, which depends, among other factors, on their dielectric properties. These properties control the interaction of the microwave radiation with matter not only through the absorbed power but also through the penetration depth. Knowledge of these properties, and how they depend on temperature during the different firing stages, is of major

[*] Corresponding Author's Email: tiago.santos@ua.pt.

In: An Introduction to Permittivity
Editor: Parveen Saini
ISBN: 979-8-88697-042-5
© 2022 Nova Science Publishers, Inc.

relevance for a better understanding, control and improvement of microwave processing, being crucial for accurate microwave heating simulations.

This work reports on the measurements of ceramic materials dielectric properties at the different firing stages of porcelain and stoneware. Measurements were extended to a variety of refractory materials, which also interact with microwave radiation and have a relevant role inside the cavities and furnaces where microwave firing of ceramic materials takes place.

Measurements were made using a resonant cavity operating in the $TE_{1,0,5}$ mode, with a resonant frequency of 2.8 GHz.

Keywords: ceramics, microwave heating, complex permittivity, dielectric loss, resonant cavity

1. Introduction

The awareness of the human environmental impact promotes the search for more efficient and green technologies and processes, and more rational use of energy, which production currently represents around 68% of the overall CO_2 emissions [1]. Among the highly energy-consuming industries, and in particular among those highly thermal energy-consuming, are the porcelain and stoneware industries, whose energy consumptions represent up to 30% of the production costs [2].

The firing of ceramic materials is conventionally promoted by fuel combustion, mainly by natural gas combustion [3, 4]. Heating processes based on firewood or coal burning, technologies used in old kilns [5–10], and heating by electrical resistances [11] also fall under the conventional heating technologies [3, 12]. In brick and tile industries, biomass is already used as the most relevant primary energy source [4]. Those energy needs are closely related to the required maximum firing temperature implemented in each industry, whose major differences are related to the contents of the raw materials used in the manufacturing process [13–17]. Stoneware industry requires firing temperatures between 1150°C and 1300°C [16, 18, 19], and porcelain industry requires firing temperatures between 1300°C and 1400°C [16, 17].

While conventional firing technologies are mainly based on convection and radiation heat transfer from the hotter kiln environment to the pieces, and also on conduction heat transfer through the contact areas of the pieces being processed with other solid materials, microwave heating has the benefit of promoting the heat generation inside the pieces themselves. The high potential of microwave radiation for materials processing is now widely recognized at the energy, economic and environmental levels. Microwave heating is usually faster and leads to lower pollutants emissions, with microwave-fired products presenting similar or even better characteristics than the conventionally fired ones [3, 10, 12, 20–22]. Microwave radiation is thus a viable alternative heating technology, used alone or combined with other heating technologies, for porcelain and stoneware firing [23–25].

Electrical energy costs discourage the use of microwave heating technology when compared to the fossil fuels (liquid hydrocarbons, natural gas and coal) costs, being these the most consumed fuels in the European Union [4]. However, increasing electricity production from renewable energy sources and a sustainable market might contribute towards a greener economy by increasingly using microwave heating for the materials processing.

In what concerns the ceramics manufacturing, and in particular stoneware and porcelain manufacturing [26, 27], it starts with the selection of the raw materials, grinding, sifting, washing, filtering and concentration of useful minerals, and the separation of impurities, including magnetic particles. It is followed by the mixing of the required proportions of raw materials, and then by the forming process of the pieces. After pieces forming, these are dried and glazed before being fired at approximately 1200°C in the stoneware case [26]. The porcelain is usually fired twice: a first firing at approximately 1000°C followed by the glass bath, and finally a second firing at approximately 1400°C [27]. It is the second firing that gives to porcelain its characteristic properties [16, 28]. Even if not so common, stoneware can also be produced through a bi-firing process. An example is the stoneware produced by the Royal Copenhagen Manufactory Company [16], which is firstly fired at 1000°C, followed by the glaze application and then subjected to a second firing at 1325°C. Ref. [18], mention the first stoneware firing at (990 ± 30)°C and the second-firing at (1180 ± 40)°C.

1.1. Microwave – Matter Interaction and Heat Generation

The shape, wall thickness, contact area with the standing base, thermal conductivity of the materials to be processed, and uniformity of heat transfer (from and to the material being heated) are of the utmost importance. The drying and firing times also depend on these parameters. Longer processing times are generally associated with pieces with reduced surface areas, high volumes, and poor thermal conductivities [16]. If these are important for conventional firing, they are even more decisive for microwave firing, as microwave heating depends essentially on the materials' intrinsic properties, and not only on the previously mentioned thermophysical properties. Microwave heating depends particularly on the complex permittivity of the material, which is frequency and temperature-dependent [12, 21, 29–32].

The permittivity of the material can be defined as

$$\varepsilon^*(\omega, T) = \varepsilon'(\omega, T) - j\varepsilon''(\omega, T) \qquad (1)$$

where ω is the angular frequency and T the temperature.

The first term is the dielectric constant, $\varepsilon'(\omega, T)$: a property that indicates the material's ability to be polarized, or, in other words, a measure of the amount of energy stored in the material in the form of an electric field. The second term of Eq. (1) is the dielectric loss factor, $\varepsilon''(\omega, T)$, which defines a measure of the capability of the material to convert microwave radiation energy into thermal energy.

Dissociating the part that governs the microwave radiation energy conversion into thermal energy from that related to heat transfer, considering a homogeneous and isotropic dielectric material, and a uniform electric field applied to the material, the power absorbed per unit volume can be expressed as [12, 21, 29–31]

$$P_a = \omega\varepsilon_0\varepsilon'' E_{rms}^2 \qquad (2)$$

where E_{rms} is the effective electric field in the material.

The dielectric properties control how the microwave radiation interacts with the materials, not only through the absorbed power but also through the penetration depth, D_p. This latter property is defined as the distance measured into the material for which the power radiation is e^{-1} (37%) of the surface-incident power radiation [12, 29, 31]

$$D_p = \frac{1}{2\alpha} = \frac{\lambda_0}{2\pi\sqrt{2\varepsilon'}}\left[(1 + (tg\delta)^2)^{1/2} - 1\right]^{-1/2} \qquad (3)$$

where λ_0 is the wavelength of the electromagnetic radiation at frequency f in vacuum.

If $\varepsilon'' \ll \varepsilon'$, D_p can be approximated (with a maximum error of 10%) [31] by

$$D_p \approx \lambda_0 \sqrt{\varepsilon'}/(2\pi\varepsilon'') \quad (4)$$

A balance is required between the absorption potential, Eq. (2), and the ability of the electromagnetic radiation to penetrate the material, Eq. (4). Materials with a loss factor lower than 10^{-2} are essentially transparent to microwave radiation and do not absorb microwave energy. Intermediate losses, between 10^{-2} and 5, lead to microwave radiation absorption and internal heat generation in the material. High frequencies and materials with high losses are essentially opaque to the electromagnetic radiation, and behave as reflective materials, for which only the surface is heated by the incident microwave radiation. This is the case of materials such as metals, with high electrical conductivity and high dielectric loss factors, $\varepsilon'' > 5$ [30, 33], which cannot be volumetrically heated using microwave radiation.

Figure 1 presents the results of a simulation of the electric field inside a microwave furnace cavity with 12 test samples (expresso coffe cups). The walls of the furnace are composed of refractory material that act as a thermal insulation, considered transparent to the microwave radiation (ε' = 1.7 and ε'' = 0.01).

It should be noted that the electric field is not uniform inside the microwave furnace cavity, varying in both space and time [34]; thus, the way the microwave radiation is absorbed by the material is more complex than expressed by Eq. (2), which is an oversimplification of how microwave radiation is absorbed by a material.

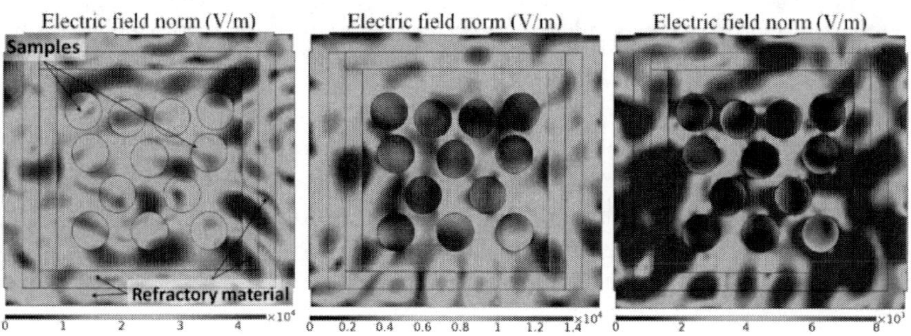

a) ε'' = 0.1 b) ε'' = 5.0 c) ε'' = 20.0

Figure 1. Simulation of the influence of the loss factor, ε', ' on the distribution of the electric field inside a microwave furnace cavity with 12 pieces with ε' = 2.09. (a) ε'' = 0.1; (b) ε'' = 5.0 and (c) ε'' = 20.0. Frequency = 2.46 GHz. (Simulations performed using COMSOL Multiphysics).

The heat transfer in the microwave heated material is governed by the heat conduction equation [36–38] as

$$\rho C_P \frac{\partial T}{\partial t} = \nabla . (k \nabla T) + P_a \tag{5}$$

which includes the volumetric heat generation rate [31, 35–37], where ρ is the density of the material, C_p its specific heat and k is the thermal conductivity.

The microwave heating rate is obtained essentially as [35]

$$\Delta T / \Delta t \approx P_a / \rho C_p \tag{6}$$

Since ceramic materials are usually good thermal insulators, with thermal conductivities lower than 1 W/(m.K) [3], the conduction heat transfer through the material is not effective. Naturally, thermal gradients arise in the material, inducing materials with non-uniform properties, especially if the heating is too fast, which might cause the appearance of cracks and even fracture due to thermally originated mechanical stresses [39, 40]. Temperature uniformity over the volume of the pieces is an essential factor, which is not achievable neither by conventional or even by pure microwave heating.

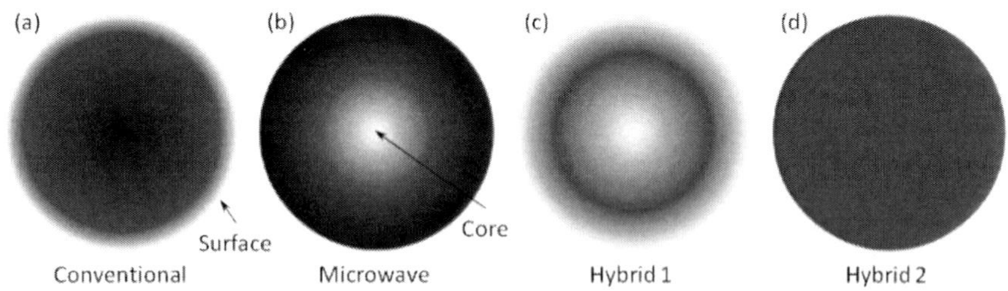

Figure 2. Temperature profiles shematization of a piece heated by (a) Conventional heating, (b) Microwave heating of a material that is not opaque to microwave radiation, (c) and (d) Hybrid (microwave + susceptor) or assisted (conventional + microwave) heating. The illustration in (d) represents the ideal case of a more homogeneous temperature/heating.

Having in mind the heat transfer associated with conventional heating, the inverse temperature profile occurs with microwave heating, that is, the piece is heated from the inside out [40–44]. Applying a hybrid or microwave-assisted heating, more homogeneous heating can be achieved, as schematized in Figure 2, allowing faster, more uniform and less energy-consuming heating processes [40, 41, 43–49].

1.2. Case Studies of Microwave Firing Clay-Based Ceramics

Few researches focus on microwave processing of clay-based ceramic materials such as faience, stoneware and porcelain. Ref. [50] mention the microwave processing of enamelled porcelain using the mono-firing process. The faster microwave-fired samples present properties equal or even better than those of the conventionally fired ones, presenting a more developed glass phase and a better finish, with brighter and smoother surfaces. In [51], when temperatures of 1100°C are required by conventionally heating, microwave fired porcelain pellets show the formation of the glass phase for lower temperatures (between 900°C and 1000°C). According to the authors, microwave heating enhances the porcelain densification at temperatures 75°C lower than those required by conventional firing. The formation of mullite is reinforced by microwave radiation, presenting crystals of smaller sizes and higher length-width ratios when compared to the mullite crystals formed by conventional firing. Similar results are presented in [23, 52, 53], where batches of stoneware mugs and porcelain cups were fired in a microwave multimode furnace.

The work developed by Hamlyn and Bowden [54] is possibly the first that focuses on the microwave firing of more traditional (clay-based) ceramic materials. The test samples are faience tiles with dimensions of 0.5 cm x 5 cm x 5 cm. is the authors' state that above 800°C the temperature in the material increases sharply, highlighting the possibility of microwave firing mechanisms changing for higher temperatures. Individualizing the primary faience raw-constituents, it was found that when kaolin (a clay raw constituent) is microwave heated, its temperature reaches 450°C in about 70 min. Above this temperature, the heating rate is greatly reduced with the formation of a plateau that lasts for about 60 min. These processes are associated with the dehydroxylation of kaolinite, $Al_2Si_2O_5(OH)_4$, the primary mineral in kaolin [53, 54].

According to the literature, the dehydroxylation of kaolinite (kaolinite → metakaolinite transformation) occurs in the temperature range of 450°C – 600°C, depending on the natural disorder state of its structure [28, 55, 56].

Returning to the work developed in Ref. [54], the feldspar heats up slowly up to 950°C in approximately 240 min, reaching temperatures higher than 1200°C in the last 10 min of the heating process. The feldspar presents higher microwave absorption for temperatures above the eutectic temperature of the ternary K_2O-Al_2O_3-SiO_2 system [28, 57–59]. This statement is consistent with that of W.W.Ho, mentioned in Ref. [60], claiming that the temperature dependence of the dielectric loss of polycrystalline ceramics is attributed to the softening of the amorphous glassy phases present at the grain boundaries, which become conductive. The authors in Ref. [54] concluded that earthenware is susceptible to microwaves only at high temperatures, and that the microwave absorption mechanisms are not trivial, varying significantly and being highly dependent on the raw materials.

To some extent, and contrary to what is mentioned in Ref. [54], Reinosa et al. [61], reporte a sharp temperature increases above 650°C, with heating rates up to 500°C/min (unexpected, according to the authors), showing exactly the opposite effect/heating behaviour reported on kaolin microwave firing [54]. The dielectric loss factor, measured up to 1300°C [61], proves this behaviour (this subject will be addressed in Sections 1.3 and 3). By the obtained data analysis [61], results also evidence a firing temperature ↔ dielectric loss response delay; as stated in Ref. [34], where the simulation of a 2D body subject to microwave heating shows a delay between the change on the electromagnetic field pattern and the respective change on the heating behaviour.

The dielectric loss factors tend to slowly increase with temperature, with an abrupt increase above the critical temperature (T_c) [30, 33, 61, 62]. This increase of the microwave energy absorption is associated with the thermal runaway phenomenon [10, 12, 31, 33]. If the electromagnetic field is not homogeneous it leads to abrupt and localized temperature increases, with the formation of hot spots and heightened local thermal gradients in the material. The thermal gradients originate stresses that will cause deformation and increased the risk of cracking.

It should be noted that the parameters and factors that originate this phenomenon are the same that provide fast microwave heating processes, and that enable the use of microwave heating for materials processing [12, 46, 63–65]. These aspects, intrinsic to the nature of microwave radiation interaction with materials, must be controlled to ensure the viability of the usage of this heating technology for materials processing. Additionally, it is noteworthy of mentioning that the frequency dependence means that it is impossible to heat all types of materials using a single frequency

[12, 30, 66, 67]. A deeper knowledge of the materials dielectric properties is essential for a more efficient control of the heating process and also for a better understanding of the differences between microwave and conventionally fired products.

1.3. Dielectric Properties Measurement

Microwave heating depends essentially on the ability of the material to generate heat from the "absorbed" electromagnetic radiation, which is related with its dielectric properties. Measurement of these properties is particularly challenging and expensive, and not easy to achieve for enlarged temperature ranges. Specific instrumentation is required when the measurement equipment is subject to high temperatures, and there are major difficulties related to the nonhomogeneous heating and sample temperature measurement [33, 60, 68, 69]. Measurement techniques are essentially based on reflection or transmission methods [21, 32, 33, 69, 70], divided in four main measurement groups: lumped circuit, transmission line, resonant cavity, and free-space measurement techniques [32, 60, 69, 71–73]. The most suitable measurement technique depends on the frequency, the dielectric properties of the target material, target material type, shape and size, contact or contactless measurement method, and the acceptable accuracy [32, 69, 74]. Examples of the different techniques, and respective advantages and disadvantages, are summarized in Refs. [69] and [73]. The sample preparation is one of the reported issues, as mentioned in Refs. [60] and [74]. For frequencies above 1 GHz, resonant cavities are generally employed [68, 69]. Very small samples can be used when using the resonant cavity method. Quoting Ref. [32] "The cavity perturbation method requires a very small sample such that the fields in the cavity are only slightly disturbed...," which can be more homogeneously heated during shorter time periods. The accurate high-temperature measurement is crucial, however are difficult to achieve [74]. The sample volume is of major relevance for the determination of the dielectric properties, more accurately the product of the sample volume and its dielectric properties. For materials with very small ε'' and ε' close to 1, relatively large sized samples are required to carry out the dielectric properties measurement. Even so, much smaller than the samples sizes of 10 cm required when the free-space measurement technique is employed [69, 74] for the frequency of 35 GHz [74]. Using a WR340 resonant cavity system, composed of a rotation sample holder, a CO_2 laser and a pyrometer, [68] shows that homogeneous heating is possible up to 1000°C with a temperature difference below 3%. This apparatus allows the measurements of the dielectric constant and dielectric loss factor with uncertainties of 5% and 10%, respectively. One drawback of this method is the need of sample shape corrections [60, 68, 75], which can induce a measurement error up to 20% [68]. This is related to the change of the electric field inside the cavity due to the insertion of the sample, which depends on the involved volumes [60, 68], and particularly on the sample volume changes during heating. During the samples heating and measurement, the cavity walls are heated and the volume of the cavity also changes, resulting on a resonant frequency shift. Although this frequency shift is small, it can be of the same order of that associated to the sample introduction into the cavity [68]. One way to minimize these errors and uncertainties is to use a cooling apparatus coupled to the cavity [60].

In Ref. [76] a single frequency high-temperature microwave dielectrometer is presented, by which the material is heated and its dielectric properties simultaneously measured. Zeolite and mullite complex permittivity measurements at 915 MHz, from 590°C up to 780°C and from 800°C up to 1200°C, respectively, are presented. In Ref. [60] the authors report the complex permittivity measurement of alumina and zirconia up to 1200°C using two different techniques: a cylindrical resonant cavity working in the $TM_{0,n,0}$ mode, and an open ended coaxial probe. Data measured at frequencies of 0.61 GHz and 2.19 GHz, using both technologies, present a good agreement, with an overall error of 8%, associated with the measurements obtained by the coaxial probe technique, and an overall error of 5%, associated with the cavity technique. One of the advantages of the coaxial probe technique is the continuous frequency measurement, from 0.2 GHz up to 4.0 GHz, in this particular case. The resonant cavity technique allows accurate measurements and simple operation advantages over the other techniques [72, 77]. In Ref. [78] a heat-resistant metal alloy coaxial probe and furnace is used to measure the permittivity of a variety of ceramic samples (titanium boride-based materials, alumina, macor and zirconia) up to 800°C, at 2.45 GHz.

By implementing the standard cavity perturbation technique, in rectangular WR 430 and WR 90 waveguides, in [79] is reported the measurement of the complex permittivity of thermoplastic composites from room temperature up to 170°C, at frequencies of 2.45 GHz and 9.4 GHz. In Ref. [80] the authors present a dual-mode cylindrical cavity, allowing the measurement of the complex permittivity during microwave heating using two different microwave sources. The microwave cavity was designed to operate in the $TE_{1,1,1}$ mode for heating and in the $TM_{0,1,0}$ mode for measuring. The resonant frequencies of the two modes are 2.432 GHz and 2.187 GHz, respectively. The determination of the permittivity is performed by an improved cavity perturbation method, with an error below 5%. Two case studies are presented, namely, a macor sample measured up to 1100°C and a Duran glass sample measured up to 1000°C. Using this technique, the complex permittivity of a ceramic frit at temperatures up to 1200°C has been analysed by [81], and the dielectric loss factor of kaolin fired up to 1300°C has been referred in Ref. [61]. Measurements consider both heating and cooling, showing distinctive curves. During heating it is observed a very sharp dielectric loss increase due to the kaolinite dehydroxylation, starting at ~475°C, up to 583°C, followed by some structural transformations with smaller changes of the dielectric loss factor, which are related to mullite and the vitreous glass (amorphous SiO_2) formation. Studies reported in Refs. [80, 81] and [61], show an abrupt increase of the materials' permittivity above their critical temperatures (T_c).

Dielectric property measurements for enlarged temperature ranges, with the materials experiencing several transformations during heating, is not an easy task, both in terms of the used measurement techniques and of the required equipment.

In the present case, as the products under study, stoneware and porcelain, undergo several crystallochemical transformations during firing, permittivity measurements must consider these transformations at the several firing stages. Nonetheless, as the resonant cavity and the used system are not equipped for measurements with temperature, only the permittivity measurements at room temperature will be reported for samples corresponding to the several firing stages. As a volume-dependent technique is used, volumes that are not instantaneously measured will induce errors difficult to quantify due to the samples shrinkage during heating. In the near future, an apparatus with an operation principle similar to that reported in Refs. [60] and in [70] will be considered. The resonant cavity walls should be at room temperature during the temperature measurements, and a sample holder and system that minimize the heat transfer to the cavity between measurements must be considered. The introduction of the sample for permittivity measurement, and its temperature measurement and monitoring should be automatic and almost instantaneous to minimize the sample heat losses to the surrounding environment, sample holder and cavity, and avoid thermal gradients in the samples.

2. Experimental Method

Porcelain and stoneware undergo several transformations during the manufacturing process. Thus, permittivity measurements were made taking into account the several stages of stoneware and porcelain manufacturing. In the case of porcelain, measurements were made in wet paste, dried paste, paste fired once at 1020°C (referred to as biscuit) and bi-fired (biscuit fired at 1380°C, referred to as unglazed porcelain). For stoneware, measurements were made in wet, dried, and dried paste fired at 1180°C (referred to as stoneware).

Moreover, several ceramic materials intended to be used as thermal insulation in the microwave furnace were also characterized. Those refractory materials are: Cerachem Blanket [82], Saffil [83], Firebricks JM [84], Rath [85], Kapyrok GR1600 [86], Promasil [87] and Promalight [88], and Silicon Carbide (SiC). The SiC is a well-known microwave absorber, referred to as a microwave susceptor, a material with a high capacity to generate heat from the absorbed microwave radiation. It can be used to enhance the microwave heating process of materials such as porcelain and stoneware as they are relatively transparent to microwave radiation at low temperatures. At the beginning of the heating process, the samples are heated mainly by the heat released from the susceptors. When the samples reach their critical temperature, they start absorbing the microwave radiation more effectively, with inner heat generation.

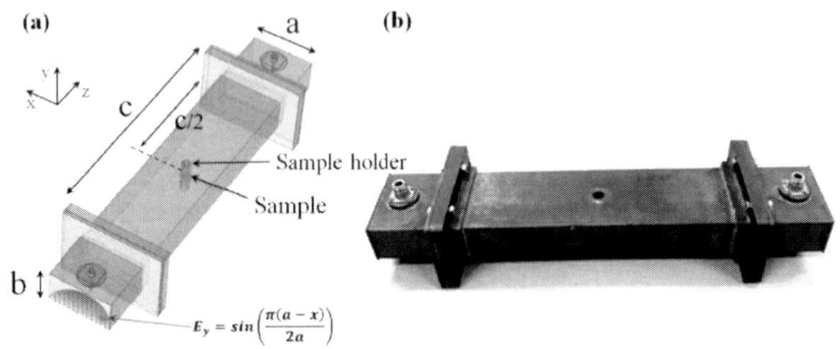

Figure 3. (a) Schematic representation of the cavity; and (b) The 2.8 GHz used cavity, with the hole for the sample at the geometric centre.

The refractory materials, used as thermal insulators, should behave differently from SiC, as they must be essentially transparent to microwave radiation throughout the whole heating process.

The microwave resonator used for the permittivity measurements is a standard rectangular metallic waveguide short-circuited at both ends, as shown in Figure 3. It operates in the $TE_{1,0,5}$ mode with a resonant frequency of 2.8 GHz.

Figure 3(a) shows a schematic representation of the cavity and Figure 3(b) an image of the used cavity.

The design of a resonant cavity can be achieved using the following expression [21, 68, 89]:

$$f = \frac{1}{2\sqrt{\mu\varepsilon}}\sqrt{\left(\frac{m}{a}\right)^2 + \left(\frac{n}{b}\right)^2 + \left(\frac{p}{c}\right)^2} \quad (7)$$

where *a*, *b* and *c* are the cavity dimensions along the *x*, *y* and *z* axes, respectively, *m*, *n* and *p* are the number of half wavelengths in each coordinate direction, and μ is the permeability of the medium.

For a typical $TE_{1,0,p}$ mode, since $n = 0$, the preceding expression reduces to:

$$f = \frac{1}{2\sqrt{\mu\varepsilon}}\sqrt{\left(\frac{m}{a}\right)^2 + \left(\frac{p}{c}\right)^2} \quad (8)$$

Therefore, the resonance frequency for the $TE_{1,0,p}$ mode will depend only on the permeability, μ, and permittivity, ε, of the medium, and on the dimensions *a* and *c* of the cavity. Since the studied ceramics are non-magnetic materials, and the cavity is almost filled with air, it is acceptable to assume $1/2\sqrt{\mu\varepsilon} = 1{,}5 \times 10^8$ (SI).

The *a* and *b* dimensions correspond to commercially available rectangular cross section ducts, in this case 86.35 mm and 43.23 mm, respectively, and *m* and *n* were set to obtain the $TE_{1,0}$ mode, ensuring the best signal and less power loss. Other modes present higher power losses; however, they are harder to control, and it is more difficult to evaluate the maximum electric field. The electric field of a wave polarized in the $TE_{1,0}$ mode is known analytically as $E(t) = Re[\left(0, \sin\left(\frac{\pi(a-x)}{2a}\right)e^{(j\omega t)}, 0\right]$ [31,89]. According to Eq. (2), for c = 340.50 mm, f = 2.8 GHz is attained.

For the cavity used in this study, $p = 5$ was set, ensuring that the respective volume (V) is much higher than the volume of the sample (v), and thus ensuring that $(\varepsilon' - 1)v \ll V$ and $\varepsilon''v \ll V$, enabling the conditions to apply the small perturbation theory [90]. Nevertheless, the difference between volumes V and v should be small enough to guarantee the best resolution and sensibility of the measurements. It should also be taken into consideration that p must always be an odd number. For a cavity with the uneven number of half wavelengths in the propagation direction (z axis), the electric field will present a maximum at the geometric centre of the cavity, where a hole, strategically placed, allows the introduction of the samples in its interior [32, 68], as shown in the simulation results presented in Figure 4.

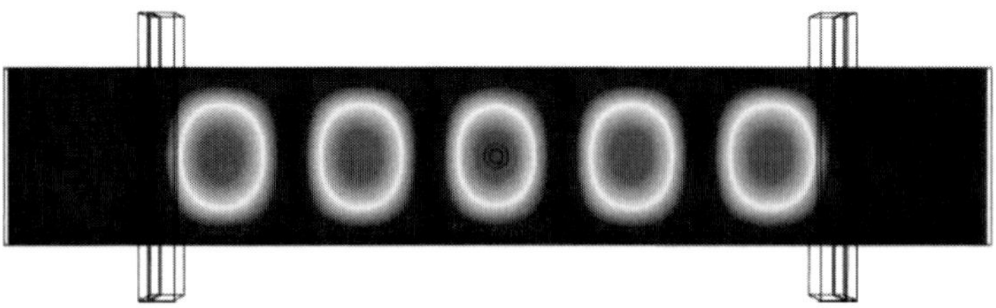

Figure 4. Normalized component of the simulated electric field excited with 1 mW at 2.8 GHz. (Simulation performed using COMSOL Multiphysics).

The fundamental concept of the small perturbation theory is that the presence of a dielectric sample in the resonant cavity shifts the resonant frequency by Δf for a lower frequency, and decreases the quality factor of the cavity, Q, expressed by a Full Width at Half Maximum (FWHM) increase [33,68], as shown in Figure 5.

Assuming that a small cylindrical sample is placed parallel to the electric field, where E_z is maximum, and considering only the first-order perturbation on the electric field caused by the sample, the dielectric constant ε' and the dielectric loss factor ε'' can be obtained as [60]:

$$\varepsilon' = K\frac{\Delta f}{f_0}\frac{V}{v} + 1 \tag{9}$$

$$\varepsilon'' = \frac{K}{2}\Delta\left(\frac{1}{Q}\right)\frac{V}{v} \tag{10}$$

where K is a constant related to the depolarization factor, which depends on geometric parameters.

Concerning the shift on the cavity resonant frequency (Δf) and the change on its quality factor $\Delta(1/Q)$ due to the sample insertion, it is expressed as

$$\Delta f = f_0 - f_l \tag{11}$$

$$\Delta\left(\frac{1}{Q}\right) = \frac{1}{Q_l} - \frac{1}{Q_0} \tag{12}$$

where indices 0 and l refer the empty and loaded cavity, respectively.

The quality factor can be obtained using the equation

$$Q = \frac{f}{\Delta f_{h/2}} \tag{13}$$

where $\Delta f_{h/2}$ is the width to half height ratio, FWHM, of the frequency-response curve [31].

Using a sample with a known dielectric constant, in this case polytetrafluorethylene (PTFE), the constant K is obtained, and thus also ε' and ε''. The sample holder, consisting of a cylindrical tube, is also made of PTFE.

The excitation of the electric field inside the resonant cavity, satisfying the boundary conditions, was considered by using transitions, small irises and shielded coaxial cables. For the permittivity measurements, an HP 8753D Network Analyzer was used and an excitation power of 1 mW applied.

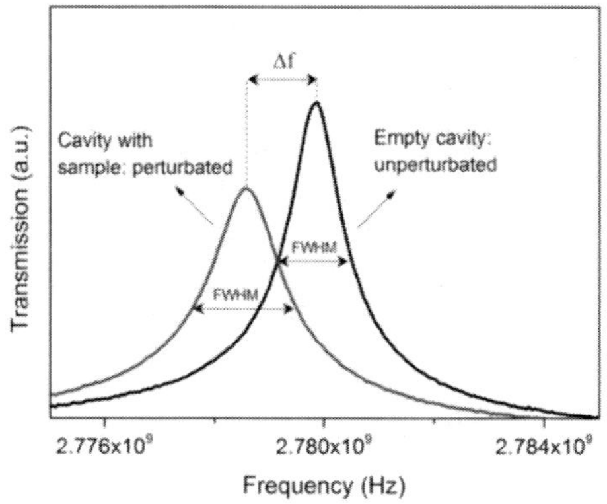

Figure 5. Transmission of the cavity, empty and with a sample (non-perturbated and perturbated signal, respectively).

3. Results and Discussion

The applicability of the small perturbation theory was tested by studying the linearity of the resonant cavity. Measurements of Δf and $\Delta(1/Q)$, as function of the perturbating volume, were carried out using micro tubes with distilled water.

Figure 6 shows the frequency response of the 2.8 GHz cavity with the different empty glass micro tubes. Since the micro tubes have a small volume and have a low dielectric loss the observed perturbations are quite small.

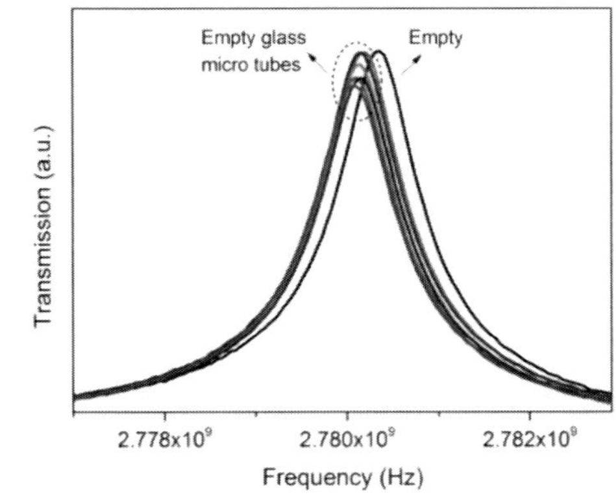

Figure 6. Transmission of the cavity with empty glass micro tubes.

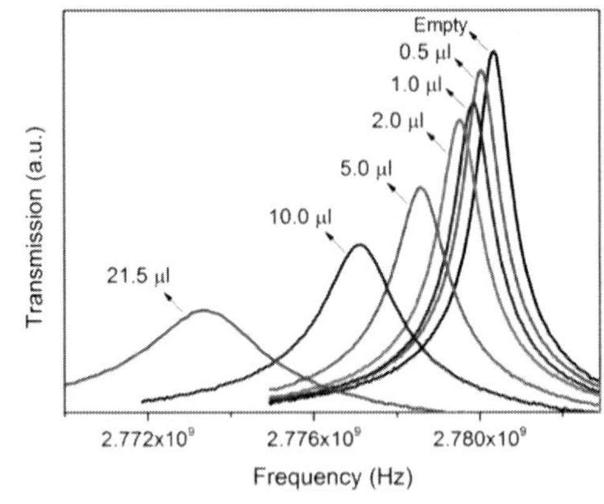

Figure 7. Transmission of the cavity with micro tubes containing different volumes of distilled water.

The different perturbations for each micro tube containing different volumes of distilled water are presented in Figure 7. As expected, the highest variations on the signal amplitude, central frequency shift and quality factor, occur for the micro tubes containing larger volumes of distilled water.

Table 1 summarizes the quality factor and the central frequency obtained for the different micro tubes, empty and filled with distilled water.

The obtained values of $\Delta(1/Q)$ and $\Delta f/f_0$ as a function of the volume of distilled water are presented in Figures 8 and 9, respectively. The linear fit of the experimental data in Figure 8 is given by the expression $\Delta(1/Q) = 5.4 \times 10^{-5}\, v + 4.4 \times 10^{-5}$, with a correlation factor, r^2, of 0.978, thus confirming linearity between the volume of water inside the cavity and $\Delta(1/Q)$. Considering the experimental data in Figure 9, the expression of the linear fit is $\Delta f/f_0 = 1.1 \times 10^{-4}\, v - 4.7 \times 10^{-6}$, with a correlation factor of 0.999, which also confirms linearity between the volume of water inside the cavity and $\Delta f/f_0$. It is thus concluded that the cavity remains in the linear regime for volumes of distilled water up to 21.5 μl, indicating that this technique can be used for variations of $\Delta f/f_0$ at least until 0.24% and $\Delta(1/Q)$ of about 1.17×10^{-3}.

To perform the dielectric characterization of the ceramic materials, small cylindrical samples were used, with diameters of 4.45 - 5.00 mm and heights of 7.70 - 13.00 mm.

Figure 10 shows the transmission signal for three situations: (1) empty cavity, (2) cavity with PTFE, and (3) cavity with SiC. In the case of the cavity with PTFE, the perturbation originated in the electric field is not as significant as the perturbation caused by the introduction of the SiC sample. This can be seen from the resonant peak shift for lower frequencies and its broadening.

Table 1. Quality factor and central frequency obtained adjusting a Lorentzian curve to the experimental data

v (μl)	Q_0 (10^3)	f_0 (GHz)	Q (10^3)	f (GHz)	$\Delta(1/Q)$ (10^{-5})	$\Delta f/f_0$ (10^{-5})
0.5	2.14	2.7802	2.07	2.7800	1.66	5.04
1.0	2.15	2.7801	1.99	2.7798	3.77	11.3
2.0	2.09	2.7801	1.87	2.7795	5.62	21.9
5.0	2.12	2.7801	1.58	2.7786	16.3	54.7
10.0	2.17	2.7801	1.16	2.7771	40.0	110
21.5	1.80	2.7881	0.58	2.7729	117	240

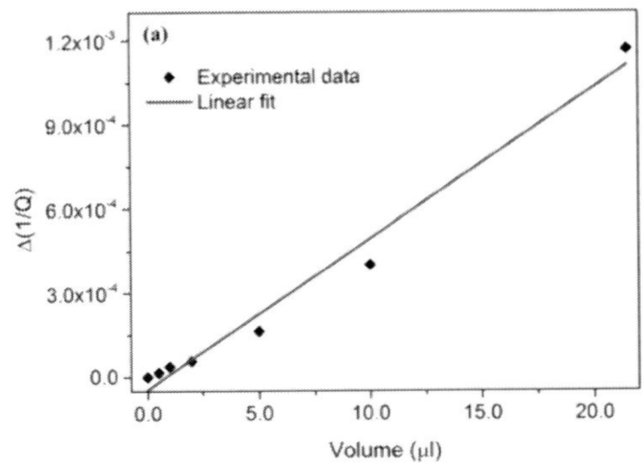

Figure 8. $\Delta(1/Q)$ for different volumes of distilled water.

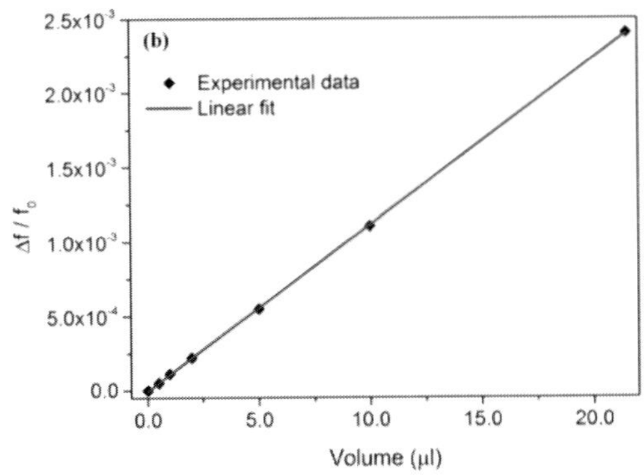

Figure 9. $\Delta f/f_0$ for different volumes of distilled water.

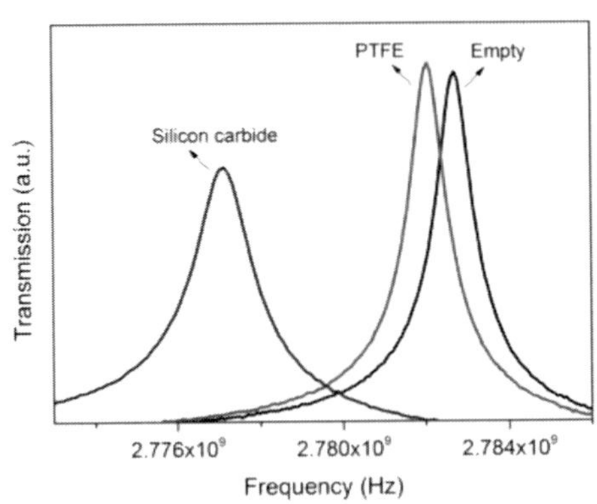

Figure 10. Transmission of the resonant cavity under three different situations: (1) empty cavity, (2) cavity with polytetrafluorethylene (PTFE), and (3) cavity with silicon carbide (SiC).

Table 2 summarizes the obtained dielectric properties, ε' and ε'' of the analyzed samples, and also $\Delta(1/Q)$ and $\Delta f/f_0$, to evaluate if the variations are within reasonable limits of the cavity linearity.

Table 2. $\Delta(1/Q)$, $\Delta f/f_0$, ε' and ε'' of the analyzed samples

Sample	**$\Delta(1/Q)$ (10^{-3})**	**$\Delta f/f_0$ (%)**	ε'	ε''
Silicon carbide	0.30	0.20	11.31	0.77
Wet porcelain paste	0.87	0.38	11.32	1.22
Dried porcelain paste	0.28	0.10	4.06	0.44
Biscuit	0.07	0.06	3.17	0.11
Unglazed porcelain	0.07	0.11	4.79	0.12
Wet stoneware paste	1.22	0.33	10.02	1.65
Dried stoneware paste	0.25	0.09	3.71	0.35
Stoneware	0.09	0.09	4.11	0.15
PROMALIGHT®	0.01	0.02	1.61	0.02
PROMASIL®	0.02	0.03	1.86	0.03
JM 26	---	0.04	2.17	---

From Table 2, at room temperature, ε' = 11.31 and ε'' = 0.77 were obtained for SiC, results that are consistent with those reported by others [45, 62, 91, 92].

However, some studies refer quite different complex permittivities for SiC, even at room temperature.

Table 3 summarizes some case studies.

Table 3. SiC complex permittivities reported in the literature

Temperature (°C)	**Frequency (GHz)**	ε'	ε''	**Ref.**
25	~ 1.4	~ 5.5	~ 0.5	[62]
750		~ 6.6	~ 0.4	
1500		~ 9.0	~ 2.1	
room	~ 2.45	~ 6.8	~ 1	[91]
700		~ 10	~ 3	
1200		~ 17	~ 8.3	
22	2.45	12.15	1.7	[45]
695		26.66	28	
room		11 to 21	1.5 to 7	[92]
30		11.45	2.0	[93]
70		11.87	1.35	
20		30	11	[31]
200		105	110	[30]
room	8.2	~ 14.2	~ 5,1	[94]

Performing an adjustment to the experimental data of Ref. [62], the critical temperature (T_c) of SiC seems to be between 1050°C and 1100°C. By performing the same for data from Ref. [91], T_c is found to be approximately 600°C. One of the possible reasons for the discrepancies observed on the reported values of ε' and ε'' of SiC is related to the specificities of each SiC sample. These differences observed for the case of SiC emphasize the need for an effective experimental characterization of each sample (SiC and other materials to be used in the construction of a microwave furnace, as well as those to be processed). The data found in the literature should be used as a guidance.

Analyzing $\Delta(1/Q)$ and $\Delta f/f_0$, for the porcelain paste sample, $\Delta f/f_0$ exceeds the 0.24% limit. For the stoneware paste, besides the $\Delta f/f_0$ limit, $\Delta(1/Q)$ also exceeds the 1.17×10^{-3} limit (Table 2). In general, if the measured values fall out of the predetermined limits, the sample volume should be reduced until the results fall within the linear regime [95]. However, in this particular case, since the cavity has a linear regime for measurements up to 21.5 μl volume of distilled water, higher volumes should be tested.

For smaller samples, and particularly for samples with smaller diameters [68], the perturbations on the electric field are smaller as well as the measurement errors. Causes of errors are the influence of the cavity holes, which can rise up to 8% [68], due to the sample introduction and removal, and to the sample alignment relative to the maximum electric field.

Nonetheless, comparing the dielectric constant and the dielectric loss factor of the porcelain paste with that of the dried porcelain paste, the decrease on the dielectric loss factor is attributed to the reduction of the water content. Similar considerations apply to stoneware.

Except for the wet samples, both porcelain and stoneware have relatively low dielectric loss factors, with a decreasing trend from wet paste samples for the fired samples. The dielectric constant decreases from the wet to the dried paste, increasing to fired samples. Both ε' and ε'' of stoneware samples show values and a trend close to that of the porcelain processing stages [wet → dried → biscuit → porcelain (unglazed)].

Comparing data with the ones in the literature, both unglazed porcelain and stoneware, have dielectric loss factors one order of magnitude higher than the "porcelain (aluminum oxide)," with $\varepsilon' = 8.9$ and $\varepsilon'' = 0.01$, measured at room temperature and at 3 GHz [33], and dielectric properties of the same order of magnitude of quartz-based porcelain at 30°C, $\varepsilon' = 6.0$ and $\varepsilon'' = 0.23$ [96].

At 1 MHz, and at room temperature, Ref. [97] presents porcelain insulators evidencing $\varepsilon' \sim 3.9$ and $\varepsilon'' \sim 0.06$.

At the frequency of 1 kHz and at the temperature of 32°C, technical porcelain with a composition of 30% kaolin, 45% potash feldspar and 25% quartz, shows values of $\varepsilon' \sim 22.3$ and $\varepsilon'' \sim 0.7$ [98]. Also, at the frequency of 1 kHz, an electro-porcelain with a composition of 37% kaolin, 35% potash feldspar and 28% quartz evidences $\varepsilon' = 21.22$ and $\varepsilon'' \sim 0.13$ at the temperature of 20°C [99].

When comparing with the main raw minerals used in the porcelain and stoneware manufacturing, it may be observed that the measured dielectric loss factors of these samples (Table 4) are higher than those of feldspar [31], of high purity SiO_2 [33], of kaolin [61], and of lower content minerals such as muscovite [31, 33] and talc [31], being these of the same order of magnitude of those of pure kaolinite, measured at the frequency of 4 GHz [100]. At the frequency of 1 GHz and at the temperature of 15°C, fine clay presents $\varepsilon'' = 0.34$ [31].

Regarding the kaolin mentioned in Ref. [61], after the sample being fired up to 1300°C, and cooled down, the dielectric loss factor of the resulted fired product (mullite + amorphous SiO_2) increased up to ~ 0.7, higher than that measured for unglazed porcelain and stoneware fired products (Table 2). The metakaolinite, resulting from firing pure kaolinite up to 800°C, presents a dielectric loss factor at room temperature that varies between 0.02 and 0.18 [61, 100]. Mullite (measured at the frequency of 1 MHz) and glass (measured at the frequency of ~2.45 GHz) present lower dielectric loss factors [30, 31, 33]. Depending on the sampling, e.g., for SiC, muscovite and alumina, considerable differences can be observed in the complex permittivities of these materials.

Returning to the present results, Table 2, it was not possible to measure the dielectric loss factor of several refractory materials, and in particular for porous samples with lower densities. As mentioned earlier, when using the resonant cavity technique, relatively large samples are required to measure the dielectric properties of materials with very small ε'.' Therefore, larger sized samples are required for the complex permittivity characterization of these materials, possibly larger than the cavity hole allows. Since the used method is based on a reference sample, PTFE, which curve is essentially overlapping the refractory samples curves, the hypothesis is that they present, at room temperature and at the frequency of 2.8 GHz, dielectric loss factors similar to those of PTFE, which according to the literature [30, 31, 104] varies between 0.0003 and 0.0005. Some values may might even to be below the sensitivity of the measurement system. Among these materials, the JM 26 brick stands out since the loss factor of this material is estimated to increase significantly as temperature increases. This was observed empirically, placing the material in a microwave furnace adapted to withstand high temperatures. During these tests, the material was damaged due to localized and abrupt microwave absorption, presenting local vitrification and melting, as depicted in Figure 11, where local temperatures above 1350°C [84] were reached.

Table 4. Dielectric constant and dielectric loss factors of some minerals/phases present in the raw materials used in the porcelain and stoneware manufacturing, and in fired products

Mineral/phase	**Temperature (°C)**	**Frequency (GHz)**	ε'	ε''	**Ref.**
Feldspar	~ 25	2.5	2.61	0.020	[31]
High purity SiO_2		3.0	3.84	0.0026	[33]
Muscovite		2.5	1.62	0.005	[31]
		3.0	5.4	0.0016	[33]
Kaolin	room	4.0	2.11	0.13	[100]
		2.45	---	< 0.10	[61]
Metakaolinite				0.08 to 0.18	
		4.0	2.19	0.02	[100]
Talc	15	1.0	1.7	0.04	[31]
Fine clay			1.8	0.34	
High purity alumina Al_2O_3	25	2.19	~ 6	---	[60]
		2.45	8.9	0.009	[30]
	room		9.85	0.0029	[101]
	22	2.45	8.9	0.0036	[45]
		2.46	8.98	< 0.001	[102]
			8.9	0.004	
	25	3.0	4.5	< 0.02	[103]
Mullite		0.001	5.8	0.012	[30]
	800	0.915	5.56	0.6	[76]
Fused quartz (SiO_2)	25	2.45	3.8	0.0001	[30]
		2.5	3.78	< 0.001	[31]
		3.0	3.78	0.0002	[33]
Soda-silica glass, 20%		2.5	5.7	0.09	[31]

Even if presenting a low loss factor at room temperature, the JM 26 brick was one of the materials that had not been chosen for the construction of the microwave furnace.

Details on the furnace developed for porcelain multi-object processing can be found elsewhere [23, 24].

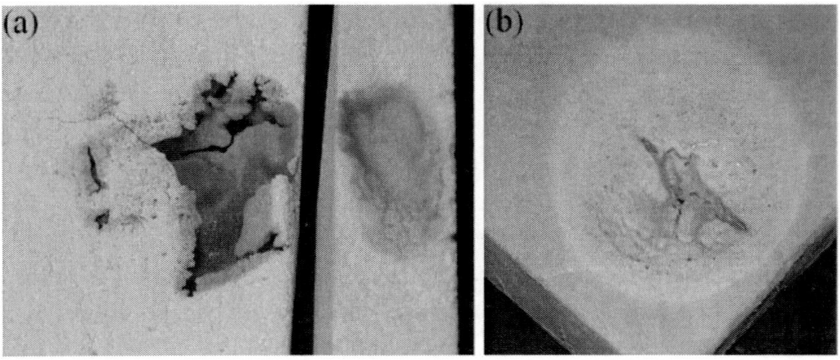

Figure 11. Images of JM Firebricks showing excessive heating and localized melting when subjected to microwave radiation: (a) without and (b) with high temperature cement covering its surface.

The phenomenon presented in Figure 11 was also observed in porcelain and stoneware non-controlled firing tests. Although these materials are poor microwave absorbers at room temperature, it was experimentally found that as the temperature increases they are more effectively heated by microwave radiation. Without heating control, and without susceptors, once reached the critical temperature it was observed that porcelain pieces were only locally heated, presenting local vitrification and melting [52], as occurred with the referred JM 26 sample. However, the main porcelain body essentially did not heat up [52]. The observed results, for porcelain, stoneware and JM firebrick, highlight the need to know the complex permittivities of the materials as functions of the processing temperature.

Conclusion

At room temperatures and for frequencies close to 2.45 GHz, porcelain and stoneware are essentially transparent to microwave radiation. SiC, as expected, presents relatively high dielectric loss coefficients, about seven times higher than those of biscuit and fired stoneware and porcelain. The wet porcelain and stoneware pastes show loss factors 2.8 and 4.7 times higher when comparing with those of the respective dried pastes. SiC, wet porcelain and stoneware pastes are relatively good microwave radiation absorbers since they have $\varepsilon'' > 10^{-2}$.

After the porcelain and stoneware lose water, although being poor microwave radiation absorbers at room temperature, it was experimentally found that these materials are more effectively heated by microwave radiation as temperature increases.

As the materials are heated, the spatial distribution of the electromagnetic field and its intensity change; consequently, to develop an accurate model it is crucial to know the complex permittivity of the materials to be microwave processed, including its dependence on temperature, since the heating rate and the way the pieces heat strongly depend on this property. Such dependence is observed not only for the materials to be microwave processed, but also for the refractory materials used in the microwave furnaces, including the SiC susceptor.

For microwave materials processing it is required to know not only the dependence of the complex permittivity as function of the temperature, but also the temperature dependence of the thermal properties during the different processing (drying and firing) stages. Knowing/understanding the temperature dependence of these properties for the different stages of the manufacturing process, allows to engineer products with increased microwave radiation absorption, by adding secondary (and non-conventionally-used) raw materials. This may enable increased volumetric heating processes, allowing improved microwave processing/firing of products that are usually processed using fossil fuels.

If this knowledge is relevant for a better understanding and control of manufacturing processes using microwave heating technology, it is crucial for accurate numerical simulations of the heating process, and therefor, for the construction and optimization of the furnace, and for microwave porcelain and stoneware firing.

Acknowledgments

We thank FEDER funds through the COMPETE 2020 Programme under CerWave: Demonstração do processo de cozedura de porcelana por gás-microondas, project POCI-01-0247-FEDER-006410 and National Funds through FCT-Portuguese Foundation for Science and Technology under the project UID/CTM/50025/2019. The authors express their sincere thanks to the Porcelanas da Costa Verde S. A., to Grestel S. A. and to Termolab High Temperature technology Lda, for providing samples for this study.

Author T. Santos acknowledges the support of CICECO-Aveiro Institute of Materials, FCT Ref. UID/CTM/50011/2019, financed by national funds through the FCT/MCTES.

Author S. Devesa acknowledges the post-doctoral grant under the project "SUSpENsE - CENTRO-01-0145-FEDER-000006."

Author V. A. F. Costa acknowledges the Portuguese Foundation for Science and Technology for the financial support provided through project UID/EMS/00481/2013-FCT, and CENTRO-01-0145-FEDER-022083.

References

[1] International Energy Agency. 2016. "CO_2 Emissions from Fuel Combustion: Highlights." *International Energy Agency*. Accessed January 11, 2020.

[2] EU-MERCI. 2012. "*Analysis of the Industrial Sectors in the European Union.*" Accessed January 11, 2020. http://www.eumerci-portal.eu/documents/20182/38527/3+-+Greece.pdf.

[3] Wroe, R. 1998. "Microwave-Assisted Firing of Ceramics." Paper presented at the *IEE Seminar New Developments in Ceramics Manufacturing*, London, UK, November 25. Accessed January 02, 2020. doi:10.1049/ic:19980856.

[4] CEPS, Economisti Associati and Ecorys. 2017. "*Cumulative Cost Assessment (CCA) of the EU Ceramics Industry*," Accessed January 11, 2020.

[5] Durantino, P. C. 1857. *I Tre Libri Dell'Arte Del Vasajo, Nei Quali Si Tratta Non Solo la Pratica, Ma Brevemente Tutti i Secreti Di Essa Cosa Che Persino al Di' D'oggi È Stata Sempre Tenuta ascosta*. Num. 387. Roma: Dallo Stabilimento tipografico. [*The Three Books of the Vasajo Art, In Which Not Only the Practice is Dealt With, But Briefly All the Secrets of it, Something that Even Today Has Always Been Kept Hidden*. Vol. 387. Rome: From the printing plant]. Accessed January 09, 2020. https://books.google.it/books?id=r2UGAAAAQAAJ& printsec=frontcover&dq=Cipriano+Piccolpasso&as_brr=1&ei=_-kKSMz_ JIOOywSXyeiaCA#v=onepage&q&f=false.

[6] Rhodes, D. 1981. *Kilns: Design, Construction, and Operation*. 2nd ed. Radnor: Chilton Book Co.

[7] All Japan Tours. 2020. "*Togei Messe Mashiko: Top Museums of Art in Japan.*" Accessed January 01. http://unisauce.com/japan-attraction/cultural-attractions/top-museum-of-art/view.php?p=AO112182110.

[8] Castro, A., N. A. de Paula, J. B. Torres, T. Curado, and A. Teixeira. 2017. "Evidências de Produção Oleira Nos Séculos XVI e XVII No Largo Das Olarias, Mouraria (Lisboa)." [Evidence of pottery production in the 16th and 17th centuries in Largo das Olarias, Mouraria (Lisbon)]. *Paper presented at the II Congress of the Association of Portuguese Archaeologists*, Lisbon, Portugal, November 22–26. Accessed December 21, 2020. https://novaresearch.unl.pt/en/publications/evid%C3%AAncias-de-produ%C3%A7%C3%A3o-oleira-nos-s%C3%A9culos-xvi-e-xvii-no-largo-das.

[9] Zhang, L., and H. Dong. 2018. "*China Hour - Porcelain Story.*" China International Television Corporation.

[10] Sutton, W. H. 1992. "Microwave Processing of Ceramics - an Overview." In *Microwave Processing of Materials III*, edited by Magdy F. Iskander Ronald L. Beatty, Willard H. Sutton, 269:3–20. San Francisco: Materials Research Society.

[11] CERAMIFOR Kilns & Equipment. n.d. "*Technical Ceramics and Refractories.*" Accessed January 08, 2019. http://ceramifor.com/en/Ceramics/Industries/Technical-Ceramics-Refractories-D30.

[12] Clark, D. E., and W. H. Sutton. 1996. "Microwave Processing of Materials." *Annual Review of Materials Science* 26(1):299–331. Accessed February 03, 2020. doi:10.1146/annurev.ms.26.080196.001503.

[13] Andreola, F., L. Barbieri, A. Corradi, I. Lancellotti, and T. Manfredini. 2002. "Utilisation of Municipal Incinerator Grate Slag Manufacturing Porcelainized Stoneware Tiles Manufacturing." *Journal of the European Ceramic Society* 22(9–10):1457–62. Accessed April 02, 2020. doi:10.1016/S0955-2219(01)00460-5.

[14] Martín-Márquez, J., Jesús Ma. R., and M. Romero. 2010. "Effect of Microstructure on Mechanical Properties of Porcelain Stoneware." *Journal of the European Ceramic Society* 30(15):3063–69. Accessed March 22, 2020. doi:10.1016/j.jeurceramsoc.2010.07.015.

[15] Sánchez, E., J. García-Ten, V. Sanz, and A. Moreno. 2010. "Porcelain Tile: Almost 30 Years of Steady Scientific-Technological Evolution." *Ceramics International* 36(3):831–45. Accessed February 19, 2020. doi:10.1016/j. ceramint.2009.11.016.

[16] Ryan, W., and C. Radford. 1987. *Whitewares Production, Testing, and Quality Control: Including Materials, Body Formulations, and Manufacturing Processes*. Oxford: Pergamon Press.

[17] Royal Copenhagen. n.d. "*General Information about Porcelain - Report.*" Accessed February 11, 2020. https://www.royalcopenhagen.com/mediaweb/General-Information-about-Porcelain.pdf?context=cmN8cm9vdHw3NjgxNzF8YXBwbGljYXRpb24vcGRmfGg1Zi9oZDMvODgwMTIzNjMyMDI4Ni5wZGZ8NzA5YzNhOTFjNmM4YTk0NTM4M2RlM2E3ODJlOWI1Mzk5YzllNzkyZDBlMWQ0YzA1NWRjM2JjNWZlMzA3M2UyYw.

[18] Matceramica - Manufacture ceramic. 2015. "*ISI 084.DQA – Especificações Técnicas Do Produto Grês.*" 1–2. [ISI 084.DQA - Stoneware Product Technical Specifications. 1–2].

[19] Fonseca, A. T. da. 2000. *Tecnologia do Processamento Cerâmico*. Lisboa: Universidade Aberta. [*Ceramic Processing Technology. Lisbon: Open University*].

[20] Fall, M., V. Yakovlev, C. Sahi, I. Baranova, J. Bowers, and G. Esquenazi. 2012. "Energy Efficient Microwave Hybrid Processing of Lime for Cement, Steel, and Glass Industries." 1–37. Ceralink Inc. *DE-EE0003472 - Final Technical Report. Energy Efficiency and Renewable Energy Investions and Innovations*, February 10, 2019.

[21] Thostenson, E. T., and T. -W. Chou. 1999. "Microwave Processing: Fundamentals and Applications." *Composites Part A: Applied Science and Manufacturing* 30(9):1055–71. Accessed January 16, 2020. doi:10.1016/S1359-835X(99)00020-2.

[22] Leparoux, S., S. Vaucher, and O. Beffort. 2003. "Assessment of Microwave Heating for Sintering of Al/SiC and for in-Situ Synthesis of TiC." *Advanced*

Engineering Materials 5(6):449–53. Accessed March 14, 2020. doi:10.1002/adem.200320136.

[23] Santos, T., L. C. Costa, L. Hennetier, M. A. Valente, J. Monteiro, and J. Sousa. 2013. "Microwave Processing of Porcelain Tableware Using a Multiple Generator Configuration." *Applied Thermal Engineering* 50(1):677–82. Accessed November 20, 2019. doi:10.1016/j.applthermaleng.2012.07.010.

[24] Santos, T., L. Hennetier, V. A. F. Costa, and L. C. Costa. 2015. "Using Microwave Radiation for Porcelain Tableware Sintering." *Paper presented at the Progress in Electromagnetics Research Symposium*, Prague, Czech Republic, July 6–9.

[25] Thuault, A., E. Savary, J. Bazin, and S. Marinel. 2014. "Microwave Sintering of Large Size Pieces with Complex Shape." *Journal of Materials Processing Technology* 214(2):470–76. Accessed January 21, 2020. doi:10.1016/j.jmatprotec.2013.09.030.

[26] Grestel. 2020. "*About Stoneware*" Accessed February 07. https://grestel.pt/en/about-stoneware.

[27] Costa Verde. 2020. "*Porcelanas da Costa Verde SA*" Accessed February 07. https://costa-verde.com/en/costaverde/.

[28] Carty, W. M., and U. Senapati. 1998. "Porcelain - Raw Materials, Processing, Phase Evolution, and Mechanical Behavior." *Journal of the American Ceramic Society* 81(1):3–20. Accessed January 07, 2020. doi:10.1111/j.1151-2916.1998.tb02290.x.

[29] Metaxas, A. C. 1991. "Microwave Heating." *Power Engineering Journal* 5(5):237. Accessed February 01, 2020. doi:10.1049/pe:19910047.

[30] Clark, D. E., and D. C. Folz. 1997. "*What Is Microwave Processing*?." Accessed January 28, 2020. https://www.researchgate.net/publication/290488291_Introduction_What_is_Microwave_Processing.

[31] Meredith, R. 1998. *Engineers' Handbook of Industrial Microwave Heating*. London: The Institution of Electrical Engineers.

[32] Keysight Technologies. n.d. "*Basics of Measuring the Dielectric Properties of Materials*." Accessed February 28, 2020. www.keysight.com.

[33] Metaxas, A. C., and R. J. Meredith. 1988. *Industrial Microwave Heating*. London: The Institution of Engineering and Technology.

[34] Santos, T., M. A. Valente, J. Monteiro, J. Sousa, and L. C. Costa. 2011. "Electromagnetic and Thermal History during Microwave Heating." *Applied Thermal Engineering* 31(16):3255–61. Accessed November 05, 2019. doi:10.1016/j.applthermaleng.2011.06.006.

[35] Oghbaei, M., and O. Mirzaee. 2010. "Microwave versus Conventional Sintering: A Review of Fundamentals, Advantages and Applications." *Journal of Alloys and Compounds* 494(1–2):175–89. Accessed March 03, 2020. doi:10.1016/J.JALLCOM.2010.01.068.

[36] Rybakov, K. I., E. A. Olevsky, and E. V. Krikun. 2013. "Microwave Sintering: Fundamentals and Modeling." *Journal of the American Ceramic Society* 96(4):1003–20. Accessed February 29, 2020. doi:10.1111/jace.12278.

[37] Jack P. H. 2010. *Heat Transfer. Series in Mechanical Engineering*. 10th ed. New York: McGraw-Hill, Inc.

[38] Darwish, M., L. Mangani, and F. Moukalled. 2017. "General Fully Implicit Discretization of the Diffusion Term for the Finite Volume Method." *Numerical Heat Transfer, Part B: Fundamentals* 71(6):506–32. Accessed April 16, 2020. doi:10.1080/10407790.2017.1330060.

[39] Carbolite, Inc. n.d. "*MAT – Microwave Assist Technology Furnace: MRF 16/22 Datasheet.*"

[40] Brandon, J. R., J. Samuels, and W. R. Hodgkins. 1992. "Microwave Sintering of Oxide Ceramics." In *Microwave Processing of Materials III*, edited by Magdy F. Iskander Ronald L. Beatty, Willard H. Sutton, 269:237–43. San Francisco: Materials Research Society.

[41] Wroe, R. 1996. "Microwave-Assisted Firing of Ceramics." *Power Engineering Journal* 10(4):181–83. Accessed January 31, 2020. doi:10.1049/pe:19960410.

[42] Janney, M. A., H. D. Kimrey, and J. O. Kiggans. 1992. "Microwave Processing of Ceramics: Guidelines Used at the Oak Ridge National Laboratoy." In *Microwave Processing of Materials III*, edited by Magdy F. Iskander Ronald L. Beatty, Willard H. Sutton, 269:173–85. San Francisco: Materials Research Society.

[43] Barmatz, M., and H. W. Jackson. 1992. "Steady State Temperature Profile in a Sphere Heated by Microwaves." In *Microwave Processing of Materials III*, edited by Magdy F. Iskander Ronald L. Beatty, Willard H. Sutton, 269:97–103. San Francisco: Materials Research Society.

[44] Carbolite, Inc. n.d. "*Microwave Assist Laboratoy Box Furnace.*" Accessed February 02, 2020. https://www.yumpu.com/en/document/read/40908823/microwave-assist-laboratory-box-furnaces-directories.

[45] Clark, D. E., D. C. Folz, R. L. Schulz, Z. Fathi, and A. D. Cozzi. 1993. "Recent Developments in the Microwave Processing of Ceramics." *MRS Bulletin* 18(11):41–46. Accessed January 13, 2020. doi:10.1017/S0883769400038525.

[46] Madhuchhanda, B., and T. Basak. 2016. "A Review on the Susceptor Assisted Microwave Processing of Materials." *Energy* 97:306–38. Accessed January 23, 2020. doi:10.1016/J.ENERGY.2015.11.034.

[47] Tayler, G. V. A, M. Anderson, and M. Hamlyn. 2018. "Heavy Clay Body Properties for Hybrid Microwave Firing." In *Ceramic Transactions Series*, edited by Kumar Sridharan, Amar Bhalla, Navin Jose Manjooran, Morsi Mahmoud, Andrei Jitianu, J. Singh, Henry Colorado, Surojit Gupta, and Jason Langhorn, 265:59–68. Hoboken: John Wiley & Sons. Accessed March 09, 2020. doi:10.1002/9781119543299.ch7.

[48] Fiona C. R. W. 2002. *Method of processing ceramic materials and a microwave furnace therefore.* US 6,172,346 B1, issued 2002. Accessed February 06, 2020.

[49] Binner, J. G. P., and T. E. Cross. 1992. "Microwave Processing in the UK." In *Microwave Processing of Materials III*, edited by Magdy F. Iskander Ronald L. Beatty, Willard H. Sutton, 269:33–38. San Francisco: Materials Research Society.

[50] Fernandes, J. P. C., R. F. K. Gunnewiek, P. M. Souto, and R. H. G. A. Kiminami. 2014. "Monoqueima de Porcelanas Esmaltadas Em Forno de Micro-Ondas." *Cerâmica* 59(352):545–50. ["One-step microwave sintering of porcelains." *Ceramic* 59 (352):545–50]. Accessed February 06, 2020. doi:10.1590/s0366-69132013000400009.

[51] Lerdprom, W., E. Zapata-solvas, D. D. Jayaseelan, A. Borrell, M. D. Salvador, and W. E. Lee. 2017. "Impact of Microwave Processing on Porcelain Microstructure." *Ceramics International* 43(16):13765–71. Accessed February 16, 2020. doi:10.1016/j.ceramint.2017.07.090.

[52] Santos, T., V. A. F. Costa, L. C. Costa. 2018. "Ovalization and Other Porcelain Flaws When Fired Using Microwave Technology." *Paper presented at the Proceedings of the 1st Iberic Conference on Theoretical and Experimental Mechanics and Materials/11th National Congress on Experimental Mechanics,* Porto, Portugal, November 4–7.

[53] Santos, T., N. F. Santos, C. S. F. Gomes, L. Hennetier, V. A. Costa, and L. C. Costa. 2019. "Features of Microwave Fired Utilitarian Stoneware." Paper presented at the *17th International Conference on Microwave and High Frequency Heating*, Valencia, Spain, September 9–12. Accessed February 01, 2020. doi:10.4995/Ampere2019.2019.9913.

[54] Hamlyn, M. G., and A. L. Bowden. 1990. "Microwave Processing of Earthenware Ceramics." *American Ceramic Society Bulletin* 69(3):368–72.

[55] Brindley, G. W., and M. Nakahira. 1957. "The Role of Water Vapour in the Dehydroxylation of Clay Minerals." *Clay Minerals Bulletin* 3(17):117–19. Accessed December 17, 2020. http://www.minersoc.org/pages/Archive-CM/Volume_3/3-17-114.pdf.

[56] Gualtieri, A., M. Bellotto, G. Artioli, and S. M. Clark. 1995. "Kinetic Study of the Kaolinite-Mullite Reaction Sequence. Part II: Mullite Formation." *Physics and Chemistry of Minerals* 22(4):215–22. Accessed February 03, 2020. doi:10.1007/BF00202254.

[57] Schüller, K. H. 1979. "Porcelain." Edited by Verlag Schmidt GmbH Freiburgi. Brg. *Ceramics Monographs - A Handbook of Ceramics*.

[58] Klein, A. A. 1916. "*Constitution and Microstructure of Porcelain*." Department of commerce - United States of America.

[59] Ernest, M. Levin, Carl R. Robbins and Howard F. McMurdie. 1964. *Phase Diagrams for Ceramists*. 3rd ed. Columbus: American Chemical Society.

[60] Arai, M., J. G. P. Binner, and T. E. Cross. 1996. "Comparison of Techniques for Measuring High-Temperature Microwave Complex Permittivity: Measurements on an Alumina/Zircona System." *Journal of Microwave Power and Electromagnetic Energy* 31(1):12–18. Accessed March 24, 2020. doi:10.1080/08327823.1996.11688287.

[61] Reinosa, J. J., B. García-Baños, J. M. Catalá-Civera, and J. F. Fernández. 2019. "A Step Ahead on Efficient Microwave Heating for Kaolinite." *Applied Clay Science* 168:237–43. Accessed May 17, 2020. doi:10.1016/j.clay.2018.11.001.

[62] Baeraky, T. A. 2002. "Microwave Measurements of the Dielectric Properties of Silicon Carbide at High Temperature." *Egypt. J. Sol.* 25(2):263–73.

[63] Zhao, C., J. Vleugels, C. Groffils, P. J. Luypaert, and O. Van Der Biest. 2000. "Hybrid Sintering with a Tubular Susceptor in a Cylindrical Single-Mode Microwave Furnace." *Acta Materialia* 48(14):3795–3801. Accessed February 28, 2020. doi:10.1016/S1359-6454(00)00160-9.

[64] Madhuchhanda, B., and T. Basak. 2017. "Susceptor-Assisted Enhanced Microwave Processing of Ceramics - A Review." *Critical Reviews in Solid State and Materials Sciences* 42(6):433–69. Accessed January 30, 2020. doi:10.1080/10408436.2016.1192987.

[65] Keyson, D., E. Longo, J. S. Vasconcelos, J. A. Varela, S. Éber, and A. DerMaderosian. 2006. "Síntese e Processamento de Cerâmicas Em Forno de Microondas Doméstico." *Cerâmica* 52(321):50–56. ["Synthesis and ceramics processing by domestic microwave oven." *Ceramic* 52(321):50–56]. Accessed April 02, 2020. doi:10.1590/S0366-69132006000100007.

[66] Bible, D. W., R. J. Lauf, and C. A. Everleigh. 1992. "Multikilowatt Variable Frequency Microwave Furnace." In *Microwave Processing of Materials III*, edited by Magdy F. Iskander Ronald L. Beatty, Willard H. Sutton, 269:77–81. San Francisco: Materials Research Society.

[67] Hartnett, J. P., T. F. Irvine, Y. I. Cho, G. A. Greene. 1999. *Advances in Heat Transfer*. 1st. Ed. San Diego: Academic Press. Accessed April 14, 2019. https://books.google.pt/books?id=RXcHuXnbntAC&pg=PA6&lpg=PA6&dq=variable+frequency+microwave+furnaces&source=bl&ots=hD2FWi1Qz2&sig=ACfU3U0IOIn--cncnVpLpnju8D0WNQxhfw&hl=pt-PT&sa=X&ved=2ahUKEwiOirWJ-c_hAhWt4IUKHdoLD944FBDoATABegQICBAB#v=onepage&q=variable%25.

[68] Pohl, V., D. Fricke, and A. Muehlbauer. 1995. "Correction Procedures for the Measurement of Permittivities with the Cavity Perturbation Method." *Journal of Microwave Power and Electromagnetic Energy* 30(1):10–26. Accessed April 10, 2020. doi:10.1080/08327823.1995.11688253.

[69] Venkatesh, M. S., and G. S. V. Raghavan. 2005. "An Overview of Dielectric Properties Measuring Techniques." *Canadian Biosystems Engineering / Le Genie Des Biosystems Au Canada* 47.

[70] Chen, L. F., C. K. Ong, C. P. Neo, V. V. Varadan, and V. K. Varadan. 2004. *Microwave Electronics: Measurement and Materials Characterization*. 1st ed. New York: John Wiley & Sons, Ltd.

[71] Araneta, J. C., M. E. Brodwin, and G. A. Kriegsmann. 1984. "High-Temperature Microwave Characterization of Dielectric Rods." *IEEE Transactions on Microwave Theory and Techniques* 32(10):1328–35. Accessed March 14, 2020. doi:10.1109/TMTT.1984.1132844.

[72] Sheen, J. 2005. "Study of Microwave Dielectric Properties Measurements by Various Resonance Techniques." *Measurement: Journal of the International Measurement Confederation* 37(2):123–30. Accessed March 14, 2020. doi:10.1016/j.measurement.2004.11.006.

[73] Andrade, O. M., M. F. Iskander, and S. Bringhurst. 1992. "High Temperature Broadband Dielectric Properties Measurement Techniques." In *Microwave Processing of Materials III*, edited by Magdy F. Iskander Ronald L. Beatty, Willard H. Sutton, 269:527–39. San Francisco: Materials Research Society.

[74] Tinga, W. R. 1992. "Rapid High Temperature Measurement of Microwave Dieletric Properties." In *Microwave Processing of Materials III*, edited by Magdy

F. Iskander Ronald L. Beatty, Willard H. Sutton, 269:505–16. San Francisco: Materials Research Society.

[75] Adams, F., M. de Jong, and R. Hutcheon. 1992. "Sample Shape Correction Factors for Cavity Perturbation Measurements." *Journal of Microwave Power and Electromagnetic Energy* 27(3):131–35. Accessed March 15, 2020. doi:10.1080/08327823.1992.11688181.

[76] Tian, B. Q., W. R. Tinga, and W. Xi. 1992. "Single Frequency Microwave Dieletric Heating and Measuring." In *Microwave Processing of Materials III*, edited by Magdy F. Iskander Ronald L. Beatty, Willard H. Sutton, 269:83–89. San Francisco: Materials Research Society.

[77] Guo, Z., P. Guo, G. Su, D. Zhai, F. Cheng, and F. Li. 2019. "High-Temperature Permittivity and Microwave Pretreatment Characteristics of Nickel Containing Sludge from Battery Production." *Processes* 7(5):1–13. Accessed March 14, 2020. doi:10.3390/pr7050257.

[78] Arai, M., J. G. P. Binner, G. E. Carr, and T. E. Cross. 1992. "High Temperature Dielectric Measurements on Ceramics." In *Microwave Processing of Materials III*, edited by Magdy F. Iskander Ronald L. Beatty, Willard H. Sutton, 269:611–16. San Francisco: Materials Research Society.

[79] Jackson, Mitchell L., and Curtis H. Stern. 1992. "Modeling the Microwave Frequency Dieletric Properties of Thermoplastic Composite Materials." In *Microwave Processing of Materials III*, edited by Magdy F. Iskander Ronald L. Beatty, Willard H. Sutton, 269:601–6. San Francisco: Materials Research Society.

[80] Catala-Civera, J. M., A. J. Canos, P. Plaza-Gonzalez, J. D. Gutierrez, B. Garcia-Banos, and F. L. Penaranda-Foix. 2015. "Dynamic Measurement of Dielectric Properties of Materials at High Temperature During Microwave Heating in a Dual Mode Cylindrical Cavity." *IEEE Transactions on Microwave Theory and Techniques* 63(9):2905–14. Accessed March 09, 2020. doi:10.1109/ TMTT.2015. 2453263.

[81] Garcia-Baños, B., J. J. Reinosa, J. F. Fernández, and P. J. Palza-González. 2015. "Dielectric Properties of Ceramic Frits up to 1200°C and Correlation with Thermal Analyses." *Paper presented at the 15th International Conference on Microwave and High Frequency Heating AMPERE 2015*, Krakow, Poland, September 14–17.

[82] Morgan Thermal Ceramics. 2018. "*Cerachem - Product Information Sheet.*" 1–2. Accessed March 01, 2020. http://www.morganthermalceramics.com/media/ 3993/cera_blankets-data-sheet-english.pdf.

[83] Unifrax. 2018. "*Saffil Blanket - Product Information Sheet.*" 1–2. Accessed March 01, 2020. https://www.unifrax.com/wp-content/uploads/2018/08/Saffil-Blanket-EN.pdf.

[84] hermal Ceramics. n.d. "*Product Datasheets*" Accessed March 01, 2020. http://www.morganthermalceramics.com/en-gb/datasheets/product-datasheets/? page=1&searchTerm=JM28.

[85] RATH. n.d. "*Rath Altra KVR Grades - Product Information Sheet.*" 12. Accessed March 01, 2020. http://www.nurzat.com.tr/assets/files/rath/products/ overview.pdf.

[86] C. M. I., Kapyrok. n.d. "*Kapyrok Product Information.*"1.

[87] Promat. n.d. "*Promasil - Product Technical Data Sheet.*" 1–2.
[88] Promat. n.d. "*PROMALIGHT Technical Data Sheet.*" 1–2.
[89] COMSOL AB. 2008. *RF Module - User's Guide*. Edited by COMSOL AB.
[90] Devesa, S., S. S. Teixeira, M. P. F. Graça, and L. C. Costa. 2017. "Design, Characterization and Test Performances of a Resonant Cavity for Complex Permittivity Measurements Using the Small Perturbation Technique." In *Electrical Measurements: Introduction, Concepts and Applications*, 69–89. Nova Science Publisher.
[91] Annino, G., A. Cintio, R. D'Ambrosio, and A. Lazzeri. 2019. "Dielectric Characterization of Si, SiC, SiC/SiC, and C/SiC Samples." In *Book of Abstracts, AMPERE 2019 - 17th International Conference on Microwave and High Frequency Heating*, edited by Beatriz García Baños, 189. Valencia: Editorial Universitat Politècnica de València.
[92] Kuang, J., and W. Cao. 2013. "Silicon Carbide Whiskers: Preparation and High Dielectric Permittivity." *Journal of the American Ceramic Society* 96(9):2877–80. Accessed April 29, 2020. doi:10.1111/jace.12393.
[93] Fayos-fernández, J., I. Pérez-conesa, J. Monzó-cabrera, S. del Pino-de León, and J. C. Albaladejo-González. 2018. "Temperature-Dependent Complex Permittivity of Several Electromagnetic Susceptors at 2.45 GHz." *Ampere Newsletter*, no. 95:2–8.
[94] Jin, H., M. Cao, W. Zhou, and S. Agathopoulos. 2010. "Microwave Synthesis of Al-Doped SiC Powders and Study of Their Dielectric Properties." *Materials Research Bulletin* 45(2):247–50. Accessed May 08 doi: 10.1016/j.materresbull. 2009.09.015.
[95] Rubinger, C. P. L., and L. C. Costa. 2007. "Building a Resonant Cavity for the Measurement of Microwave Dielectric Permittivity of High Loss Materials." *Microwave and Optical Technology Letters* 49(7):1687–90. Accessed April 29, 2020. doi:10.1002/mop.22506.
[96] Satapathy, L. N. 2009. "Microwave Assisted Sintering of High Voltage Porcelain Material and Its Characterization." *Journal of Ceramic Processing Research* 10(5):637–42.
[97] Belhouchet, K., A. Bayadi, H. Belhouchet, and M. Romero. 2019. "Improvement of Mechanical and Dielectric Properties of Porcelain Insulators Using Economic Raw Materials." *Boletin de La Sociedad Espanola de Ceramica y Vidrio* 58(1):28–37. Accessed April 19, 2020. doi:10.1016/j.bsecv.2018.05.004.
[98] Kasrani, S., A. Harabi, S. E. Barama, L. Foughali, M. T. Benhassine, and D. M. Aldhayan. 2016. "Sinterização e propriedades dielétricas de uma porcelana técnica preparada a partir de matérias-primas naturais de baixo custos." *Cerâmica* 62 (364):405–12. ["Sintering and Dielectric Properties of a Technical Porcelain Prepared from Economical Natural Raw Materials" *Ceramic* 62(364):405–12]. Accessed April 19, 2020. doi:10.1590/0366-69132016623641994.
[99] Kitouni, S. 2014. "Dielectric Properties of Triaxial Porcelain Prepared Using Raw Native Materials Without Any Additions." *Balkan Journal of Electrical and Computer Engineering* 2(3):128–31. Accessed April 19, 2020. doi:10.17694/ bajece.40949.

[100] Orzechowski, K., T. Słonka, and J. Głowinski. 2006. "Dielectric Properties of Intercalated Kaolinite." *Journal of Physics and Chemistry of Solids* 67(5–6):915–19. Accessed May 12, 2020. doi:10.1016/j.jpcs.2006.03.001.

[101] Guan, Y., and Y. Nikawa. 2007. "Measurement of Temperature-Dependent Complex Permittivity for Materials Using Cylindrical Resonator under Microwave Irradiation." *Electronics and Communications in Japan, Part II: Electronics (English Translation of Denshi Tsushin Gakkai Ronbunshi)* 90(11):1–8. Accessed April 08, 2020. doi:10.1002/ecjb.20346.

[102] Hutcheon, R. M., M. S. de Jong, F. P. Adams, P. G. Lucuta, J. E. McGregor, and L. Bahen. 1992. "RF and Microwave Dielectric Measurements to 1400°C and Dielectric Loss Mechanisms." In *Microwave Processing of Materials III*, edited by Magdy F. Iskander Ronald L. Beatty, Willard H. Sutton, 269:541–51. San Francisco: Materials Research Society.

[103] Xi, Weiguo, and W. R. Tinga. 1992. "Microwave Heating and Characterization of Machinable Ceramics." In *Microwave Processing of Materials III*, edited by Magdy F. Iskander Ronald L. Beatty, Willard H. Sutton, 269:569–77. San Francisco: Materials Research Society.

[104] Balmus, S. B., G. N. Pascariu, F. Creanga, I. Dumitru, and D. D. Sandu. 2006. "The Cavity Perturbation Method for the Measurement of the Relative Dielectric Permittivity in the Microwave Range." *Journal of Optoelectronics and Advanced Materials* 8(3):971–77.

Chapter 4

Dielectric Permittivity and Models of Dielectric Relaxation

Georgios Giannoukos*, **Vasileios Chioktour and Athanasios Kakarountas**

Department of Computer Science and Biomedical Informatics,
University of Thessaly, Lamia, Greece

Abstract

In this chapter a study on dielectrics is made. More specifically, the electric dipole in homogeneous and inhomogeneous electric fields is studied as well as polar and non-polar molecules. The four types of polarization of dielectric materials, namely electronic, atomic, bipolar and interfacial polarization, are also analyzed. This is followed by a review of dielectric permittivity and its dependence on frequency, temperature, pressure and humidity. Subsequently a capacitor with parallel plates as well as RC circuits in series and in parallel is studied. Finally, the equations of the relaxation model of Debye, Cole-Cole, Cole-Davidson and Havriliak–Negami are presented.

Introduction

The terms dielectric and insulator are often used synonymously, although we must note that the term dielectric is broader than the term insulator. In any case, these are materials with a very large electrical resistance and with an energy gap greater than 3eV.

---

* Corresponding Author's Email: g.giannoukos@gmail.com.

In: An Introduction to Permittivity
Editor: Parveen Saini
ISBN: 979-8-88697-042-5
© 2022 Nova Science Publishers, Inc.

A dielectric is a material which is not a conductor but whose basic electrical property is its ability to polarize in an electric field. Likewise an insulator is a material which has low conductivity and which creates obstruction to the flow of current and can't be polarized in an electrical field. Examples of dielectric materials are rubber, glass, porcelain, plastics, etc (Serway1990).

Electric Dipole

The electric dipole is used to describe many natural phenomena such as in the operation of an antenna and in the explanation of the forces between molecules. In chemistry, the compound molecules, such as the water molecule (H_2O), behave like a dipole in many cases.

The electric dipole is defined as a system of two opposite electric charges ±q located at a specific distance from each other. One characteristic of a dipole is that it has zero total charge but it has an important property, its electric torque which is a vector quantity that is symbolized by the letter p and is defined by the equation:

$$\vec{p} = q\vec{L} \tag{1}$$

and has the following features:

- direction from negative to positive charge and
- magnitude given by the multiplication of the absolute value of the charge with the distance L between the two charges of the dipole.

$$p = qL \tag{2}$$

We understand that the orientation of the dipole in space is determined by its electric torque, the behaviour and orientation of the dipole depending on the type of electric field, that is whether the field is homogeneous or heterogeneous (Stenzel 2015).

In complex continuous electric charge distributions the electric torque can be calculated with the following integral:

$$\vec{p} = \int \rho_{(\vec{r}')}\vec{r}'\, dr' \tag{3}$$

where $\rho_{(r')}$ is the electric density at a distance r′ (Bonin and Kresin 1997).

When an electric dipole is placed in a homogeneous electric field, its electric torque p forms an angle φ with the intensity E of the field and an electric force is exerted on each charge according to the equation

$$F = Eq \tag{4}$$

The electric charge +q receives a force F+ which is balanced by the field strength, while the electric charge -q receives a force F- which is opposite to it. These two forces are a pair of forces, opposite in equal magnitude so mechanical torque appears in the dipole which is a vector quantity and is defined by the equation:

$$\vec{\tau} = \vec{F} x \, \vec{d} \tag{5}$$

and has the following features:

- direction perpendicular to the plane formed by the direction of the forces and the distance between the two charges,
- direction determined by the right-hand rule and
- magnitude given by the following equation:

$$\tau = FL \sin\varphi => \tau = EqL \sin\varphi \;\; => \tau = Ep \sin\varphi \tag{6}$$

where φ is the angle made between p and E.

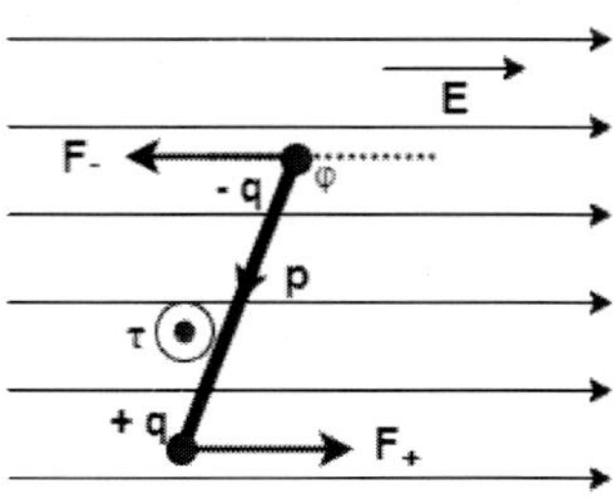

Figure 1. Electric dipole in a homogeneous electric field.

Mechanical torque forces the dipole to rotate and it can be observed that the angle φ, which tends to be zero, changes so that the electric vector torque points in the same direction as the electric intensity E of the electric field.

When the dipole is in the same direction as the dynamic lines, the mechanical torque is zeroed and the dipole is balanced.

During the rotational movement of the dipole at an elementary angle dφ, elementary work is produced:

$$dW = \tau\, d\varphi \tag{7}$$

The produced work W for each angle change dφ is given by the equation:

$$W = \int_{\varphi_0}^{\varphi} \tau d\,\varphi \;=>\; W = \int_{\varphi_0}^{\varphi} pE \sin\varphi\, d\,\varphi \;=>$$

$$W = pE[(-\cos\varphi) - (-\cos\varphi_0)] \;=>$$

$$W = pE(\cos\varphi_0 - \cos\varphi) \tag{8}$$

where φ_0 is the angle of the dipole reference position.

If the angle $\varphi_0 = 90^o$ ($\cos\varphi_0 = 0$), is defined, the reference position of the dipole shows that the electric torque is perpendicular to the intensity of the electric field, the produced work is:

$$W = pE\cos\varphi \tag{9}$$

Work W during the rotational motion of the dipole is saved in the form of potential energy:

$$U = -\vec{p}\vec{E} \tag{10}$$

where the potential energy of the dipole indicates the state of equilibrium of the dipole. When the vectors E and p are in the same direction (φ = 0°), potential energy is negative and a state of stable equilibrium prevails, while when the vectors E and p are in opposite direction (φ = 180°), potential energy is positive and a state of unstable equilibrium in the dipole prevails.

When an electric dipole is placed in an inhomogeneous electric field, the forces F+ and F- that receive the two charges +q and –q, respectively, are not the same and the dipole, in addition to being rotated by the mechanical torque, also receives a net force F which moves it into the field. In this case, the dipole performs a complex motion, mechanical torque forcing the dipole to orient itself almost parallel to the dynamic lines in that area of the electric field, while the net force moves it within the field depending on the direction of bipolar torque. More specifically, when its bipolar torque is oriented towards the field, then the dipole moves to a point of stronger field, while when the bipolar torque is opposite the field then it moves to a point of weaker field.

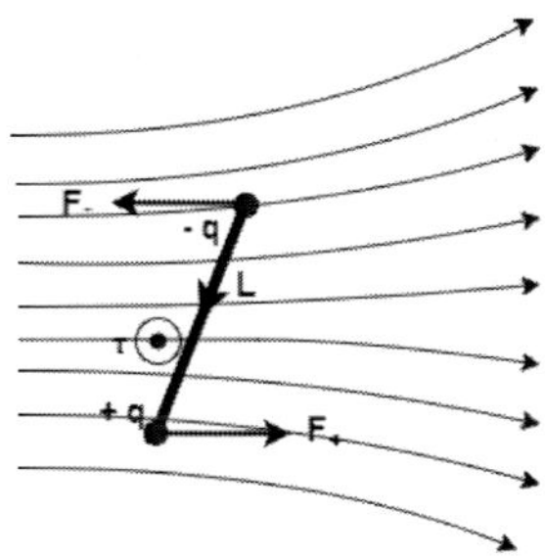

Figure 2. Electric dipole in an inhomogeneous electric field.

If the dipole is ideal (L $\rightarrow$ 0) the mechanical torque τ that it receives in a field with intensity E, is calculated by the equation:

$$\vec{\tau} = \vec{p} x \vec{E} \tag{11}$$

When the dipole is not ideal, then the mechanical torque is equal to the vector sum of the torque of the pair of forces that result from F+ and F-.

The translatory motion of the dipole depends on the net force F given by the equation:

$$\vec{F} = \overrightarrow{F_+} + \overrightarrow{F_-} = q\overrightarrow{E_+} + (-q)\overrightarrow{E_-} = q\left(\overrightarrow{E_+} - \overrightarrow{E_-}\right) \tag{12}$$

where E+, E- are the intensities at the points of the field where the charges are +q and –q, respectively.

Polar and Non-Polar Molecules

The molecules within substances are divided into polar or non-polar and present different chemical and physical behaviour. In a diatomic molecule there is the electropositive and the electronegative atom that determine the electrical torque of the dipole. Respectively in a polyatomic molecule its geometry plays an important role, more specifically each bond can be a dipole, the total bipolar moment determined by the orientation of the individual dipoles (Pauling 1960).

A molecule is polar when the centres of mass of its positive and negative charges do not coincide which occurs if the bond is formed between atoms of different elements. In this case the molecule will have an electric torque and will behave like an electric dipole when it is in a homogeneous electric field. Typical examples of polar molecules are water, hydrochloride, ethyl alcohol molecules, etc.

To be more specific, the water molecule (H_2O) behaves as an electric dipole. Although its molecule is electrically neutral, we know that the molecule causes a permanent shift of electric charges due to the nature of the chemical hydrogen-oxygen bond and as a result polarization occurs. Due to this phenomenon, the water is an excellent solvent for ionic compounds, such as NaCl, and owing to this excellent property of water to behave as an electric dipole, the dissolution of various substances in the water is vital for chemistry and biology (Debye 1929).

In contrast, a molecule is non-polar when the centres of mass of the positive and negative charges coincide so that it does not show electric torque, in which case a bond is formed between atoms of the same elements and in completely symmetrical molecules. When these molecules, which do not have a permanent electric torque, are found in an electric field, they will acquire an electric torque and will be converted into polar molecules. This occurs since the centre of mass of the positive and negative charges shift and they balance in new positions that no longer coincide. All monoatomic and diatomic molecules between atoms of the same element are non-polar. Typical examples are hydrogen, iodine, benzene, ethane molecules, etc (Tykodi 1989).

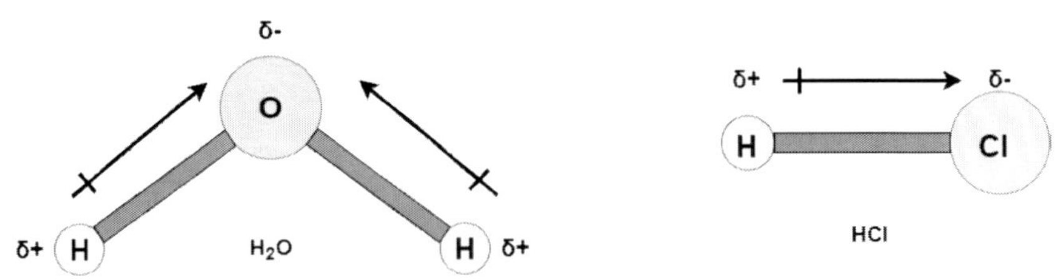

Figure 3. Polar molecules of water (H_2O) and hydrochloride (HCl).

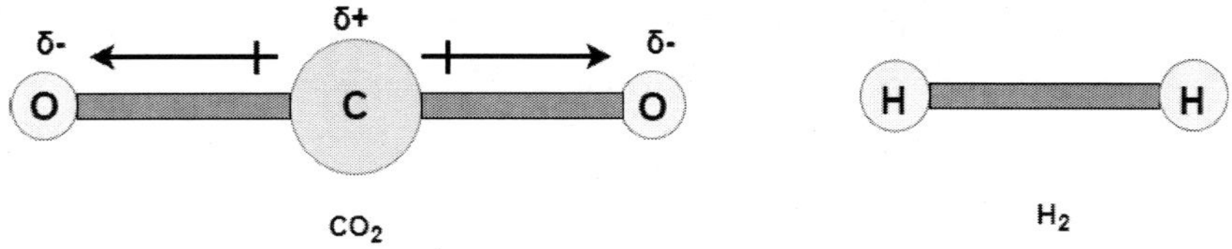

Figure 4. Non-polar molecules of carbon dioxide (CO_2) and hydrogen (H_2).

Any material that does not have free electric carriers inside is called a dielectric material or insulator. When the dielectric material is not in electric field, its bipolar molecules, due to their erratic thermal motion, have a random orientation in the space occupied by the dielectric material, resulting in a total absence of electric torque. This is on account of the fact that the electric torques of the bipolar molecules are, in general, mutually exclusive. However, when the material is placed in an electric field, the bipolar molecules of an area are oriented so that their electric torques become balanced in the same direction as the intensity of the electric field of that region. Thus each region of the dielectric material acquires electric torque. Of course, full orientation of the bipolar molecules of an area does not occur because in this process the thermal motion of the molecules reacts. The lower the temperature of the dielectric, the easier it is to orient its molecules from the electric field and at absolute zero temperature their orientation is complete (Hill, Vaughan, Price, and Davis 1969).

Polarization of Dielectric Material

Generally when a material is placed inside an external electric field, then all its charged particles will receive forces so they tend to move along the field. Each charged particle of the material contributes to the polarization of the material depending on their degree of agility. Dielectric polarization $\vec{P}$ for a dielectric material is a vector quantity which is defined as the quotient of the electric torque dp of an elementary volume of a part of the material to this elementary volume dV (Boettcher 1973).

$$\vec{P} = \frac{d\vec{p}}{dV} \tag{13}$$

The dielectric polarization $\vec{P}$ has the same direction as the electric torque of an elementary part of the material. If the bipolar molecules of the dielectric material are uniformly oriented, in the case of an isotropic dielectric then, the dielectric polarization at each point is the same and is equal to:

$$\vec{P} = \frac{\vec{p}}{V} \tag{14}$$

where p is the total electric torque of the dielectric material and V is its volume.

We can say that the dielectric polarization expresses the electric torque per unit volume of the dielectric (Boettcher and Bordewijk 1980).

Dielectric Polarization

There are different mechanisms of polarization, the main ones being electronic, ionic, bipolar, and atomic (Hofmann 2015).

- Electronic polarization occurs on each atom or molecule because the center of mass of the electron cloud that surrounds the atom shifts due to the EF. This effect is extremely fast as the electrons are very light and can be observed by optical frequencies. This shift depends on the intensity of the field and the attractive forces between the electrons and the nucleus. The time constant of this process is in the ultraviolet region. Due to electronic polarization, permittivity ε_∞ is determined at high frequencies.

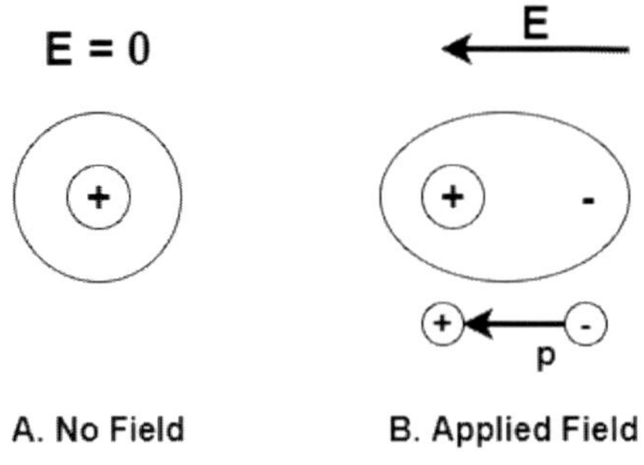

Figure 5. Electronic polarization.

- Atomic or ionic polarization refers to materials whose molecules form ions. When an EF is applied then cations and anions become displaced in opposite directions. Apart from electric polarization, elastic displacements of the electrical charges (nuclei and electrons) will also occur. This means that these types of molecules are polar compounds that can be polarized up to infrared frequencies.

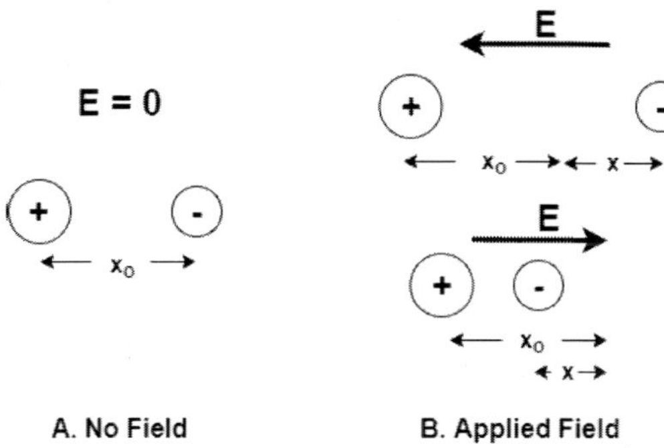

Figure 6. Ionic polarization.

- Bipolar polarization refers to materials that contain molecules with permanent dipole moments. Under the influence of EF, the dipoles are oriented in the direction of the EF. Orientation polarization depends on temperature as opposed to electronic and ionic polarization. The ability of polar molecules to orient themselves is offset by their thermal motion. The higher the temperature the lower the degree of the orientation.

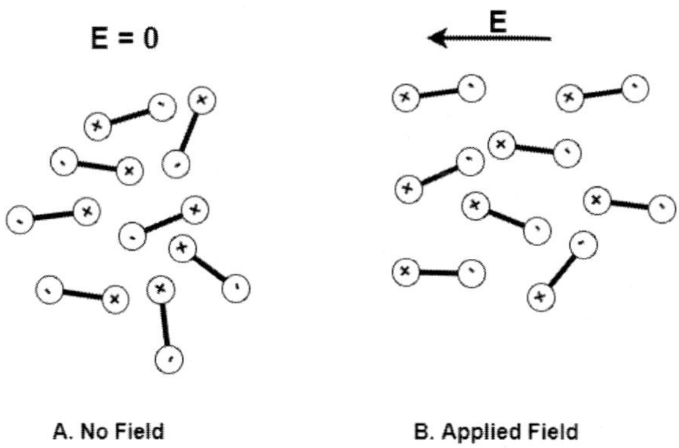

Figure 7. Bipolar polarization.

- Interfacial (or space charge) polarization refers to materials that have different phases (inhomogeneous) so when an electric field is applied, the space charges move to the limits of each phase known as interfaces. Eventually they are oriented to increase the polarization of the material, an effect known as Maxwell-Wagner-Sillars polarization (MWS polarization) (Hedving 1977). During the manufacture of the material, free charges are created which move within the material under the influence of an electric field, whereupon continuous (dc) electrical conductivity occurs. Cargo carriers behave like large dipoles in areas of different conductivity. This phenomenon is often very slow and is generally perceived in the power frequency range and below ($f \leq 50Hz$).

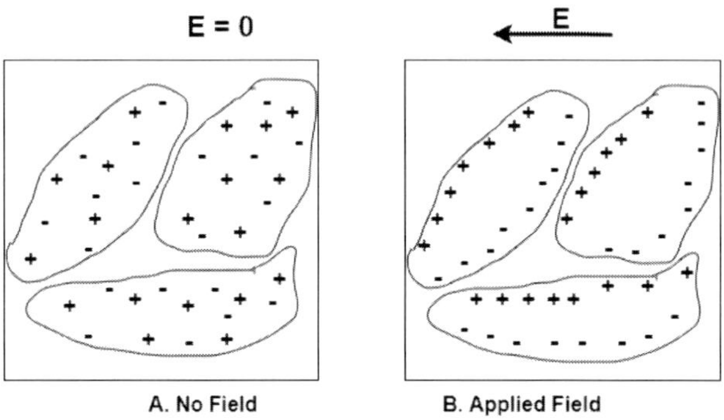

Figure 8. Interfacial polarization.

In conclusion, dielectric polarization is the result of the relative displacement of positive and negative charges in a material. During this process the EF is not able to force the charge to escape from the material which would cause electrical conductivity.

Each of the dielectric polarization mechanisms is associated with a characteristic resonance frequency (sharp increase of dielectric permittivity at a given frequency) and relaxation frequency (gradual decrease in dielectric permittivity with increasing frequency). As the frequency increases, the slower dielectric polarization mechanisms are eliminated leaving only the fastest to contribute to the phenomena of energy storage (ε_r'). Similarly, the rate of loss (ε_r'') is incremented at each critical frequency. Resonance is usually associated with electronic polarization (Dakin 2006).

The mechanisms of dielectric polarization depend on frequency and are shown in the figure below.

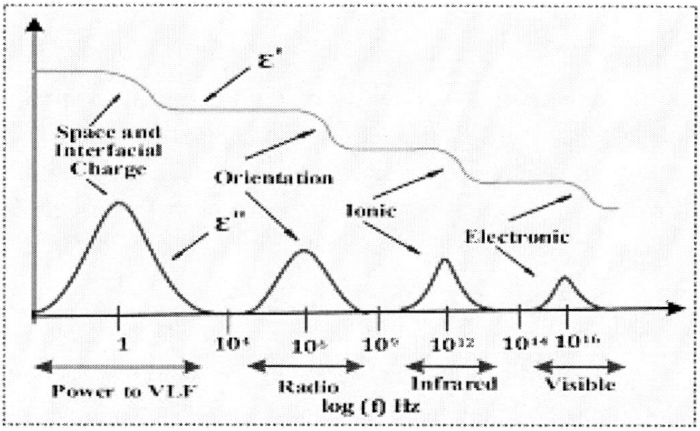

Figure 9. Dielectric polarization mechanisms (Morsalin, Phung, Danikas, and Mawad 2018).

Dielectric Permittivity

When an electric field is applied to a material then the electric charges of the material are moved and shifted. The movement of the electric charge is characterized by the electrical conductivity σ while the displacement of the electric charge is characterized by the dielectric constant ε, which shows how the electric field affects the polarization of the material. The vector of the polarization P of a material is the electric torque of the dipoles per unit volume and the relation that connects the vector of the dielectric shift D with the intensity of the applied electric field E and the polarization P are given by the following (Jonscher 1983):

$$D = \varepsilon_0 \cdot E + P \tag{15}$$

where ε_o is the vacuum permittivity.

The vectors of the dielectric displacement D and the intensity of the electric field E can be interconnected by a ratio constant called relative permittivity and is denoted by ε_r.

$$D = \varepsilon_0 \cdot \varepsilon_r \cdot E \tag{16}$$

From equations (15) and (16) we get the following relationship between vectors P and E:

$$P = \varepsilon_0 \cdot (\varepsilon_r - 1)\, E = \chi \cdot \varepsilon_0 \cdot E \tag{17}$$

where χ is the dielectric susceptibility of the material which expresses the polarization created by an electric field in a particular material.

The following equation expresses the total electric current density J resulting from the application of an electric field to the material:

$$J = \sigma \cdot E + \varepsilon_0 \frac{\partial E}{\partial t} + \frac{\partial P}{\partial t} \tag{18}$$

As already mentioned, there are different types of polarization mechanisms some of which are faster than others. Therefore, when a constant electric field is applied, the total polarization of the material varies. This is also shown in the figure below, in which P_∞ is polarization due to fast mechanisms and P_s is the total polarization of the material after an extended length of time. Instantaneous polarization is given by the following equation (Jonscher 1983; Frubing 2001):

$$P_{(t)} = P_\infty + (P_S - P_\infty) \cdot g\,(t - t_0)\,, if\ t \geq t_0 \tag{19}$$

where $g_{(t)}$ is a monotonous ascending function that satisfies the following condition:

$$g_{(t)} = 0\,, if\ t \leq t_0$$

$$g_{(t)} = 1\,, if\ t \rightarrow \infty$$

$$g_{(t)} \geq 0\,, \frac{dg}{dt} \geq 0, if\ t \geq t_0 \tag{20}$$

From equations (17) and (19), the total polarization due to a constant electric field E_0 can be expressed by the following equation:

$$P_{(t)} = \{\varepsilon_0(\varepsilon_\infty - 1) + (\varepsilon_S - \varepsilon_\infty) \cdot g\,(t - t_0)\} \cdot E_0 \tag{21}$$

where ε_s and ε_∞ are the static and high frequency dielectric constants of the material, respectively. Another form of the above equation is the following:

$$P_{(t)} = \varepsilon_0(\varepsilon_\infty - 1) \cdot E_{(t)} + \varepsilon_0 \int_{-\infty}^{\tau} f_{(t-\tau)} \cdot E_{(\tau)} d\tau \tag{22}$$

where $f_{(t)}$ is a monotone decreasing function known as a dielectric response function. The first part of equation (22) corresponds to the rapid polarization processes in the material.

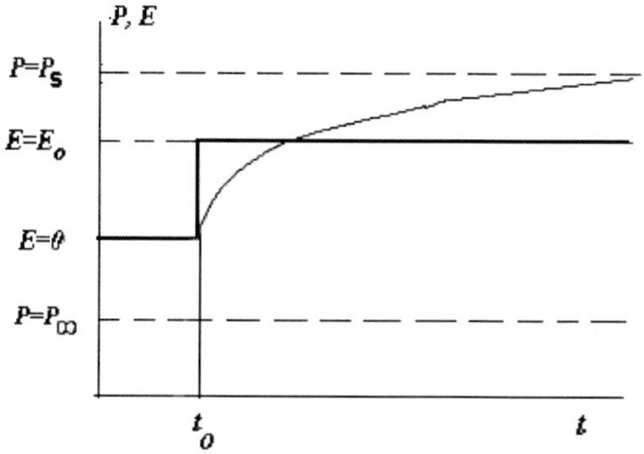

Figure 10. Time dependence of polarization due to constant electric field strength E0 (Jonscher 1983).

The total current density $J_{(t)}$ due to a constant electric field results from equations (18) and (22):

$$J_{(t)} = \sigma \cdot E_{(t)} + \varepsilon_0\left[\varepsilon_\infty \cdot \delta_{(t)} + f_{(t)}\right] \cdot E_{(t)} \tag{23}$$

The term $\sigma \cdot E_{(t)}$ corresponds to the current density due to the conductivity of the material, the term $\varepsilon_0 \cdot \varepsilon_\infty \cdot \delta_{(t)} \cdot E_{(t)}$ to the current density resulting from the rapid polarization processes and the term $\varepsilon_0 \cdot f_{(t)} \cdot E_{(t)}$ to the current density due to the slow polarization processes.

The Dielectric Response in a Frequency Range

If the electric field strength is exponentially variable, according to equation (Jonscher 1983):

$$E_{(t)} = E_0 \, exp(j\omega t) \tag{24}$$

and by substituting equation (24) with (22), the electric displacement D(t) can be expressed by the following equation:

$$D_{(t)} = \varepsilon_0 \cdot \{\varepsilon_\infty + \int_0^\infty f_{(t)} \cdot exp(-j\omega t)dt\} E_0 \cdot exp(j\omega t) \tag{25}$$

The term $\int_0^\infty f_{(t)} \cdot exp(-j\omega t)dt$ is equivalent to the Fourier transform of the function f(t), which is defined as the frequency-dependent complex dielectric susceptibility $\chi_{(\omega)}$:

$$\chi_{(\omega)} = \chi'_{(\omega)} - j\,\chi''_{(\omega)} = \int_0^\infty f_{(t)} \cdot exp(-j\omega t)dt \tag{26}$$

where $\chi'_{(\omega)}$ and $\chi''_{(\omega)}$ are the real and imaginary part of the complex susceptibility, respectively. Therefore equation (25) is written:

$$D_{(t)} = \varepsilon_0 \cdot \{\varepsilon_\infty + \chi_{(\omega)}\} E_0 \cdot exp(j\omega t) \tag{27}$$

We can conclude, therefore, that in the frequency domain the current density $J_{(\omega)}$ generated in a dielectric material by an externally applied electric field $E_{(\omega)}$, is given by the equation:

$$J_{(\omega)} = \sigma \cdot E_{(\omega)} + j \cdot \omega \cdot [\varepsilon_0 \cdot \varepsilon_\infty + \varepsilon_0 \cdot \{\chi'_{(\omega)} - j \cdot \chi''_{(\omega)}\}] \cdot E_{(\omega)} \Rightarrow$$

$$J_{(\omega)} = j \cdot \omega \cdot \varepsilon_0 \cdot \left[\varepsilon_\infty + \chi'_{(\omega)} - j\left\{\frac{\sigma}{\varepsilon_0 \cdot \omega} + \chi''_{(\omega)}\right\}\right] \cdot E_{(\omega)} \Rightarrow$$

$$J_{(\omega)} = j \cdot \omega \cdot \varepsilon_0 \cdot [\varepsilon'_{(\omega)} - j \cdot \varepsilon''_{(\omega)}] \cdot E_{(\omega)} \tag{28}$$

where

$$\varepsilon'_{(\omega)} = \varepsilon_\infty + \chi'_{(\omega)} \tag{29}$$

$$\varepsilon''_{(\omega)} = \frac{\sigma}{\varepsilon_0 \cdot \omega} + \chi''_{(\omega)} \tag{30}$$

Equation (29) corresponds to the real part, while equation (30) corresponds to the imaginary part of the complex permittivity, which is given by the following equation:

$$\varepsilon^*_{(\omega)} = \varepsilon'_{(\omega)} - j \cdot \varepsilon''_{(\omega)} \tag{31}$$

The real part $\varepsilon'_{(\omega)}$ of the complex permittivity is related to the amount of energy of the electric field stored in the material. The imaginary part $\varepsilon''_{(\omega)}$ indicates the loss of energy of the electric field in the material. The term $\frac{\sigma}{\varepsilon_0 \cdot \omega}$ is associated with ohmic losses due to the free movement of electrical charges in the material (Hedvig 1977; Giannoukos, Min and Rang 2017).

The quotient of the imaginary to the real part is known as the loss tangent and is given by the equation:

$$\tan\delta = \frac{\varepsilon''}{\varepsilon'} \tag{32}$$

This quotient shows the relationship between the rate of energy loss and the rate at which energy is stored in the material.

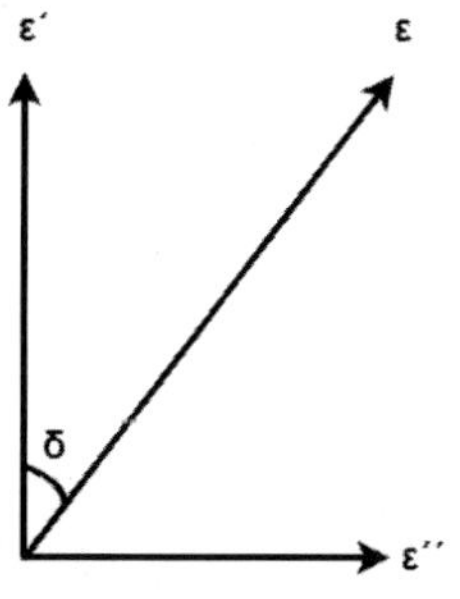

Figure 11. Diagram of the real part vs the imaginary part of the complex permittivity.

Dependence of Permittivity on Temperature, Pressure and Humidity

Permittivity depends not only on the frequency but also on temperature and pressure. The ideal Debye response depends exponentially on temperature according to the equation:

$$\chi_{(\omega)} \propto \frac{1}{1+i \cdot \omega \cdot \tau_\infty \cdot exp\left(\frac{W}{kT}\right)} \tag{33}$$

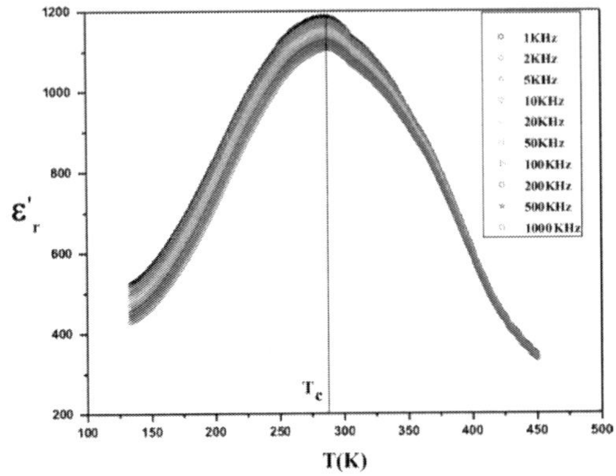

Figure 12. Temperature dependence of the permittivity ε'r for a ceramic with composition Ba 0.95 Na 0.05 Ti 0.95 (Chihaoui, Chaker, Chiheb and Khemakhem 2017).

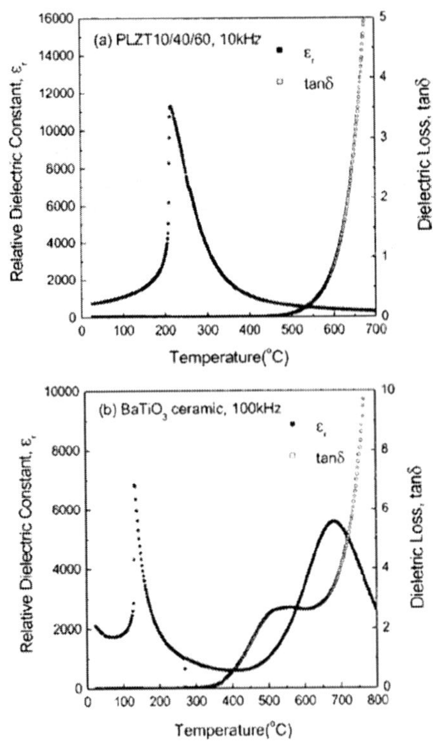

Figure 13. Temperature dependence of the dielectric constant in (a) PLZT 10/40/60 ceramics and (b) BaTiO3 ceramics (Kang, Choi and Park 2003).

The subsequent diagrams show the dependence between permittivity and temperature for various materials such as ceramics and an amorphous polymer.

If the response is not of the Debye type then there is no way to correlate the temperature dependence with the frequency but it has been found that frequency does not change significantly with temperature (Zhong and Xu 1996).

Pressure also affects permittivity in a manner similar to that of temperature.

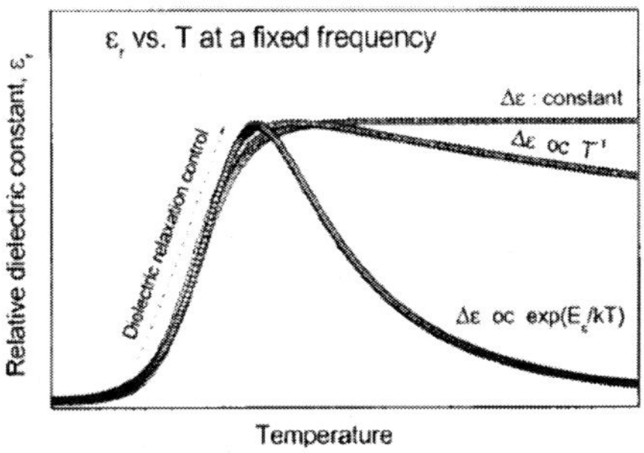

Figure 14. Diagram of the temperature dependent behavior of the dielectric relaxation (Kang, Choi and Park 2003).

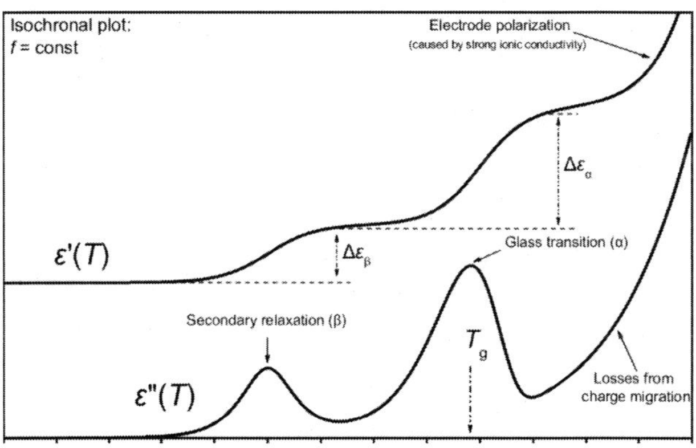

Figure 15. Behavior of temperature dependence of ε' and ε" for an amorphous polymer (Vassilikou-Dova and Kalogeras 2008).

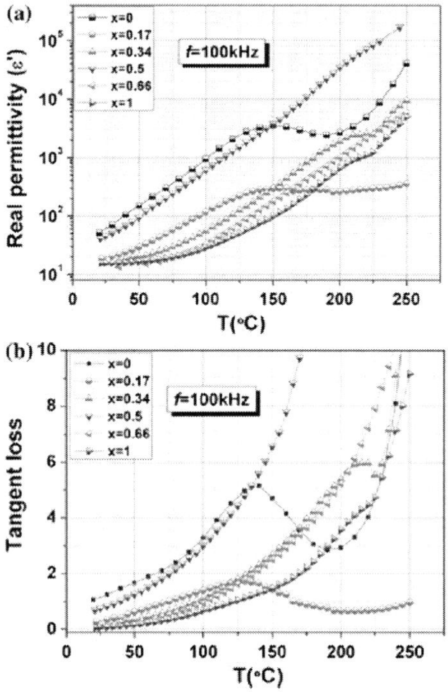

Figure 16. Temperature dependence of (a) ε′and (b) tangent loss for Mg x Ni 1-x Fe 2 O 4 ceramics (Mocanu, Airimioaei, Curecheriu, Tudorache, Tascu, Iordan, Palamaru and Mitoseriu 2014).

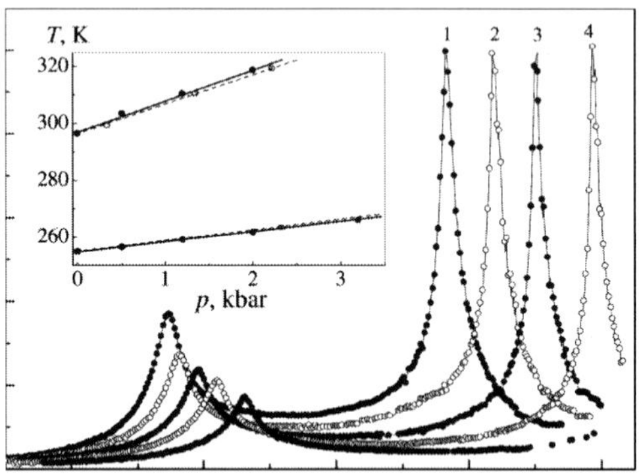

Figure 17. Temperature dependence of the dielectric permittivity of Rochelle salt at different values of hydrostatic pressure p (Slivka, Kedyulich, Iqbal Khan, Moina, Romanyuk and Guivan 2005).

The diagram below shows how the electrical susceptibility of a material changes at two different temperatures for different pressure values. The result is that pressure and temperature have a corresponding effect on electrical susceptibility.

Also another factor that affects permittivity is humidity. The following diagrams show how the $\varepsilon''_{(\omega)}$ curve changes for different percentages of sand moisture.

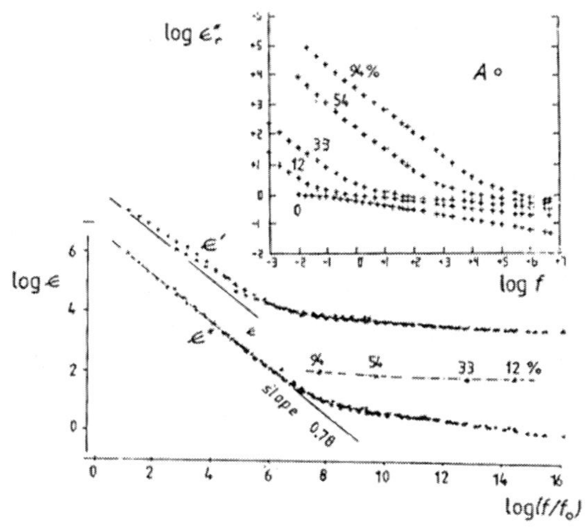

Figure 18. Results of dielectric measurements in sand samples with different humidity (Jonscher 1983).

Parallel Plates with a Dielectric

The most common dielectric characterization device is a capacitor with parallel plates between which the dielectric material under study is located.

The capacity of a pair of parallel charged plates is increased by introducing a dielectric material. The capacitance is inversely proportional to the electric field, while the presence of the dielectric reduces the field between the plates due to the existence of a field with opposite direction within the material because of the polarization of the material.

The capacity of a pair of parallel charged plates increases when a dielectric material is introduced (Stanford and Tanner 1985). It is known that the dielectric is characterized by the dielectric permittivity ε_r and the capacity is multiplied by this factor. An overview of the capacity is:

$$C = \frac{\varepsilon_r \cdot \varepsilon_0 \cdot A}{d} \tag{34}$$

if there is air between the plates then:

$$C = \frac{\varepsilon_0 \cdot A}{d} \tag{35}$$

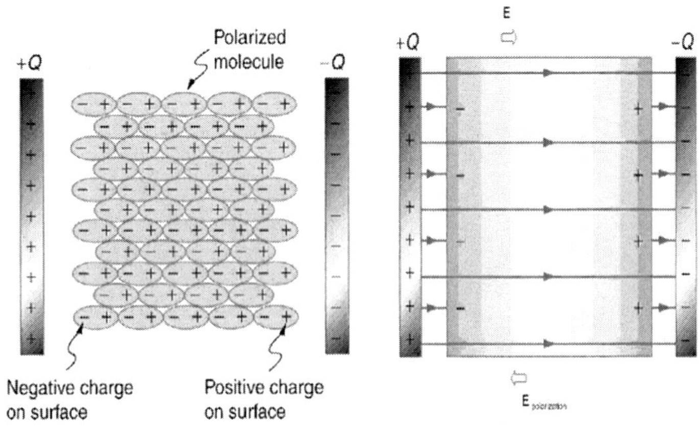

Figure 19. Parallel plates charged (Retrieved from https://uncw.instructure.com/courses/17048/modules/items/158531).

The electric field when there is air or a dielectric material between the plates is:

$$E = \frac{\sigma}{\varepsilon} = \frac{V}{d} \tag{36}$$

and

$$E_{effective} = E - E_{polarization} = \frac{\sigma}{\varepsilon_r \cdot \varepsilon_0}, \text{ respectively} \tag{37}$$

where V is the applied voltage and d the distance between the plates and σ is the surface charge density in each plate.

The relative dielectric permittivity ε_r shows the amount of reduction of the active field in relation to the initial field. This reduction of the field results in the ability for more energy to be stored in the capacitor having the same voltage between the plates (Kaiser 1990).

Along with energy storage, non-ideal capacitor is characterized by power consumption (losses). In the dielectric there are two types of losses:

i) ohmic losses: there is a difference between the volume resistance (R_v) and surface resistance (R_s). The total resistance is:

$$R_{total} = \frac{R_V \cdot R_S}{R_V + R_S} \quad (38)$$

The ohmic losses are greater when the temperature of the insulator is increased.

ii) Dielectric losses: They are the result of a change of polarity of the dielectric due to changing the direction of the field in each period. This change causes oscillations that produce heat due to friction so part of the electrical energy turns to heat. The extent of loss is:

$$P = V_{rms}^2 \cdot 2\pi f \cdot C \cdot \tan\delta \quad (39)$$

where $\tan\delta$ is the loss factor and is a measure of dielectric losses depending on the frequency (Froehlich 1958).

RC Ciruit

The electrical equivalent circuit which is used to simulate a parallel plate capacitor may consist of a resistor and an ideal capacitor connected either in series or parallel.

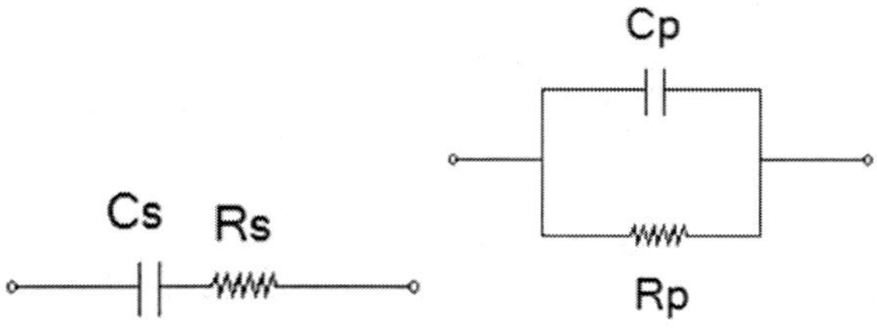

Figure 20. The RC series and parallel circuit respectively.

The parallel equivalent circuit is usually used. The vector diagrams of voltages and currents for the two equivalent circuits are, respectively (Figure 21).

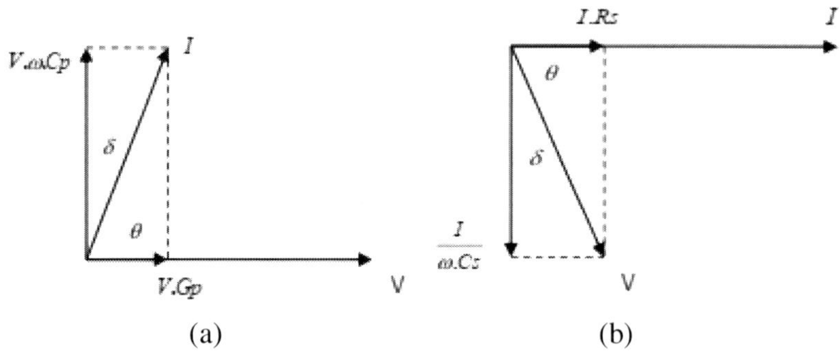

Figure 21. Vector diagram of voltages (a) and currents (b) of the series and the parallel circuit, respectively.

These two circuits describe the same system which is why the result regarding the relationship between voltage and current must be the same. In this case the angles θ, δ in both diagrams and the complex impedance must tally.

Subsequently, the necessary definitions are given to determine the analytical equations which connect the elements of the serial equivalent circuit with the complex dielectric permittivity and the loss tangent obtained from the parallel equivalent circuit (*Kaplan and White 2003).*

The loss tangent from the parallel circuit is

$$tan\,\delta = \frac{1}{\omega \cdot C_p . R_p} = \frac{\varepsilon_r''}{\varepsilon_r'} \tag{40}$$

and from the in series circuit is

$$tan\,\delta = \omega \cdot R_s \cdot C_s \tag{41}$$

The complex impedance of the parallel circuit is

$$Z_p = \frac{1}{\frac{1}{R_p} + j \cdot \omega \cdot R_p} \tag{42}$$

and from the in series circuit is

$$Z_s = R_s + \frac{1}{j \cdot \omega \cdot C_s} \tag{43}$$

Given that $Z_p = Z_s$ this means the real parts must be equal as must the imaginary parts, and by solving the equation system we get *(Dorf and Svoboda 2015)*

$$C_P = \frac{C_S}{1+tan^2\,\delta} \Rightarrow C_P = \frac{1}{1+\omega^2 \cdot R_S{}^2 \cdot C_S{}^2} \tag{44}$$

and for the resistors

$$\frac{R_P}{R_S} = \frac{1+tan^2\,\delta}{tan^2\,\delta} = 1 + \frac{1}{tan^2\,\delta} \tag{45}$$

So the real and the imaginary part of dielectric permittivity are

$$\varepsilon_r{}' = \frac{C_P}{C_O} \Rightarrow \varepsilon_r{}' = \frac{C_S}{C_O \cdot (1+\omega^2 \cdot R_S{}^2 \cdot C_S{}^2)},$$

$$\varepsilon_r{}'' = \varepsilon_r{}'\,tan\,\delta = \varepsilon_r{}' \cdot \omega \cdot R_S \cdot C_S = \frac{\omega \cdot R_S \cdot C_S{}^2}{C_O(1+\omega^2 \cdot R_S{}^2 \cdot C_S{}^2)} \tag{46}$$

but

$$\varepsilon_r = \varepsilon_r{}' - j\varepsilon_r{}'' \tag{47}$$

And finally the complex permittivity is

$$\varepsilon_r = \frac{C_S}{C_O(1+\omega^2 \cdot R_S{}^2 \cdot C_S{}^2)} - j\frac{\omega \cdot R_S \cdot C_S{}^2}{C_O(1+\omega^2 \cdot R_S{}^2 \cdot C_S{}^2)} \tag{48}$$

The last equation describes the complex dielectric permittivity with the help of a serial equivalent circuit. In practice, the result which is obtained for the material being tested does not depend on which equivalent circuit was used. The parallel circuit is more commonly used because the calculation of ε_r' and ε_r'' is simpler.

Debye Relaxation Model

The real part of dielectric permittivity ε′ varies between a maximum value ε_s, corresponding to static fields or very low frequencies and a minimum value ε_∞, corresponding to very high frequencies. The imaginary part ε′′ is related to losses and has a maximum at the resonance frequency (Havriliak and Negami 1966; Runt and Fitzgerald 1997).

The ε′ and ε′′are given by Debye's equations (49):

$$\varepsilon' = \varepsilon_\infty + \frac{\varepsilon_S - \varepsilon_\infty}{1+\omega^2\cdot\tau^2}, \varepsilon'' = \omega\cdot\tau\cdot\frac{\varepsilon_S - \varepsilon_\infty}{1+\omega^2\cdot\tau^2}$$

$$\frac{\varepsilon_{(\omega)} - \varepsilon_\infty}{\varepsilon_S - \varepsilon_\infty} = \frac{1}{1+j\cdot\omega\cdot\tau} \Rightarrow \varepsilon_{(\omega)} = \varepsilon_\infty + \frac{\varepsilon_S - \varepsilon_\infty}{1+j\cdot\omega\cdot\tau}$$

$$\tan(\delta) = \frac{\varepsilon''}{\varepsilon'} = \frac{(\varepsilon_S - \varepsilon_\infty)\cdot\omega\cdot\tau}{\varepsilon_S + \varepsilon_\infty\cdot\omega^2\cdot\tau^2} \quad (49)$$

where τ is the relaxation time of the material.

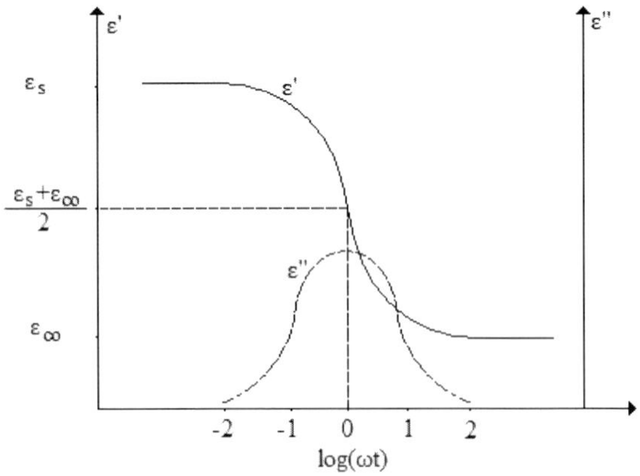

Figure 22. Real and imaginary part of the dielectric permittivity for a Debye mechanism (Levy 1968).

From the above equations we get

$$\varepsilon - \varepsilon_\infty = (\varepsilon' - j\cdot\varepsilon'') - \varepsilon_\infty = \varepsilon' - \varepsilon_\infty - j\cdot\varepsilon'' \Rightarrow$$

$$\varepsilon - \varepsilon_\infty = \frac{\varepsilon_S - \varepsilon_\infty}{1+\omega^2 \cdot \tau^2} - j \cdot \omega \cdot \tau \cdot \frac{\varepsilon_S - \varepsilon_\infty}{1+\omega^2 \cdot \tau^2} \Rightarrow$$
$$\varepsilon'^2 + \varepsilon''^2 = \varepsilon'(\varepsilon_S - \varepsilon_\infty) - \varepsilon_S \cdot \varepsilon_\infty \quad (50)$$

Cole-Cole Relaxation Model

The first empirical expressions for $\varepsilon^*(\omega)$ was given by K. S. Cole and R. H. Cole

$$\varepsilon^*(\omega) = \varepsilon_\infty + \frac{(\varepsilon_S - \varepsilon_\infty)}{1+(i\omega\tau)^{1-\alpha}}, \quad (51)$$

$where\ 0 \leq a \leq 1$

The exponent parameter α describes different spectral shapes, where

$$\varepsilon' = \varepsilon_\infty + (\varepsilon_s - \varepsilon_\infty)\frac{1+(\omega\cdot\tau)^{1-\alpha}\, sin\left(\frac{1}{2}\alpha\pi\right)}{1+2(\omega\cdot\tau)^{1-\alpha}\, sin\left(\frac{1}{2}\alpha\pi\right)+(\omega\cdot\tau)^{2(1-\alpha)}} \quad (52)$$

and

$$\varepsilon'' = (\varepsilon_s - \varepsilon_\infty)\frac{(\omega\cdot\tau)^{1-\alpha}\, cos\left(\frac{1}{2}\alpha\pi\right)}{1+2(\omega\cdot\tau)^{1-\alpha}\, sin\left(\frac{1}{2}\alpha\pi\right)+(\omega\cdot\tau)^{2(1-\alpha)}} \quad (53)$$

The equations can be reduced to the Debye expression when α = 0 (Cole and Cole 1941; Cole and Cole 1942).

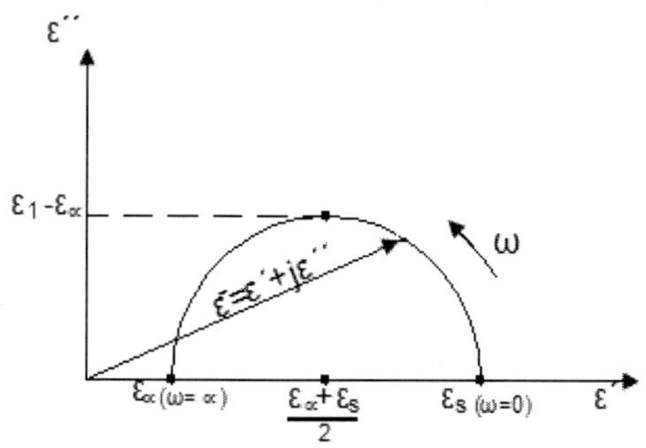

Figure 23. Cole-Cole graph (Levy 1968).

Cole-Davidson Relaxation Model

$$\varepsilon^*(\omega) = \varepsilon_\infty + \frac{(\varepsilon_S - \varepsilon_\infty)}{(1 + i \cdot \omega \cdot \tau)^\beta} \tag{54}$$

where

$$1+\text{i}\cdot\omega \cdot \tau = e^{i\varphi}\sqrt{1 + \omega^2 \cdot \tau^2} = \frac{e^{i\varphi}}{\cos\varphi} \tag{55}$$

where $\tan\varphi = \omega \cdot \tau \Rightarrow \varphi = \tan^{-1}(\omega \cdot \tau)$.

This can be reduced to the Debye equation for β=1 (Davidson and Cole 1950).

Real and imaginary parts are as follows

$$\varepsilon' = \varepsilon_\infty + (\varepsilon_s - \varepsilon_\infty) \cos\varphi^\beta \cos(\beta \cdot \varphi) \tag{56}$$

$$\varepsilon'' = (\varepsilon_s - \varepsilon_\infty) \cos\varphi^\beta \sin(\beta \cdot \varphi) \tag{57}$$

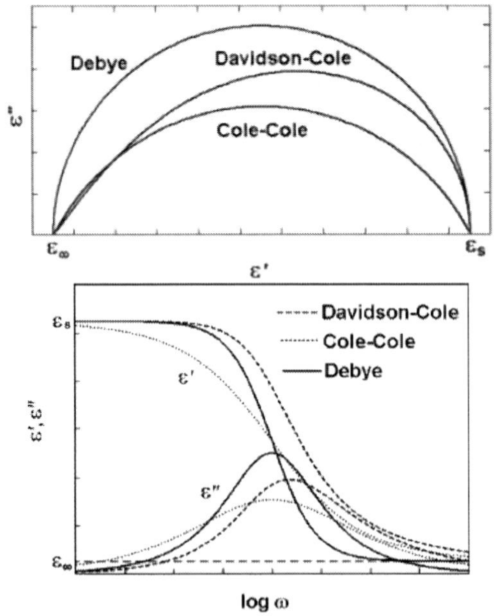

Figure 24. Comparison of Debye, Davidson-Cole and Cole-Cole relaxation models (Lukichev 2012; Lukichev 2014).

The Havriliak–Negami Equation

This model is an empirical modification of the Debye relaxation

$$\varepsilon^*(\omega) = \varepsilon_\infty + \frac{(\varepsilon_S - \varepsilon_\infty)}{(1+(i\cdot\omega\cdot\tau)^\alpha)^\beta} \tag{58}$$

where

$$\varepsilon' = \varepsilon_\infty + (\varepsilon_s - \varepsilon_\infty)\frac{1+(\omega\cdot\tau)^{1-\alpha} sin\left(\frac{1}{2}\alpha\pi\right)}{1+2(\omega\cdot\tau)^{1-\alpha} sin\left(\frac{1}{2}\alpha\pi\right)+(\omega\cdot\tau)^{2(1-\alpha)}} \tag{59}$$

$$\varepsilon'' = (\varepsilon_s - \varepsilon_\infty)\frac{sin(\beta\cdot\varphi)}{\{1+2(\omega\cdot\tau)^{1-\alpha} sin\left(\frac{1}{2}\alpha\pi\right)+(\omega\cdot\tau)^{2(1-\alpha)}\}^{\beta/2}} \tag{60}$$

where $\varphi = arctan\left(\frac{(\omega\cdot\tau)^\alpha\ sin(\pi\frac{\alpha}{2})}{1+(\omega\cdot\tau)^\alpha\ cos(\pi\frac{\alpha}{2})}\right)$

For $\beta = 1$ equation (58) is reduced to the Cole-Cole equation and for α=1 to the Cole-Davidson equation ((Havriliak and Negami 1967; Navriliak and Havriliak 1997).

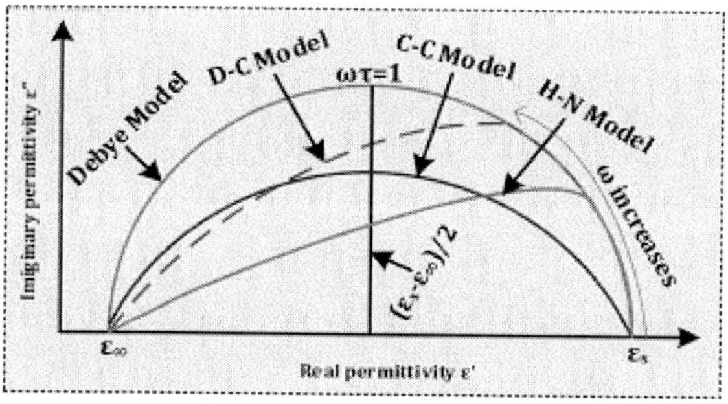

Figure 25. Comparison of Debye, Davidson-Cole, Cole-Cole and Havriliak–Negami relaxation models (Morsalin, Phung, Danikas and Mawad 2019).

Conclusion

This chapter aims to examine the characteristics of dielectrics as well as analyse the way in which relative complex permittivity is affected by frequency and to put forth various approaches to and equations for relative complex permittivity like those of Debye, Cole – Cole, Cole – Davidson and Havriliak – Negami. Furthermore, a capacitor with parallel plates in which the dielectric material under study is found, was discussed this being the most common dielectric characterization device. Then Ohmic and dielectric losses were explained and finally electronic, atomic and bipolar polarization mechanisms are discussed.

Acknowledgments

The authors thank Ms Kate Somerscales for proofreading the manuscript.

References

Boettcher, C. J. (1973). *Theory of Electric Polarization, Vol. 1: Dielectrics in Static Fields, 2nd ed.*, Amsterdam: Elsevier Science B. V.

Boettcher, C. F. and Bordewijk, P. (1980). Theory of Electric Polarisation, Vol. 2: *Dielectrics in Time-Dependent Fields, 2nd ed.,* Amsterdam: Elsevier Science B. V.

Bonin, K. and Kresin, V. (1997). *Electric-dipole polarizabilities of atoms, molecules, and clusters.* Singapore: River Edge, NJ: World Scientific.

Chihaoui, S., Chaker, H., Chaker, C., and Khemakhem, H. (2017). X-ray diffraction, dielectric and Raman studies of the Ba 1-x Na x Ti 1-x (Nb 1-y Sb y) x O3 ceramics. *Ceramics International.*

Cole, R. H. and Cole, K. S. (1941). Dispersion and Absorption in Dielectrics I. Alternating Current Characteristics. New York: *Journal of Chemical Physics.*

Cole, K. S. and Cole, R. H. (1942). Dispersion and Absorption in Dielectrics II. Direct Current Characteristics. New York: *Journal of Chemical Physics.*

Dakin, T. W. (2006). Conduction and polarization mechanisms and trends in dielectric. *Electrical Insulation Magazine, IEEE.*

Davidson, D. W. and Cole, R. H. (1950). *Dielectric relaxation of glycerine.* New York: Journal of Chemical Physics.

Debye, P. (1929). *Polar molecules. Chemical Cataloque Company*, reprinted in New York: Dover Publications.

Dorf C. R. and Svoboda A. J. (2015). Introduction to Electric Circuits, (9rd ed.). John Wiley.

Froehlich, H. (1958). *Theory of Dielectrics. Dielectric Constant and Dielectric Loss, 2nd ed.*, Oxford: Clarendon Press.

Frubing P. (2001). *Dielectric spectroscopy.* Potsdam: University of Potsdam, Institute of Physics.

Giannoukos, G., Min, M., and Rang, T. (2017). Relative complex permittivity and its dependence on frequency. *World Journal of Engineering.*

Havriliak, S. and Negami, S. (1966). A complex plane analysis of α-dispersions in some polymer systems. *Journal of Polymer Science Part C: Polymer Symposia*, Wiley Periodicals, Inc., A Wiley Company.

Havriliak, S.and Negami, S. (1967). *A complex plane representation of dielectric and mechanical relaxation processes in some polymers*. Polymer, Elsevier.

Hedving, P. (1977). *Dielectric spectroscopy of polymers*. John Wiley & Sons.

Hill, N. E., Vaughan W. E., Price A. H., and Davis M. (1969). *Dielectric Properties and Molecular Behavior*. London: Van Nostrand.

Hofmann, P. (2015). *Solid state physics: an introduction*. John Wiley & Sons.

Jonscher, A. K. (1983). *Dielectric relaxation in solids*. London: Chelsea Dielectrics Press.

Kaiser C. J. (1990). *The capacitor handbook*. London: Van Nostrand Reinhold.

Kang, B., Choi, S., Park, C.ul Hong. (2003). Diffuse dielectric anomaly in perovskite-type ferroelectric oxides in the temperature range of 400-700°C. *Journal of Applied Physics*.

Kaplan, D., and White, C. (2003). RC circuits. In *Hands-On Electronics: A Practical Introduction to Analog and Digital Circuits*. Cambridge: Cambridge University Press.

Levy, R. A. (1968). *Principles of Solid State Physics*. Academic Press.

Lukichev, A. A. (2012). Graphical method for the Debye-like relaxation spectra analysis. *Journal of Non-Crystalline Solids*, Amsterdam: Elsevier Science B. V.

Lukichev, A. A. (2014). Relaxation function for the non-Debye relaxation spectra description. *Chemical Physics*.

Mocanu, Z., Airimioaei, M., Curecheriu, L., Tudorache, F., Tascu, S., Iordan, A., Palamaru, M. N. and Mitoseriu, L. (2014). Investigation of the functional properties of Mg x Ni 1−x Fe2O4 ceramics. *Journal of Materials Science*.

Morsalin, S., Phung, T. B., Danikas, M. and Mawad, D. (2019). Diagnostic challenges in dielectric loss assessment and interpretation: a review. *IET Science, Measurement & Technology*.

Navriliak, S. and Havriliak, S. J. (1997). Dielectric and Mechanical Relaxation in Materials – Analysis, Interpretation and Application to Polymers. Munich: Hanser.

Pauling, L. (1960). *The Nature of the Chemical Bond (3rd ed.)*. Oxford: Oxford University Press.

Runt, J. P. and Fitzgerald, J. J. (1997). *Dielectric Spectroscopy of Polymeric Materials, Fundamentals and Applications*. Washington: American Chemical Society.

Serway, R. (1990). *Physics for Scientists and Engineers*. Philadelphia: Saunders College Publishing.

Slivka, A., Kedyulich, V., Iqbal Khan, M., Moina, A., Romanyuk, M. and Guivan, M. (2005). The effect of external factors on dielectric permittivity of Rochelle salt: Humidity, annealing, stresses, electric field. *Condensed Matter Physics*.

Stanford A. L. and Tanner J. M. (1985). *Physics for Students of Science and Engineering*. Elsevier Inc.

Stenzel, O. (2015). *The physics of thin film optical spectra*. Springer.

Tykodi J. (1989). Identifying polar and nonpolar molecules. *Journal of Chemical Education*

Vassilikou-Dova, A. and Kalogeras, I. M. (2008). Dielectric analysis (DEA). *Thermal Analysis of Polymers: Fundamentals and Applications*.
Zhong, L. and Xu, C. (1996). Studying on temperature Cole-Cole diagrams by dielectric temperature spectrum. *Proceedings of Conference on Electrical Insulation and Dielectric Phenomena,* IEEE.

Chapter 5

Physical Characterization of the $(Bi_{1-x}Cr_x)NbO_4$ System

S. Devesa[1,2,*], P. S. P. da Silva[1], M. P. Graça[2], L. C. Costa[2] and J. A. Paixão[1]

[1]CFisUC, Department of Physics, University of Coimbra, Coimbra, Portugal
[2]I3N and Physics Department, University of Aveiro, Aveiro, Portugal

Abstract

In the last decades, there has been considerable interest in low temperature co-fired ceramic materials showing high dielectric constant and low dielectric losses. Such materials offer the opportunity to enhance the performance and/or allow a reduction in the physical volume of capacitive components. Bismuth-based materials are promising low-firing temperature ceramics, namely the bismuth niobate, $BiNbO_4$, which has been a subject of several studies, aiming the improvement of its structural, morphological and dielectric properties. In this work, $(Bi_{1-x}Cr_x)NbO_4$ system, prepared using the sol-gel method, was analysed. The structure of all samples was studied by X-ray diffraction and the morphology of the grains was investigated by scanning electron microscopy.

The dielectric characterization, in the microwave region, was made at room temperature, using the small perturbation theory, with a resonant cavity operating in $TE_{1,0,11}$ mode, at the frequency of 5 GHz. The dielectric properties, such as dielectric constant (ε') and dielectric losses (ε''), were described as a function of composition and treatment temperature.

* Corresponding Author's Email: susana_devesa@hotmail.com.

In: An Introduction to Permittivity
Editor: Parveen Saini
ISBN: 979-8-88697-042-5
© 2022 Nova Science Publishers, Inc.

Keywords: bismuth niobate, chromium, sol-gel, microwave dielectric properties

5.1. Introduction

In the last decades, the low temperature co-fired ceramic (LTCC) technology has attracted considerable interest due to its important role in the fabrication of modern electronic devices, since it allows the production of three-dimensional ceramic modules with low dielectric loss and embedded low-cost metal electrodes. In LTCC, a sintering temperature lower than the melting points of the inner metals and suitable microwave dielectric properties are the required characteristics for practical applications [1-3].

The requirements that a ceramic material must have to be used in microwave applications are high dielectric constant, for miniaturization; low dielectric loss (or high quality factor), for enhancing the signal-to-noise ratio; near-zero temperature coefficient of resonant frequency, for temperature stability [1,4-6].

Bismuth niobate, $BiNbO_4$, has been evaluated as a material with potential for dielectric applications and that can be co-fired at low temperatures [7-9].

At ambient pressure, $BiNbO_4$ possesses two polymorphs structures, an orthorhombic (α) phase and a triclinic (β) phase. Recently, a new phase of $BiNbO_4$, referred as HP-$BiNbO_4$, has been obtained above 3 GPa and 800ºC [10, 11].

Since 1992, when Kagata et al. [12] reported, for the first time, the microwave dielectric properties of bismuth niobate, using as sintering aids CuO and V_2O_5, to obtain higher densification and enhanced dielectric properties, several researchers attempted to improve the physical and the dielectric properties of $BiNbO_4$ ceramics, either by the addition of sintering aids (oxides) [13-19], or by the substitution of bismuth or niobium by other cations [20-23], or even by the combination of the two techniques [7, 9, 24-28].

Despite all the research already performed, searching for novel microwave dielectric ceramics with low sintering temperatures and promising properties is still an important issue, since these materials are a key factor in the modern communication systems, an industry that revealed, in the last decades, one of the most remarkable developments [13, 29-31].

The solid state reaction has been the most common method to process bismuth niobate [30, 32, 33], which means that its advantages and disadvantages were already reported in the literature. $BiNbO_4$ powders prepared by solid state reaction results in high crystallinity, but not always in pure and homogeneous phases, because some metastable phases can be obtained under the same conditions [34]. Furthermore, this method may cause other drawbacks such as large grain growth, segregation of components and loss of stoichiometry due to the volatilization of bismuth at high temperature [35].

The sol-gel method, through the citrate route, is considered a simple way to obtain stable precursors and stoichiometric fine powders and has been adopted in the production of various oxides. This method, despite being time consuming, is considered promising for achieving high homogeneity, since the formation of citrate complexes allows mixing the component cations at an atomic scale, thus they can immediately react with each other at relatively low temperatures [4, 34, 35].

In this work, $(Bi_{1-x}Cr_x)NbO_4$ ($0.00 \leq x \leq 0.75$) ceramics were designed, and heat-treated at 3 different temperatures, 500, 650 and 800°C.

In previous works, the substitution of Bi^{3+} by other cations was achieved only when the ionic radius of the cation to include was relatively smaller than the bismuth ionic radius. The substitution of Bi^{3+} by iron [21] was achieved, but in the case of europium, which ionic radius is very similar to the bismuth, the substitution did not occurred [22].

The Cr^{3+} ion, presenting a radius of 0.62 Å [36], is a suitable candidate to accomplish this substitution. Moreover, the use of a material such as chromium is aligned with important goals, such as the sustainable consumption of natural resources, since its average crustal abundance is similar to commonly used industrial metals such as nickel or zinc [37, 38], and the use of non-toxic materials, since the Cr^{3+} is considered relatively non-toxic, in contrast with other oxidation states [39].

The microwave dielectric properties of the prepared samples were measured at 5 GHz, using the resonant cavity method.

X-Ray diffraction (XRD) and scanning electron microscopy (SEM) were used to find the influence of the microstructures in the microwave dielectric properties.

5.2. Methods

The $(Bi_{1-x}Cr_x)NbO_4$ (x = 0.00, 0.25, 0.50, 0.75) powders were prepared using the sol-gel method, through the citrate route. Bismuth nitrate

($Bi(NO_3)_3 \cdot 5H_2O$), niobium chloride ($NbCl_5$), chromium nitrate ($Cr(NO_3)_3.9H_2O$), were used as starting materials and citric acid and ethylene glycol as chelating agent and reaction medium, respectively. All reagents used in the whole procedure were of analytical grade. A suspension containing stoichiometric amounts of starting materials was previously prepared in a minor amount of hydrogen peroxide and dispersed in a mixture of citric acid and ethylene glycol in a molar ratio 1:3. To promote the solubility, the suspension was stirred, for 7 days, until a clear colloidal suspension was obtained.

To establish the period of stirring, the viscosity of the mixture with x = 0.00 was monitored with a *AND Vibro Viscosimeter*. The viscosity varied between 2.61 and 3.30 mPa.s, and, as shown in figure 2, it tends to stabilize after 7 days of stirring.

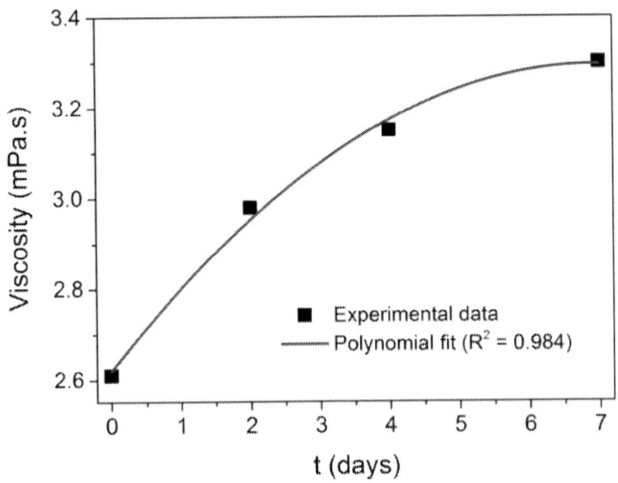

Figure 1. Viscosity of the mixture with x = 0.00 as a function of the stirring time.

To evaporate the solvent, the obtained gels were dried. The mixture with x = 0.00 was dried at 400ºC, for 48h, and the mixtures with $0.25 \leq x \leq 0.75$ were dried at 500°C for 6 h to evaporate the solvent. The obtained powders were pressed into cylinders, and finally heat-treated at 500, 650 and 800ºC, using a dwell time of 4h, with a heating rate of 5ºC/min.

The flow chart for the preparation of the samples by the sol-gel method is shown in Figure 2.

The structural characterization was performed using X-ray diffraction (XRD). The patterns data were obtained on a Bruker D8 Advance X-Ray

Diffractometer (Cu Kα radiation, λ=1.54060 Å), equipped with a LynxEye linear detector, at 40kV and 40mA, in a Bragg-Brentano parafocusing optics configuration. Intensity data were collected by the step counting method (step 0.02° in 0.5 s) in the 2θ angle range of 10-120°.

The morphology of the obtained samples was analyzed by scanning electron microscopy (SEM). The measurements were performed on a TESCAN-Vega III instrument.

The microwave characterization was performed using a resonant cavity, operating at 5 GHz. The measurements were made at room temperature, using a HP 8753D network analyser. Figure 3 shows a schematic representation of the cavity and the frequency spectrum for this cavity in the range [4.968; 4.980] GHz, where a resonant peak can be observed. In Table 1 are presented the characteristics of the resonant cavity.

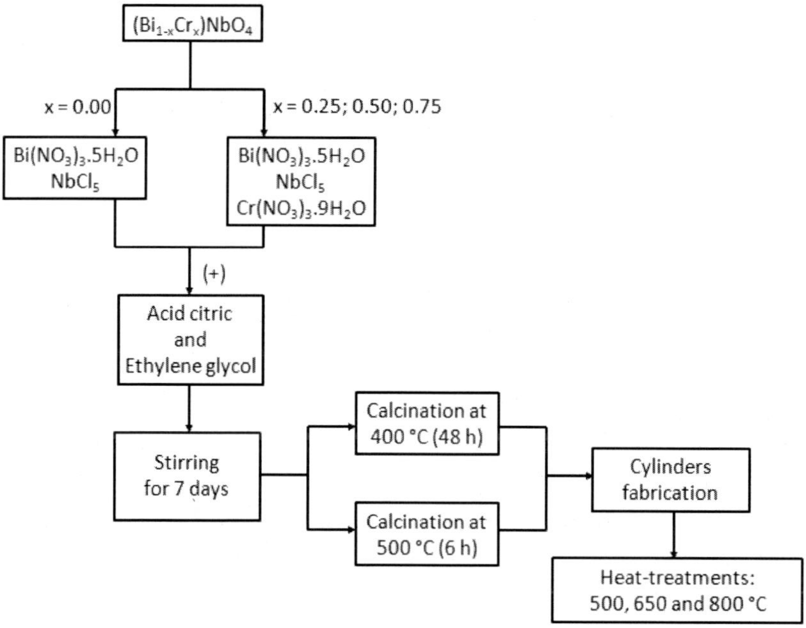

Figure 2. Schematic diagram for the preparation of $(Bi_{1-x}Cr_x)NbO_4$ samples.

The quality factor for the empty cavity, Q_0, can be obtained by the equation

$$Q_0 = \frac{f_0}{\Delta f_{h/2}} \tag{1}$$

where $\Delta f_{h/2}$ is the full width at half maximum, FWHM, of the frequency-response curve [40].

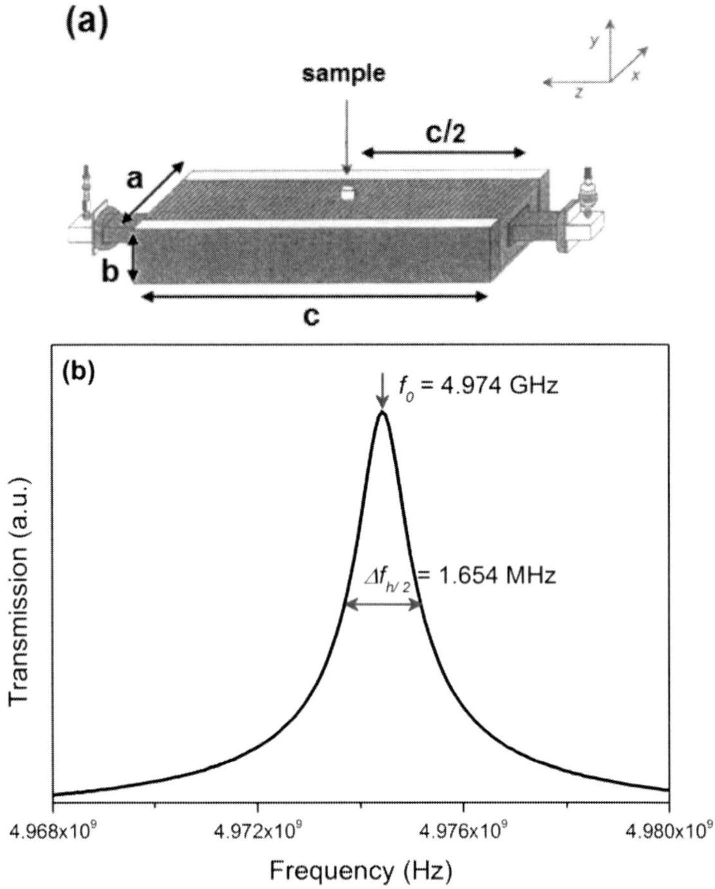

Figure 3. (a) Schematic representation of the resonant cavity, with an aperture for the sample in its geometric center [41]; (b) Cavity spectrum in the range of [4.968;4.980] GHz, emphasizing the resonant frequency and the FWHM of the frequency-response curve.

Table 1. Resonant cavity characteristics

a (mm)	47.50
b (mm)	22.20
c (mm)	425.64
Resonant mode	$TE_{1,0,11}$
Resonant frequency for the empty cavity, f_0 (GHz)	4.974
Quality factor for the empty cavity, Q_0	3007

The calculation of the complex permittivity, $\varepsilon^* = \varepsilon' - i\varepsilon''$, was made using the small perturbation theory. In this method, the resonance peak frequency and the quality factor of the cavity, with and without a sample, are used to obtain the complex dielectric permittivity of the material. The shift in the resonant frequency of the cavity, $\Delta f = f_0 - f_S$, caused by the insertion of the sample, is related to the real part of the complex permittivity, ε', and the change in the inverse of the quality factor of the cavity, $\Delta(1/Q) = 1/Q_S - 1/Q_0$, gives the imaginary part, ε''. Considering only the first order perturbation in the electric field caused by the sample and separating the real and imaginary parts of the complex permittivity, we can obtain

$$\varepsilon' = K\frac{f_0 - f_s}{f_0}\frac{V}{v} + 1 \tag{2}$$

$$\varepsilon'' = \frac{K}{2}\left(\frac{1}{Q_S} - \frac{1}{Q_0}\right)\frac{V}{v} \tag{3}$$

where K is a constant related to the depolarisation factor, which depends upon the geometric parameters and where the indexes 0 and S refer to the empty cavity and loaded with the sample, respectively [42-46].

Using a material of known complex permittivity, the value of K can be calculated. In this study was used a polytetrafluorethylene (PTFE) cylinder, with the same size and shape of the samples.

5.3. Results and Discussion

5.3.1. Structural Characterization

The crystalline phases present in the sintered powders were identified by X-ray diffraction, as shown in Figure 4.

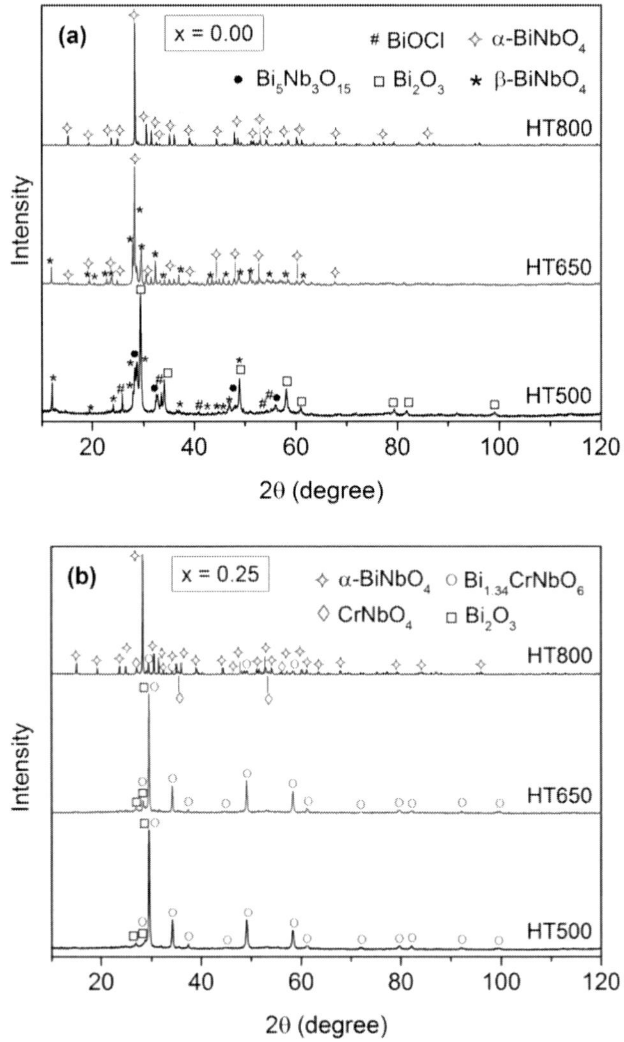

Figure 4. (Continued)

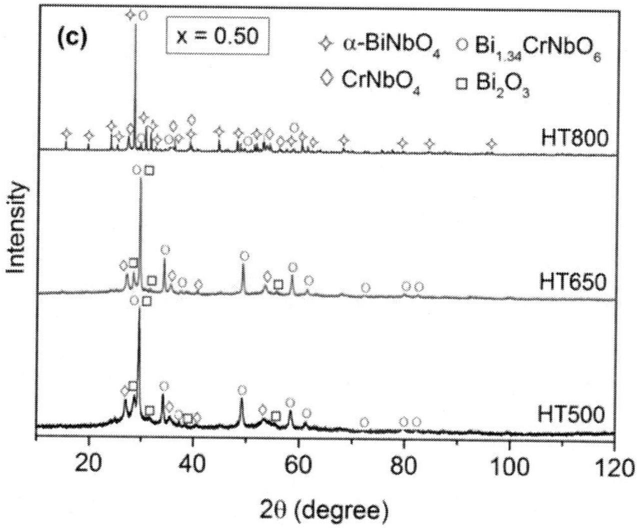

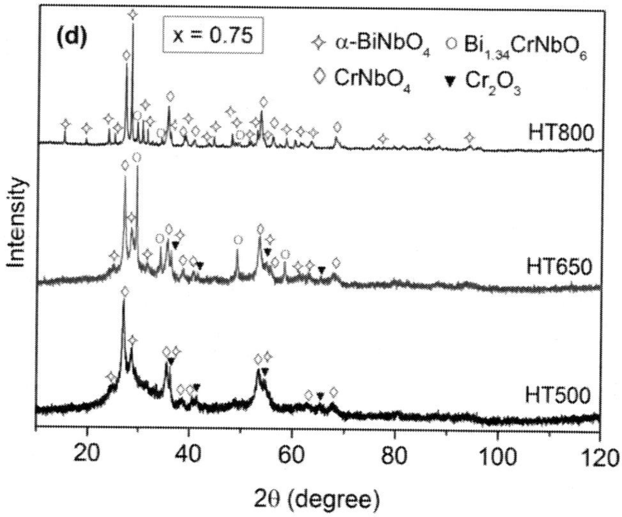

Figure 4. X-ray diffraction patterns of $(Bi_{1-x}Cr_x)NbO_4$ samples:(a) $x = 0.00$; (b) $x = 0.25$; (c) $x = 0.50$; (d) $x = 0.75$.

Analyzing the diffraction patterns shown in Figure 4(a), that concern the reference samples, in the powders treated at 500ºC, besides the triclinic bismuth niobate, β-$BiNbO_4$, three secondary phases, BiOCl, Bi_2O_3 and $Bi_5Nb_3O_{15}$, were identified. Increasing the treatment temperature to 650ºC, only bismuth niobate was present, in a combination of β- and α-$BiNbO_4$

phases. At 800ºC the diffraction patterns confirmed the presence of pure orthorhombic bismuth niobate.

This evolution is in agreement with the literature, since the low temperature β-$BiNbO_4$ was the first to be formed, with the transition to α-$BiNbO_4$ occurring at temperatures between 600 and 750ºC, or lower, if the sol-gel method is used [23].

For the samples with x = 0.25, at 500 and 650ºC, two phases were identified, Bi_2O_3 and $Bi_{1.34}CrNbO_6$, showing that, even with low treatment temperatures, it is possible to promote the substitution of bismuth by chromium. At 800ºC, besides the $Bi_{1.34}CrNbO_6$, α-$BiNbO_4$ and $CrNbO_4$ phases were identified.

Increasing the x value to 0.50, the evolution of the formed phases is similar, however, with the $CrNbO_4$ phase already present at 500ºC.

Analyzing the system with x = 0.75, it is possible to notice that the samples heat-treated at 500 and 650ºC show a lower crystallinity. Furthermore, the α-$BiNbO_4$ phase was formed at lower temperature in comparison with the reference samples, with the substitution of bismuth by chromium occurring only in the samples heat-treated at 650 and 800ºC.

Table 2. Crystalline phases of $(Bi_{1-x}Cr_x)NbO_4$ (x = 0.00, 0.25, 0.50, 0.75) samples

HTT (ºC)	x = 0.00	x = 0.25	x = 0.50	x = 0.75
500	BiOCl Bi_2O_3 $Bi_5Nb_3O_{15}$ β-$BiNbO_4$	Bi_2O_3 $Bi_{1.34}CrNbO_6$ (94.8%)	Bi_2O_3 $Bi_{1.34}CrNbO_6$ (40.5%) $CrNbO_4$	Cr_2O_3 α-$BiNbO_4$ $CrNbO_4$
650	β-$BiNbO_4$ α-$BiNbO_4$	Bi_2O_3 $Bi_{1.34}CrNbO_6$ (90.9%)	Bi_2O_3 $Bi_{1.34}CrNbO_6$ (41.4%) $CrNbO_4$	Cr_2O_3 $Bi_{1.34}CrNbO_6$ (13.6%) α-$BiNbO_4$ $CrNbO_4$
800	α-$BiNbO_4$	$Bi_{1.34}CrNbO_6$ (6.6%) α-$BiNbO_4$ $CrNbO_4$	$Bi_{1.34}CrNbO_6$ (3.5%) α-$BiNbO_4$ $CrNbO_4$	$Bi_{1.34}CrNbO_6$ (6.0%) α-$BiNbO_4$ $CrNbO_4$

The identification of the crystalline phases of the $(Bi_{1-x}Cr_x)NbO_4$ (x = 0.00, 0.25, 0.50, 0.75) samples, as well as the percentage of the $Bi_{1.34}CrNbO_6$ phase, where the substitution of Bi^{3+} by Cr^{3+} was achieved, are presented in Table 2. The percentage of $Bi_{1.34}CrNbO_6$ was obtained with the Rietveld refinement method. For the samples heat-treated at 500 and 650ºC this percentage decreases with the increase of the chromium content. On the other hand, for the samples with $x = 0.25$ and 0.50 the amount of $Bi_{1.34}CrNbO_6$ is drastically reduced with the heat-treatment of 800ºC.

5.3.2. Morphological Characterization

The SEM micrographs of the prepared samples are shown in Figure 5. Regardless of the *x* value, the samples treated at 500 and 650ºC show nanosized particles with spherical shape. The heat-treatment performed at 800ºC promoted morphological changes. In the reference sample, $x = 0.00$, it is perceptible a grain growth and the formation of rod-shaped grains, though, the surface of the sample maintained an homogeneous configuration. In the samples with chromium, $x > 0.00$, the changes promoted by this heat-treatment are more significant, with the samples showing particles with diversity of size and geometry.

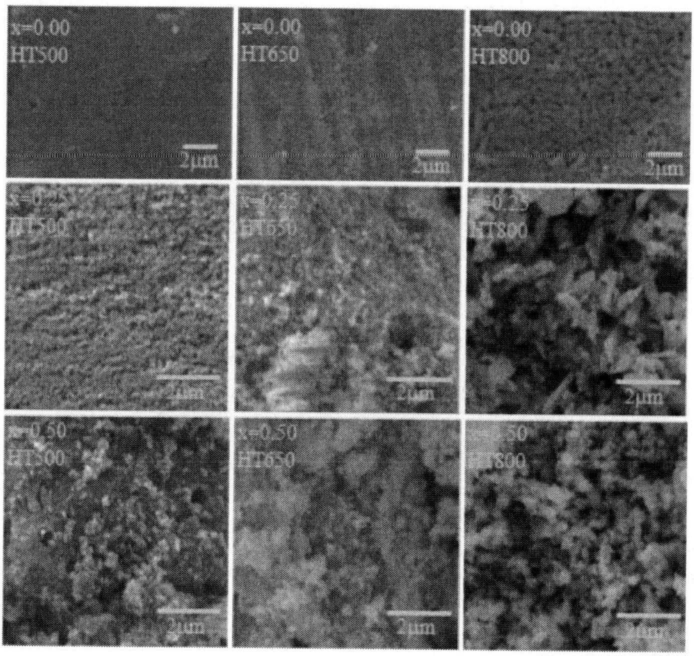

Figure 5. SEM micrographs of $(Bi_{1-x}Cr_x)NbO_4$ ($x = 0.00, 0.25, 0.50$) samples, with magnification of 20kX.

5.3.3. Dielectric Characterization

For the measurement of the dielectric properties of a material with low losses, that is, low values of ε'', the small perturbation theory can be used with good results. However, the applicability of this method, namely equations (2) and (3), depends on the existence of linearity between the measured perturbation and the volume of the inserted sample [41]. To confirm the linearity of the cavity, measurements using empty glass micro tubes and also filled with different volumes of distilled water were performed. Figure 6 shows the frequency response of the cavity with the different micro tubes, empty and filled with distilled water. As expected, the perturbation caused by the empty micro tubes is insignificant in comparison with the perturbation caused by the presence of distilled water.

The obtained values of $\Delta f/f_0$ and $\Delta(1/Q)$ as a function of the water volume are presented in Figure 7, which shows that the cavity remains in the linear regime for measurements up to variations of $\Delta f/f_0$ at least until 0.34% and $\Delta(1/Q)$ of about 2.6×10^{-3}.

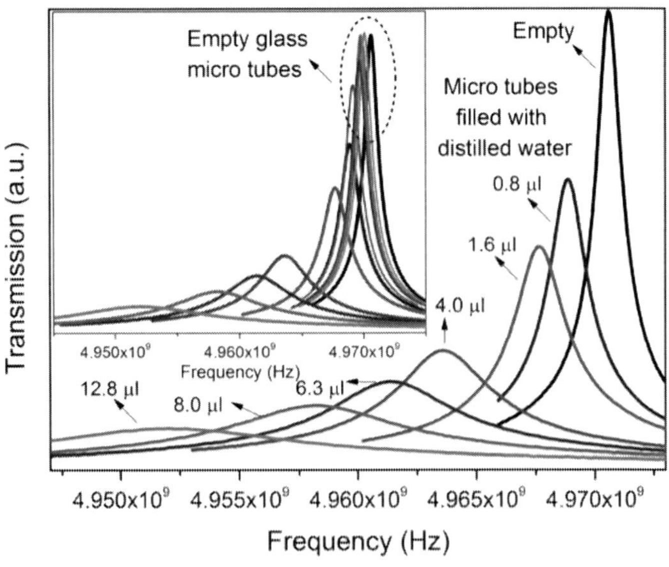

Figure 6. Transmission of the cavity, with the empty micro tubes and with the micro tubes filled with distilled water.

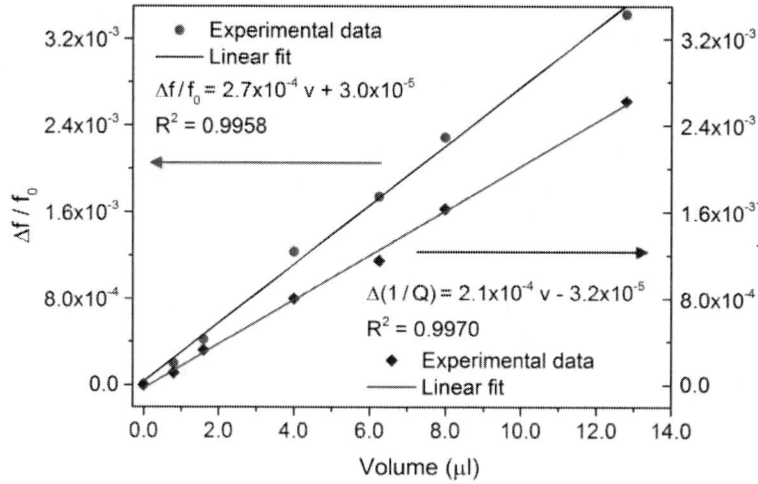

Figure 7. Calculated $\Delta f/f_0$ and $\Delta(1/Q)$ for different volumes of distilled water.

The real, ε', and the imaginary, ε'', parts of the complex permittivity, at 5 GHz, as function of the treatment temperatures are shown in Figure 8. All the $\Delta(1/Q)$ and $\Delta f/f_0$ values obtained in the measurements of these samples were in the limits of the linearity of the cavity, validating the small perturbation technique application.

Figure 8(a) shows that in the samples with chromium, and regardless of the x value, the dielectric constant decreases with the increase of the heat-treatment temperature. Furthermore, regardless of the treatment temperature, the samples with $x = 0.25$ show the highest ε'. In the samples with $x = 0.25$ and 0.50 the decrease of ε' is more pronounced with the heat-treatment of 800ºC, which can be correlated with the decrease of the $Bi_{1.34}CrNbO_6$ content, or the formation of the α-$BiNbO_4$ phase.

In the reference samples, the heat-treatment of 650ºC promoted a significant decrease in the dielectric constant, which can be due to the increase of the β-$BiNbO_4$ content, since no significant morphological changes occurred.

The dependence of the dielectric losses with the treatment temperature, presented in Figure 8(b), shows that, for all of the x values, the dielectric losses decrease with the increase of the temperature, with the samples with $x = 0.75$ having higher losses.

In the samples with chromium, the ones heat-treated at 800ºC show minimum values for the dielectric constant and dielectric losses. This behavior must be related to the structural changes and not so much to the morphological evolution, since the decrease of the dielectric constant and dielectric losses was already verified for the samples heat-treated at 650ºC.

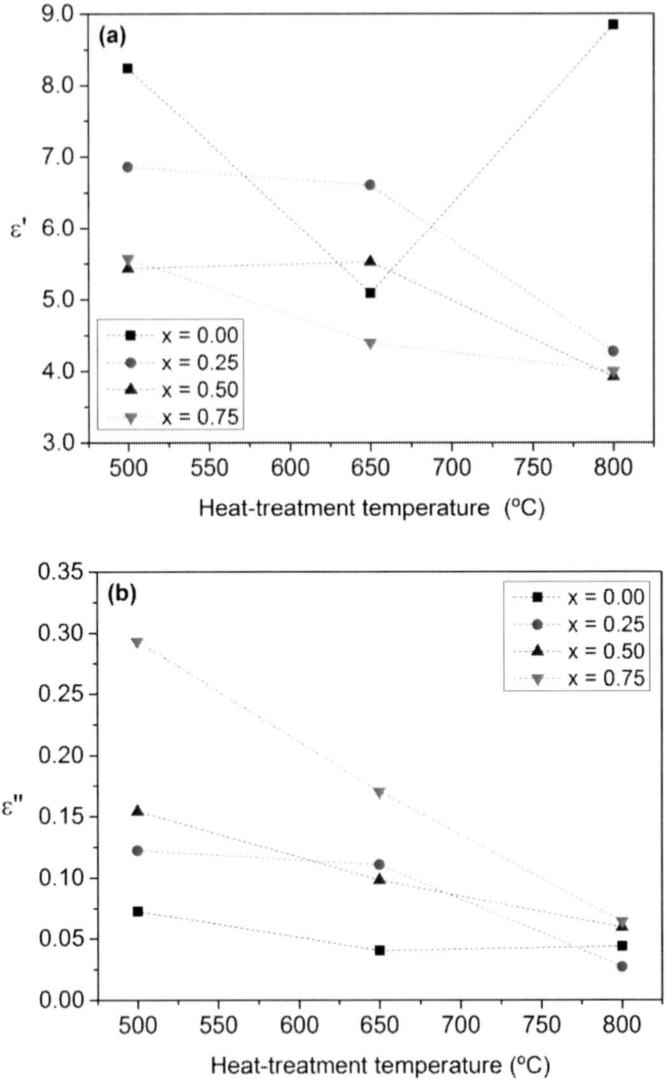

Figure 8. (a) ε' and (b) ε'' for the $(Bi_{1-x}Cr_x)NbO_4$ ($x = 0.00, 0.25, 0.50, 0.75$) samples, at room temperature.

Figure 9 shows the dielectric constant of the samples with $Bi_{1.34}CrNbO_6$, the phase formed due to the substitution of bismuth by chromium. As one can see, the dielectric constant increases almost linearly with the content of $Bi_{1.34}CrNbO_6$.

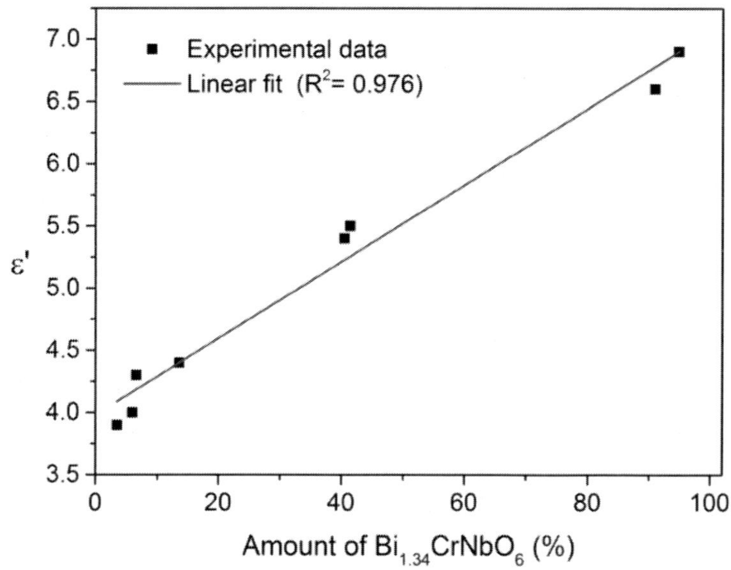

Figure 9. Dielectric constant as a function of $Bi_{1.34}CrNbO_6$ amount.

Figure 10 presents the measured transmission of the empty cavity, with PTFE and with the different samples with $x = 0.75$, where a shift in the resonant frequency, Δf, is observed. This shift can be related to the real part of the complex permittivity ε', where higher shift corresponds to higher ε', since the samples have the same dimension [47]. Therefore, these results are in agreement with the information shown in Figure 8. Furthermore, and as expected, the shift produced by the PTFE sample is less expressive, which confirms the fact that this material has low dielectric constant of 2.1 [48], lower than the samples under study.

Despite the approximations inherent to the small perturbation theory, namely the calculation of ε' and ε'' using only the first order perturbation in the electric field caused by the sample, the precision associated to the method validates its application, since the errors, $\Delta\varepsilon'$ and $\Delta\varepsilon''$, can be controlled within 10% [44, 48, 49].

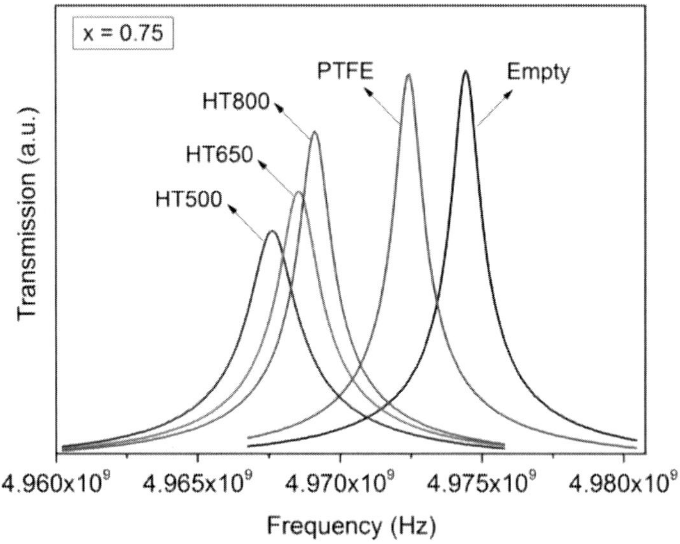

Figure 10. Transmission of the cavity, empty, with a PTFE sample and with the samples with $x = 0.75$, heat-treated at different temperatures.

Figure 11 shows the values of the dielectric constant, dielectric losses and tangent losses, $tg\ \delta = \varepsilon''/\varepsilon'$, determined at room temperature.

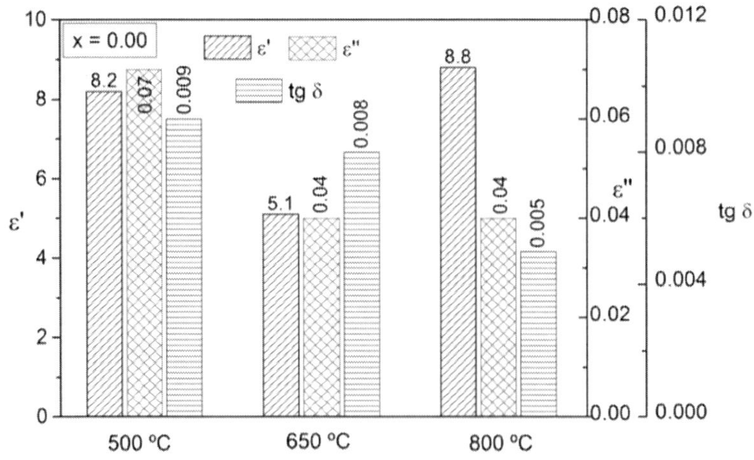

Figure 11. (Continued)

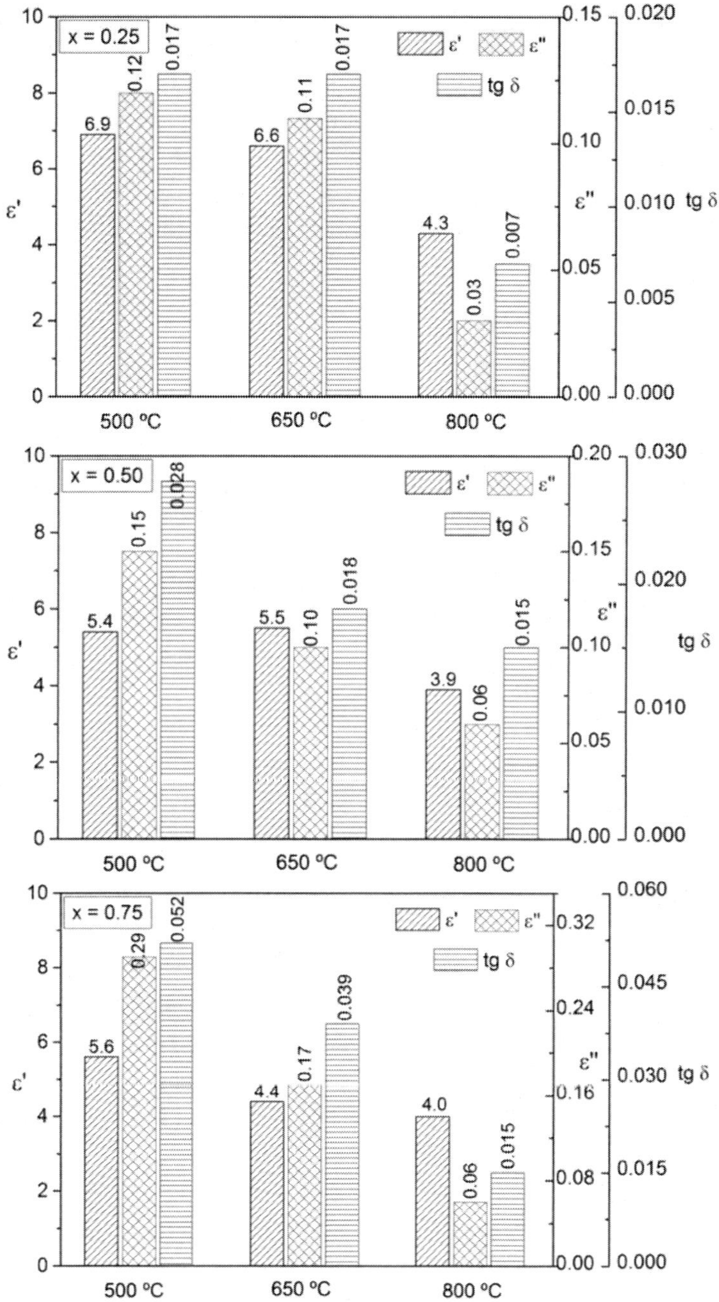

Figure 11. ε', ε'' and $tg\delta$, at room temperature.

By comparing these three parameters, it can be inferred that the most promising sample, from a dielectric properties perspective, is the reference sample treated at 800ºC. Considering only the samples with $0.25 \leq x \leq 0.75$, in other words, the samples with chromium, the most promising is the sample with $x = 0.25$, also heat-treated at 800ºC. Although this sample is not the one that presents the higher dielectric constant, it has lower losses, and a compromise between dielectric constant and dielectric losses is required.

Conclusion

The $(Bi_{1-x}Cr_x)NbO_4$ ceramic powders were prepared successfully by the sol-gel method, with the substitution of bismuth by chromium being performed with success. The non-stoichiometric phase $Bi_{1.34}CrNbO_6$ was obtained, even with low treatment temperatures. Moreover, one can conclude that the inclusion of chromium led to a reduction of the conversion temperature of low-temperature β-$BiNbO_4$ in α-$BiNbO_4$ or even inhibited the formation of the low-temperature β-$BiNbO_4$.

Concerning the dielectric characterization, the resonant cavity method proved to be very useful for dielectric measurements on ceramic materials, featuring important advantages, as the easy sample preparation, the absence of measurement contacts and the non-destructive character. Concerning the small perturbation technique, it is important to highlight the simplicity of the calculations and that all the performed measurements were in the limits of the linearity of the cavity used.

The dielectric study revealed that in the samples with chromium, both dielectric constant and dielectric losses decreased with the increase of the sintering temperature.

Moreover, in the samples with the $Bi_{1.34}CrNbO_6$ the dielectric constant showed a linear increase with the increasing amount of this crystalline phase.

Acknowledgment

Susana Devesa acknowledges the post-doctoral grant under the project "SUSpENsE - CENTRO-01-0145-FEDER-000006".

This work was partially supported by FEDER funds through the COMPETE 2020 Programme and National Funds through FCT - Portuguese

Foundation for Science and Technology under the project UID/CTM/50025/2019 and UID/FIS/04564/2020.

Access to TAIL-UC facility funded under QREN-Mais Centro Project ICT/2009/02/012/1890 is gratefully acknowledged.

References

[1] Chen J, Li C, Wang D, Xiang H, Fang L. 2017. "Preparation, crystal structure, and dielectric characterization of $Li_2W_2O_7$ ceramic at RF and microwave frequency range." *Journal of Advanced Dielectrics*, 1720001-7(01). Accessed January 2, 2020.doi: 10.1142/S2010135X17200016.

[2] Di Z, Li-Xia P, Ze-Ming Q, Biao-Bing J, Xi Y. 2014. "Novel ultra-low temperature co-fired microwave dielectric ceramic at 400 degrees and its chemical compatibility with base metal." *Scientific Reports*, 5980-4. Accessed January 2, 2020.doi: 10.1038/srep05980.

[3] Li J, Li C, Wei Z, Tang Y, Su C and Fang L. 2015. "Microwave dielectric properties of a low-firing $Ba_2BiV_3O_{11}$ ceramic." *Journal of the American Ceramic Society*, 683:686-98(3). Accessed January 2, 2020. doi: 10.1111/jace.13418.

[4] Wang N, Zhao MY, Yin ZW, Li W. 2003. "Low-temperature synthesis of β-$BiNbO_4$ powder by citrate sol-gel method." *Materials Letters*, 4009:4013-57(24-25). Accessed January 6, 2020. doi: 10.1016/S0167-577X(03)00256-8.

[5] Xia WS, Yang FY, Zhang GY, Han K, Guo DC. 2016. "New low-dielectric-loss $NiZrNb_2O_8$ ceramics for microwave application." *Journal of Alloys and Compounds*, 470:475-656. Accessed January 2, 2020. doi: 10.1016/ j.jallcom.2015.10.008.

[6] Bi J, Niu Y, Wu H. 2017. "$Li_4Mg_3Ti_2O_9$: a novel low-loss microwave dielectric ceramic for LTCC applications." *Ceramics International*, 7522:7530-43(10). Accessed January 2, 2020. doi: 10.1016/j.ceramint.2017.03.041.

[7] Butee S, Kulkarni AR, Prakash O, Aiyar RPRC, Sudheendran K, Raju KJ. 2010. "Effect of lanthanide ion substitution on RF and microwave dielectric properties of $BiNbO_4$ ceramics." *Journal of Alloys and Compounds*, 351:357-492(1-2). Accessed January 4, 2020. doi: 10.1016/j.jallcom.2009.11.093.

[8] Rao KS, Buddhudu S. 2010. "Structural, thermal and dielectric properties of $BiNbO_4$ ceramic powder." *Ferroelectrics Letters*, 101:109-37(6). Accessed January 2, 2020. doi: 10.1080/07315171.2010.527806.

[9] Shihua D, Xi Y, Yu M, Puling L. 2006. "Microwave dielectric properties of $(Bi_{1-x}R_x)NbO_4$ ceramics (R = Ce, Nd, Dy, Er)." *Journal of the European Ceramic Society*, 2003:2005-26(10-11). Accessed January 4, 2020. doi: 10.1016/ j.jeurceramsoc.2005.09.049.

[10] Liu YJ, Zhang JW, He DW, Xu C, Hu QW, Qi L, Liang AK. 2017. "Exploring the compression behavior of HP-$BiNbO_4$ under high pressure." *Chinese Physics B*, 116202-26(11). Accessed January 3, 2020. doi: 10.1088/1674-1056/26/11/116202.

[11] Xu C, He D, Liu C, Wang H, Zhang L, Wang P, Yin S. 2013. "High pressure and high temperature study the phase transitions of $BiNbO_4$." *Solid State Communications*, 21:24-156. Accessed January 3, 2020. doi: 10.1016/j.ssc.2012.11.007.

[12] Kagata H, Inoue T, Kato J, Kameyama I. 1992. "Low-fire bismuth-based dielectric ceramics for microwave use." *Japanese Journal of Applied Physics*, 3152-31(9S). Accessed January 4, 2020. doi: 10.1143/JJAP.31.3152.

[13] Tzou WC, Yang CF, Chen YC, Cheng PS. 2000. "Improvements in the sintering and microwave properties of $BiNbO_4$ microwave ceramics by V_2O_5 addition." *Journal of the European Ceramic Society*, 991:996-20(7). Accessed January 5, 2020. doi: 10.1016/S0955-2219(99)00228-9.

[14] Yang CF. 1999. "Improvement of the dielectric properties of $BiNbO_4$ ceramics by the addition of CuO–V_2O_5 mixtures." *Journal of Materials Science Letters*, 805:807-18(10). Accessed January 4, 2020. doi: 10.1023/A:1006697318870.

[15] Shihua D, Xi Y, Yong Y. 2004. "Dielectric properties of B_2O_3-doped $BiNbO_4$ ceramics." *Ceramics International*, 1195:1198-30(7). Accessed January 4, 2020. doi: 10.1016/j.ceramint.2003.12.030.

[16] Zhong C, Yuan Y, Zhang S, Pang Y, Tang B. 2010. "Low-fired $BiNbO_4$ microwave dielectric ceramics modified by CuV_2O_6 addition sintered in N_2 atmosphere." *Ceramics-Silikaty*, 103:107-54(2). Accessed January 4, 2020.

[17] Almeida JS, Fernandes TSM, Sales AJM, Silva MAS, Júnior GP, Rodrigues HO, Sombra ASB. 2011. "Study of the structural and dielectric properties of Bi_2O_3 and PbO addition on $BiNbO_4$ ceramic matrix for RF applications." *Journal of Materials Science: Materials in Electronics*, 978:987-22(8). Accessed January 4, 2020. doi: 10.1007/s10854-010-0247-z.

[18] Sales AJM, Oliveira PWS, Almeida JS, Costa MM, Rodrigues HO, Sombra ASB. 2012. "Copper concentration effect in the dielectric properties of $BiNbO_4$ for RF applications." *Journal of Alloys and Compounds*, 264:270-542. Accessed January 4, 2020. doi: 10.1016/j.jallcom.2012.07.025.

[19] Yang Y, Ding S, Yao X. 2004. "Influences of Fe_2O_3 additives on the dielectric properties of $BiNbO_4$ ceramics under different sintering atmosphere." *Ceramics International*, 1341:1345-30(7). Accessed January 4, 2020. doi: 10.1016/j.ceramint.2003.12.110.

[20] Zhou D, Wang H, Yao X, Liu Y. 2008. Microwave dielectric properties of low-firing $BiNbO_4$ ceramics with V_2O_5 substitution." *Journal of Electroceramics*, 469:472-21(1-4). Accessed January 4, 2020. doi: 10.1007/s10832-007-9223-2.

[21] Devesa S, Graça MP, Henry F, Costa LC. 2015. "Microwave dielectric properties of $(Bi_{1-x}Fe_x)NbO_4$ ceramics prepared by the sol-gel method." *Ceramics International*, 8186:8190-41(6). Accessed January 4, 2020. doi: 10.1016/j.ceramint.2015.03.038.

[22] Devesa S, Graça MPF, Henry F, Costa LC. 2017. "Structural, morphological and microwave dielectric properties of $(Bi_{1-x}Eu_x)NbO_4$ ceramics prepared by the sol-gel method." *International Journal of Materials Engineering Innovation*, 12:26-8(1). Accessed January 4, 2020. doi: 10.1504/IJMATEI.2017.10006773.

[23] Devesa S, Graça MPF, Costa LC. 2018. "Dielectric Properties of Bismuth Niobate Ceramics." *Bismuth: Advanced Applications and Defects Characterization, InTech*, 93:111. Accessed January 4, 2020. doi: 10.5772/intechopen.75100.

[24] Butee SP, Kambale KR, Upadhyay S, Bashaiah S, Raju KJ, Panda H. 2016. "Synthesis and microwave dielectric behavior of $(Bi_{1-x}Pb_x)$ NbO_4 ceramics." *Journal of Advanced Dielectrics*, 1650006-6(01). Accessed January 4, 2020. doi: 10.1142/S2010135X16500065.

[25] Huang CL, Weng MH, Yu CC. 2001. "Low firable $BiNbO_4$ based microwave dielectric ceramics." *Ceramics International*, 343:350-27(3). Accessed January 4, 2020. doi: 10.1016/S0272-8842(00)00086-9.

[26] Wang N, Zhao MY, Yin ZW, Li W. 2004. "Effects of complex substitution of La and Nd for Bi on the microwave dielectric properties of $BiNbO_4$ ceramics." *Materials Research Bulletin*, 439:448-39(3). Accessed January 4, 2020. doi: 10.1016/j.materresbull.2003.10.016.

[27] Tzou WC, Yang CF, Chen YC, Cheng PS. 2002. "Microwave dielectric characteristics of $(Bi_{1-x}Sm_x)NbO_4$ ceramics." *Ceramics International*, 105:110-28(1). Accessed January 4, 2020. doi: 10.1016/S0272-8842(01)00064-5.

[28] Wang N, Zhao MY, Yin ZW. 2003. "Effects of Ta_2O_5 on microwave dielectric properties of $BiNbO_4$ ceramics." *Materials Science and Engineering: B*, 238:242-99(1-3). Accessed January 4, 2020. doi: 10.1016/S0921-5107(02)00464-6.

[29] Sebastian MT, Jantunen H. 2008. "Low loss dielectric materials for LTCC applications: A review." *International Materials Reviews*, 57:90-53(2). Accessed January 4, 2020. doi: 10.1179/174328008X277524.

[30] Filho RC, Araújo JH, Ginani MF, d'Assunção Jr AG, Martins RA, d'Assunção AG, Mendonça LM. 2010. "Simulation and measurement of inset-fed microstrip patch antennas on $BiNbO_4$ substrates." *Microwave and Optical Technology Letters*, 1034:1036-52(5). Accessed January 6, 2020. doi: 10.1002/mop.25098.

[31] Narang SB, Bahel S. 2010. "Low loss dielectric ceramics for microwave applications: A review." *Journal of Ceramic Processing Research,* 316:321-11(3). Accessed January 5, 2020.

[32] Radha R, Gupta UN, Samuel V, Muthurajan H, Kumar HH, Ravi V. 2008. "A co-precipitation technique to prepare $BiNbO_4$ powders." *Ceramics International*, 1565:1567-34(6). Accessed January 6, 2020. doi: 10.1016/j.ceramint.2007.03.022.

[33] Zhou D, Wang H, Yao X, Wei X, Xiang F, Pang L. 2007. "Phase transformation in $BiNbO_4$ ceramics." *Applied Physics Letters*, 172910-90(17). Accessed January 6, 2020. doi: 10.1063/1.2732833.

[34] Almeida CG, Andrade HMC, Mascarenhas AJS, Silva LA. 2010. "Synthesis of nanosized β-$BiTaO_4$ by the polymeric precursor method." *Materials Letters*, 1088:1090-64(9). Accessed January 6, 2020. doi: 10.1016/j.matlet.2010.02.020.

[35] Radha R, Muthurajan H, Rao NK, Pradhan S, Gupta UN, Jha RK, Mirji SA, Ravi V. 2008. "Low temperature synthesis and characterization of $BiNbO_4$ powders." *Materials Characterization*, 1083:1087-59(8). Accessed January 6, 2020. doi: 10.1016/j.matchar.2007.08.026.

[36] Payne SA, Chase LL, Krupke LW. 1987. "Optical Properties of Cr^{3+} in Fluorite-Structure Hosts and in MgF^*_2." *The Journal of Chemical Physics*, 3455-3461-86(6). Accessed March 27, 2020. doi: 10.1063/1.452001.

[37] Haxel G. 2002. *Rare Earth Elements: Critical Resources for High Technology*. US Geological Survey.

[38] Gambogi J. 2017. *Rare Earths-Mineral Commodity Summaries*. US Geological Survey.

[39] Bateman N, Jefferson R, Thomas S, Thompson J, Vale A. 2014. *Oxford Desk Reference: Toxicology*. OUP Oxford.

[40] Meredith RJ. 1998. *Engineers' Handbook of Industrial Microwave Heating*. The Institution of Electrical Engineers. ISBN0852969163.

[41] Costa LC, Devesa S, Henry F. 2006. "A 5 GHz resonant cavity for complex permittivity measurements: design, test performances and application." In *Materials Science Forum*, Trans Tech Publications, 1561:1565-514. Accessed December 31, 2019. doi: 10.4028/www.scientific.net/MSF.514-516.1561.

[42] Birnbaum G, Franeau J. 1949. "Measurement of the dielectric constant and loss of solids and liquids by a cavity perturbation method." *Journal of Applied Physics*, 817:818-20(8). Accessed December 31, 2019. doi: 10.1063/1.1698535.

[43] Parkash A, Vaid JK, Mansingh A. 1979. "Measurement of dielectric parameters at microwave frequencies by cavity-perturbation technique." *IEEE Transactions on Microwave Theory and Techniques*, 791:795-27(9). Accessed December 31, 2019. doi: 10.1109/TMTT.1979.1129731.

[44] Henry F. 1982. "Development of microwave metrology and application to the study of hydration and water diffusion in macromolecular materials". *PhD diss.*, University of Paris.

[45] Carter RG. 2001. "Accuracy of microwave cavity perturbation measurements." *IEEE Transactions on Microwave Theory and Techniques*, 918:923-49(5). Accessed December 31, 2019. doi: 10.1109/22.920149.

[46] Devesa S, Teixeira SS, Graça MPF, Costa LC. 2017. "Design, characterization and test performances of a resonant cavity for complex permittivity measurements using the small perturbation technique." *Electrical Measurements: Introduction, Concepts and Applications*. Nova Publishers, 69-90. Accessed December 31, 2019.

[47] Devesa S, Graça MP, Costa LC. 2017. "Sol-Gel Synthesis, Structural characterization and microwave dielectric properties of bismuth Niobate modified by iron inclusion." *Recent Applications in Sol-Gel Synthesis, InTech*, 63:80. Accessed December 31, 2019. doi: 10.5772/intechopen.68211.

[48] Balmus S, Pascariu G, Creanga F, Dumitru I, Sandu DD. 2006. "The cavity perturbation method for the measurement of the relative dielectric permittivity in the microwave range." *Journal of Optoelectronics and Advanced Materials*, 971-8(3). Accessed December 31, 2019.

[49] Costa LC. 1995. "Electrical properties of glasses with some rare-earth ions." *PhD diss.*, University of Aveiro.

Chapter 6

Dielectric Behavior of Ferroelectric Liquid Crystals with Gold Nanoparticles, Water, and Ethanol for Innovative Electro-Optical Applications

Amit Choudhary[1,*] and Ashok M. Biradar[2]

[1]Physics Department, Deshbandhu College, University of Delhi, Kalkaji, New Delhi, India

[2]CSIR - National Physical Laboratory, Dr. K. S. Krishnan Marg, New Delhi, India

Abstract

We have reviewed the modification in the properties of ferroelectric liquid crystal (FLC) when an external entity is added into it. Here, three different materials, as external entities, gold nanoparticles, water, and ethanol, are added into the FLC matrix and the resulting composition is the formation of mixed system. The characterization of the mixed system has been analyzed by electro-optical and majorly by dielectric spectroscopy. The choice of these materials is specific because of the aspect that when nanoparticles are added into the FLC, the solvent of nanoparticles is mostly either water or ethanol. During the mixing process, part of the solvent may also remain into FLC and can leave to impact the properties of FLC. This compilation of the studies would lead to the understanding of identifying the impact of solvents in the FLC matrix.

---

[*] Corresponding Author's Email: amitnpl2005@gmail.com.

In: An Introduction to Permittivity

Editor: Parveen Saini

ISBN: 979-8-88697-042-5

© 2022 Nova Science Publishers, Inc.

6.1. Introduction

Ferroelectric liquid crystal (FLC) is well established for their advanced application in electro-optical devices such as fast response, bistability effect, second harmonic generation, etc. [1–3] They have been studied experimentally and analyzed with materials particularly gold nanoparticles and isotropic liquid compositions to enhance the dielectric and electro-optical properties in last two decades. This is because of the fact that even after a lot of studies, the proposed devices based on FLCs could not be developed, and hence the efforts of enhancing and tuning the basic properties have been started to remove the hurdles of the intrinsic defects. This has been started by employing nanoparticles, particularly gold nanoparticles in the majority of cases whereas in some cases, it is liquids [4–9]. Thus, monitoring the advancement in the basic properties of FLCs after incorporating these external entities is significant for the advancement in the design and development of devices based on such ferroelectric materials. Studies of the dielectric properties allow the researcher to explore the internal structure of the molecular arrangement and its scope for the application in devices.

The application of FLCs could fairly be decided by analyzing their dielectric properties and electro-optic response to the applied electric or magnetic field. For example, if one wants to use FLC for the electro-optical switching-based application, its molecules must respond at that desired frequency and hence the electro-optical switching must also be realized in the transmission of light through the sample. This kind of information can easily be realized by performing the dielectric spectroscopy from a few hertz frequencies to few MHz [10]. If the dielectric permittivity is quite high at low frequency, say 100 Hz to 500 kHz as compared to its value at higher frequencies, say 10 MHz, then it means that the high permittivity value is because of the contribution from dipolar fluctuations. Fluctuations of dipoles are because of the applied measuring electric field which is kept as low as just to be able to fluctuate the dipole and not to disturb their orientation up to great extent. That is why a weak electric field is applied for the knowledge of dipolar arrangement inside the bulk of materials. In the case of FLCs, the dipoles are associated firmly with the FLC molecules that why the information of molecular alignment can be extracted by extracting the information of dipoles. Besides this, the application of DC bias field and during this configuration of bias, the dielectric measurement reveals the information of molecular reorientation. If this is large enough to change the dielectric properties then it is quite common that the optical properties would also change. The change in optical properties will allow the selectivity of material for a particular application. Thus, dielectric spectroscopy is a very crucial tool to characterize the FLC matrix system.

In this chapter, the dielectric and electro-optical properties of FLC in combination with gold nanoparticles and liquids (common solvents of gold nanoparticles), particularly water and ethanol, are explained. In the first section, the introduction of each type such as FLC, water, and ethanol is discussed. In the later sections, some results of the modification in the basic properties of FLC by employing gold nanoparticles, water, and ethanol are presented. This is also worth discussing here why this combination of various mutually independent materials is incorporated with FLC. The studies of nanoparticles in FLC and other materials have become very common. The nanoparticles are used in the form of suspension where the most common solvents are water and ethanol. Now, there are chances that during the mixing of gold nanoparticles, the solvent could leave its separate impact on the basic properties of FLC. This curiosity of the solvent effect led to the incorporation of the independent compilation of the studies on the impact of gold nanoparticles, water, and ethanol in this chapter.

6.2. Ferroelectric Liquid Crystals

6.2.1. Origin of Tilting of FLC Molecular Alignment in Smectic C Phase

Several theories have been developed to produce a realistic model for the origin of tilted director of the smectic C (SmC) phase of LC [11], Figure 1. Initially, Wulf [12] had assumed the zig-zag shape of the molecule and considered the freezing of the free rotation of this shape around the long axes of the molecule. Later, McMillan [13] proposed this theory of freezing the core of a molecule along the transverse dipole moment of the molecules which are present in every material possessing the SmC phase. It states that the two oppositely oriented outward dipoles, which are away from the centers of the molecules, would not produce the net spontaneous dipole moment. Now, if there are two dipolar components along with longitudinal directions, the tilting of molecules can be expected in the layer to minimize the free energy of the dipolar systems. However, some molecular level studies (NMR and experimental details) have shown that molecules are freely rotating about the long molecular axis in SmC phase and have a disagreement with the assumptions of both the above theories. Further, several assumptions have been assumed to clarify the origin of the tilt [14], among them Goossens has made a detailed calculation of the intermolecular interactions between shaped molecules, the attractive potential arising from the anisotropic dispersion energy as well as permanent quadrupole moments [15–21].

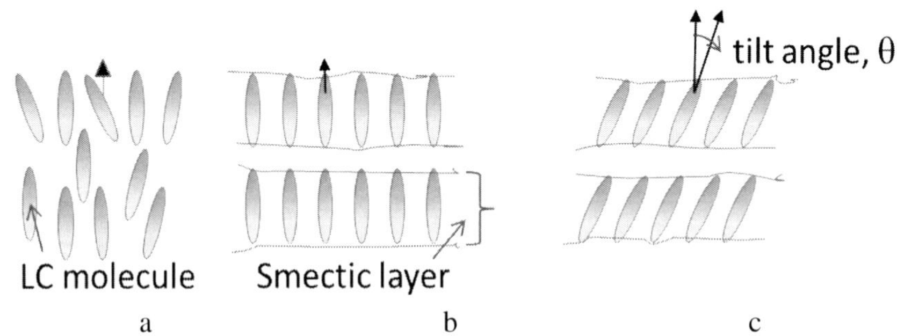

Figure 1. Arrangement of molecules in (a) nematic mesophase, (b) SmA and (c) SmC phase. 'n' represents the director of nematic phase which is in the direction of arrow (a) and (b) whereas it is tilted in (c).

6.2.2. Origin of Ferroelectricity in Chiral Smectic C Phase

The transverse dipole moment in the compound exhibiting the SmC phase is to build an effectively strong quadrupole moment due to which the titling structure of the molecule in the layered structure. Moreover, the existence of chirality in the SmC (SmC*) phase shows the strong dielectric anisotropy as a result of the symmetry reduction to C_2 from C_{2h} and dipole moment ($\mu_{\perp}$) originates perpendicular to the layer normal which results in spontaneous polarization ($P_s = \mu_{\perp}$) along the SmC* layer plane, Figure 2. However, the spontaneous polarization is the result of the contribution of dipole-dipole and dipole-quadrupole induction interaction [22]. The spontaneous polarization P_s is a vector and represents the breakdown of symmetry, i.e., there is a directional electronic shift. By convention, P_s is defined as positive as per the rule $k \times n$, where k and n are the layer normal and director unit vectors, respectively, and negative otherwise. Since n (director) precesses around z (layer normal) when going from one layer to another, P_s also precesses around z in the plane perpendicular to the director n and parallel to the smectic layers. The successive tilted smectic layers show a gradual change in the direction of tilt, such that the director precesses about the z from layer to layer, always lying on the surface of a hypothetical cone of angle 2θ. The angle around the circle of precession is known as the azimuthal angle Φ. Thus, the director n, as well as P_S, constitutes the helix of a specific pitch being along with the layer normal and defined as the distance needed to reach the same molecular orientation, Figure 2 shows the basic geometry of the SmC* phase. However, the orientation of P_S leads to zero net spontaneous polarization due to helix formation. But, the net P_S can be obtained in conditional sampling such as in a sample of thickness less than the pitch value of FLC material. Clark and Lagerwall in 1980 had proposed that if FLC material is confined between two parallel glass plates separated by a distance less than the pitch value of materials, the strong surface anchoring holds the molecules at the surface and does not allow the helix to complete. In this way, the helix could be suppressed and the device was called a surface *stabilized ferroelectric liquid crystal* (SSFLC) device [23]. It is well known that the solid ferroelectrics are characterized by a macroscopic polarization that is energetically stable in the absence of an external field; whereas bulk ferroelectric liquid crystals do not have any macroscopic polarization due to the formation of the helical structure. The switch ON and OFF state can be turned within a few tens of microseconds i.e., about 1000 times faster than the simple twisted nematic device.

The helix is inherently absent in the surface-stabilized structure due to its unwinding by elastic interaction of FLC molecules on the surfaces of substrates. The force, which keeps the helix unwinding, generates interaction, plays the role of an external applied field and is non-interacting with the electric field that has to be applied. In terms of unwinding, there is no principal difference between a cholesteric helix and a chiral smectic helix. The deformation by external electric or magnetic field energy is on the one hand whereas energy of elastic deformation in the liquid crystal is on the other which are essentially equivalent in the two cases. Thus, the surface interaction is equivalent to the above two cases. Suppose, the whole sample is helix free, then the molecular arrangement in layers becomes such that the layers arrange themselves in the way as books in the shelf and the arrangement is called '*book shelf geometry*'. Hence, Clark and Lagerwall showed that the true ferroelectricity can be achieved by the suppression of helix with surface control.

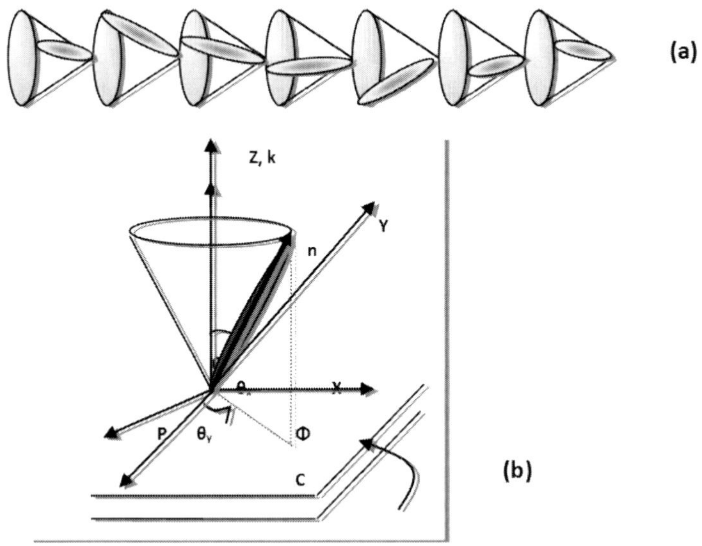

Figure 2. (a) Spiraling of the director in SmC* phase. (b) Basic geometry of the SmC* phase showing layer normal z, tilt angle θ(T), molecular director n, director C, polarization P along (k x n) and azimuthal angle Φ of P about z. Directions of both n and P vary spatially in a helical manner.

6.2.3. Symmetry Concepts of Ferroelectric Liquid Crystals

There are three concepts of symmetry that are commonly used by researchers in different fields of science and technology and used in liquid crystals as well. They are i. *molecular symmetry*, ii. *space symmetry*, and iii. *phase symmetry* [24].

6.2.3.1. Molecular Symmetry

This concept is used to define the spatial configuration of a single molecular structure. For instance, the configurational isomers are two types of stereoisomers; one is an enantiomer and the other is diastereo isomer. Enantiomers are two molecules having a non-superimposable mirror image. On the other hand, diastereoisomers are those which possess more than one asymmetric atom and non-superimposable mirror image [24]. The spatial configuration of asymmetric carbon atom is indicated by R (rectus, right) or S (sinister, left) as used for the first time by Cahn et al. in 1950 [25, 26]. The scheme is based on the priority arrangement of groups in the molecule about the asymmetric carbon atom.

6.2.3.2. Space Symmetry

The electric polarization of the medium must be invariant under symmetry operations and this requires the polar axis to be along a unique rotation axis (only one may exist) to which no perpendicular plane of reflection symmetry exists. Usually, very high symmetry of some phases, such as nematics ($D_{\infty h}$, uniaxially birefringent) and smectics (SmA: D_{∞}; SmC: C_{2h}; biaxially birefringent) is not matchable with a spontaneous macroscopic polarization vector *P*. The lower symmetry of these phases is just like the symmetry group as the letter *N, S,* or *Z,* which cannot be associated with vectors due to their mirror reflection perpendicular to the plane of the paper. However, a SmC phase composed of a chiral molecule has symmetry allowing it to be ferroelectric because now the symmetry plane has vanished [27]. The possibility of ferroelectricity in liquid crystal phase was proposed by Saupe in 1969 [28]. Meyer has proposed the ferroelectricity in liquid crystal phase by symmetry argument if the molecules in the material comprise of dipole moment in the direction $k \times n$ [1]. Now, we are in the position to discuss the symmetry which states that SmC* phase (chiral phase) shows the ferroelectric phase whereas SmC phase (non-chiral phase) does not.

Consider, the average molecular director *n* playing an important role in local symmetry consideration for example in defining the local symmetry axis. In solids any descriptive vectors fit well into the definition of polarization of the medium. However, in liquid crystals, it is likewise fundamental that in the formation of all properties, the medium is invariant against the sign reversal of the director, $n \rightarrow -n$ *symmetry operation.* On the other hand, if this invariance is not required, then ferroelectricity would have been observed in the simple liquid crystal phase also, but this is not observed in reality. Therefore, *n* is not a variable quantity that could lend itself as a direction for introducing polarization *P*, any component of it along with *n* violates the condition. The polarization vector is always locally perpendicular to the director, i.e., $P \perp n$.

Assuming that the system exhibits a net spontaneous polarization $P(x,y,z) = (P_x, P_y, P_z)$. Here, we must demand this vector to be invariant under all symmetry operations which are compatible with the symmetry of the phase under study. Let us consider the case of ordinary liquid with optically active molecules as shown in Figure 3. The system has complete rotational symmetry allowing no reflection operation. Under 180° rotation about *y*-axis, any assumed polarization vector is forced to change the sign in its x and z components, which therefore must be zero for P to remain invariant. The remaining P_y is transformed into P_x at a subsequent 90° rotation around the z-axis, and so likewise has to vanish. Thus, an isotropic liquid can be optically active but not ferroelectric. The argument is similar for the nematic phase and the SmA phase [27].

On the basis of molecular symmetry in SmC phase, there is a molecular symmetry while not in SmC*. The SmC* i.e., ferroelectric phase can be prepared in many ways for instance, by synthesizing a chiral variant of a molecule, which is known for its stable SmC phase, or by doping an SmC phase in any chiral molecular material soluble in it; symbolically C + * → C*.

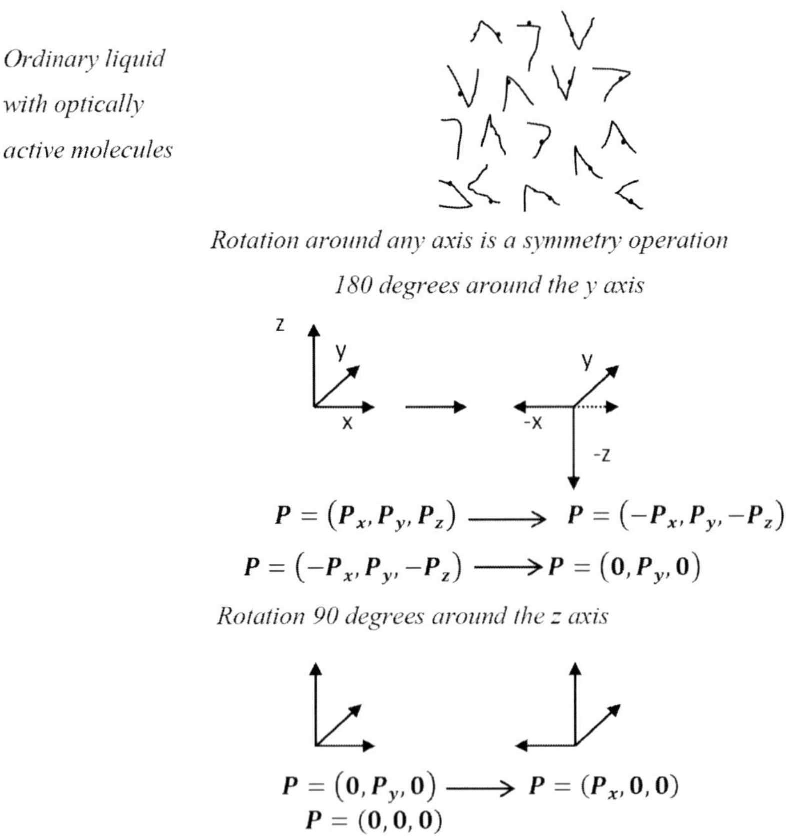

Figure 3. The isotropic liquid is showing the non-existence of spontaneous polarization.

A molecule with one or several asymmetric carbons atoms can be a dopant, which is usually the case, but it can also be a molecule without asymmetric centers, having chirality built into its integral structure. The phase transition properties of the medium have neither been affected very much by the doping nor the chirality of molecules. The lack of mirror symmetry of the medium now gives rise to the summing up of dipolar contributions in a certain direction for every molecule with appropriate dipoles. Figure 4 shows a small portion of the SmC and SmC* phase, respectively with the coordinates needed for the present purpose [27].

In case of the achiral SmC phase, the symmetry operation is the two-fold rotation along the plane of layers and perpendicular to the long molecular axis of the material. The symmetry group C_{2h}, comprising with mirror plane normal to the rotation axis and a center of symmetry, has been found. But if the molecules are chiral, the mirror plane and center of symmetry disappear as a result reducing the symmetry to a point group C_2. In this case, there is a possibility of obtaining spontaneous polarization along the rotation axis. If the molecule has a permanent dipole moment perpendicular to the long axis and free rotation around, this axis is hindered by the surrounding molecules.

In both SmC and SmC* phase a 180° rotation around the y-axis is an allowed symmetry operation [29]. This operation allows the polarization vector P to transform accordingly as

$$(P_x, P_y, P_z) \rightarrow (-P_x, P_y, -P_z)$$

Thus, the resultant spontaneous polarization i.e., by combining the polarization before and after the operation, the net polarization will be as

$P = (0, P_y, 0)$ exist.

In the SmC phase, an x-z plane is a mirror plane as shown in Figure 4. The corresponding symmetry operation transforms the vector P according to

$$(0, P_y, 0) \rightarrow (0, -P_y, 0)$$

This shows that P_y changes to $-P_y$ which results in total P zero. However, in the case of SmC* phase the molecules are chiral, i.e., they cannot be superimposed over the mirror images of themselves. Thus, the x-z plane is no longer a mirror plane and symmetry thus permits the system to exhibit a permanent polarization in the y-direction, i.e., the direction perpendicular to the direction of the tilt.

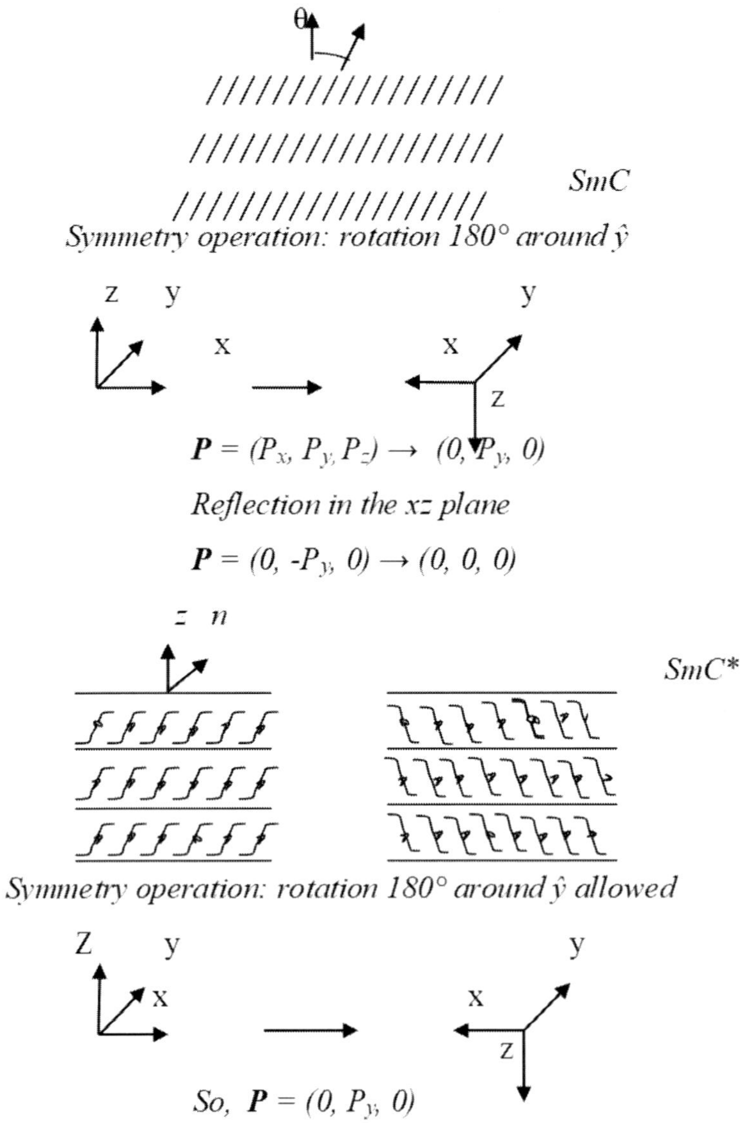

Figure 4. The anisotropic liquid is showing the existence of spontaneous polarization.

Therefore, the reduced symmetry and interactions due to chirality across the layer planes define that the molecular tilt direction will twist through successive layers, which leads to the rotation of the spontaneous polarization that is always parallel to the smectic layers as well as perpendicular to the long molecular axis and the direction of the tilt. This results in a helicoidal structure with a pitch *p*, which is the distance along the *z*-axis needed to reach the same molecular orientation. Thus, the macroscopic polarization of the Sm-C* phase is average to zero. However, the system will exhibit a local spontaneous polarization according to

$$\begin{Bmatrix} P_x = -P \sin\Phi \\ P_y = P\cos\Phi \\ P_z = 0 \end{Bmatrix}$$

Hence, due to the helicoidal nature of the molecular arrangement, the macroscopic polarization is zero in the bulk sample. The ferroelectric properties are in a way lying hidden in the helicoidal structure. Since the bulk polarization is zero and ferroelectric properties are also hidden in the bulk sample, therefore such types of ferroelectrics are called *improper ferroelectrics*. Therefore to bring the properties of proper ferroelectric, one needs either to apply critical electric field strength, $E = \pi^4 K/4\,Pp^2$, where *p* is the pitch of unwinding helix and *K* the twist elastic constant; or suppressing the helix by surface interactions i.e., SSFLC.

6.2.3.3. Phase Symmetry

The phase symmetry is related to the symmetry of the orientational ordering of the molecules in smectic phases from layer to layer [24]. For example, the helical phase of the SmC* possesses the rotation of the average molecular director around an axis and draws the shape of helix. This helix has a simple two-fold rotation symmetry about the axis perpendicular to the axis of the helix. This does not possess the superimposable mirror image of the helix, for example, deformed helix FLC (DHFLC) and cholesteric. These helical phases are optically active if the helical pitch lies in the range of the visible part of the electromagnetic spectrum. In general, the plane of polarization of the polarized light is rotated by the optically active molecules whereas in helical phases it is dependent on the helical pitch of the chiral phase.

In the present chapter, the effect of gold nanoparticles on the electro-optical and dielectric properties of FLCs has been discussed. There have been found an efficient increment in the basic properties such as optical tilt angle and operating voltage of host FLC material. The gold nanoparticles have been coated with the electrostatically protecting material coating, which avoids the agglomeration of GNPs in the dispersing medium. These dispersing mediums of GNPs are always isotropic materials. Therefore, the effect of some isotropic mediums such as water and ethanol on the properties of FLC material has also been demonstrated. One of these results, the switching outside the effective electrode region has been the major outcome of the study. There has been the observation of a new collective dielectric process in water added FLC as observed by the dielectric spectroscopy technique.

6.3. Gold Nanoparticles

Gold nanoparticles (GNPs), also called gold colloids, have been the center of research since the 17th century [30]. They have been in the limelight of research due to their fascinating properties such as the behavior of individual particles, size-related electronic, magnetic, optical properties (quantum size effect), assembly of multiple types, and their applications for catalysis and biology. They are significant due to their method of synthesis by various techniques [31–35]. The unique properties within the gold material appear when the dimension of bulk gold is reduced below the wavelength of visible radiations, the electric field of the incident visible part of radiations spectrum on the GNPs in the dielectric medium induces the dipole moment in the surface of gold nanoparticles due to the shifting of electron cloud called as surface plasmons. The dielectric properties of gold nanoparticles vary with frequencies as $\varepsilon(\omega) = \varepsilon_1(\omega) - i\varepsilon_2(\omega)$where $\varepsilon_1(\omega)$ and $\varepsilon_2(\omega)$ are the real and imaginary parts of the complex permittivity of dielectric constant. Furthermore, the dielectric constant of metal can be obtained by using Drude free electron model theory suggests that the dielectric constant as

$$\varepsilon^*(\omega) = 1 - \omega_p^2/\omega\,(\omega + i/\tau), \tag{1}$$

where $\omega_p^2 = 4\pi\,\mathrm{Ne}^2/m_o$. Here N is the density of conduction electrons, $m_o$is their effective optical mass of electron, and $\tau$is the relaxation time. Separating the real and imaginary parts, we get

$$\varepsilon_1(\omega) = 1 - \omega_p^2\,\tau^2/(1 + \omega^2\tau^2) \tag{2}$$

and

$$\varepsilon_2(\omega) = \omega_p^2\,\tau/\omega\,(1 + \omega^2\tau^2). \tag{3}$$

The induced polarization phenomenon and surface chemistry of GNP have been found to enhance the electro-optical properties of liquid crystals. The optical tilt angle of a commercial FLC material has been found to increase up to five times by adding the gold nanoparticles and the respective driving voltage reduces up to 10 times. Besides this, the memory effect has also been observed [4, 5]. Since the dispersing medium of several nanomaterials is water and ethanol, therefore it has been worth studying the dispersion of these isotropic fluids in the FLC materials. In the present chapter, we would analyze the behavior of double distilled water and ethanol in the FLC material.

Gold nanoparticles are the most commonly synthesized nanoparticles using some solvents [32, 36–38]. When the gold nanoparticles are to be added to other materials and if solvent could not be evaporated properly, then there are chances of getting the impact of solvent also in the mixture. This aim has been focused to explore the impact of the two most common solvents. They are water and ethanol and are explored for their impact on the basic properties of FLC when added into the bulk.

6.3.1. Water

Water is also called a universal solvent due to its nature of dissolving most of the materials in it. Gold nanoparticles are also most commonly synthesized in water as a solvent. It is readily understandable that the water is inorganic and FLC is organic so it becomes unusual to mix them for the study. From the experimental experience, it has been observed that the mixing of water in FLC is a very challenging task and hence, ultimately, it could not be mixed properly and uniformly but an emulsion could be formed. This emulsion is further treated under various steps of temperature for making the incursion of water molecules inside the FLC matrix which is used for the study.

6.3.2. Ethanol

Ethanol is one of the most commonly used solvents. Gold nanoparticles can also be easily synthesized by making ethanol its dispersing medium. Thus, studying ethanol in the bulk of the FLC matrix would also be interesting for the same reason as for water if it was left unevaporated during the mixing of gold nanoparticles in FLC.

6.4. Experimental Techniques

There are some specific techniques to analyze the particular property of FLC material. The following are those which have been used discussed

6.4.1. Sample Cell Fabrication

Highly conducting indium tin oxide (ITO) coated glass plates with a flatness of $\lambda/2$ have been used as electrodes on substrates and used to fabricate the sample cell. Both the electrodes of the cell were treated with adhesion promoter and later nylon 6/6 and rubbed uni-directional on both the surfaces to get homogeneous alignment [39]. The two polymer-treated substrates were assembled over each other facing the polymer surface inside. The ultra violet radiation effective adhesion polymer was applied to seal the sample cell, Figure 5. The thickness of the cells was maintained employing photolithographic technique and/or Mylar sheets. The cells were calibrated using air and toluene as standard references.

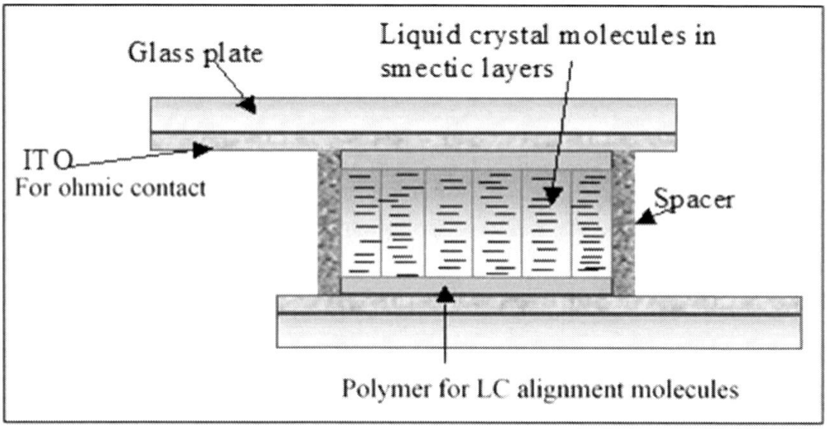

Figure 5. Structure of the sample cell used in the experiments.

6.4.2. Dielectric Spectroscopy

Dielectric spectroscopy (DS), sometimes called impedance spectroscopy, is used to measure the dielectric properties of the materials as a function of frequency, which can provide the information about the dielectric relaxation mechanisms along with the variation with time 10^{-12} s to 10^5 s by means of computer-controlled impedance analyzers [40–47]. It occupies a special place among the numerous modern methods used for the physical and chemical analyses of materials because it enables to investigate the collective relaxation processes of complex systems over an extremely wide range of characteristic times by using network analyzers as well as time-domain spectrometers [44]. At higher frequencies (above 10 GHz) one should use waveguides, which means that for each frequency a separate apparatus is necessary to perform the accurate dielectric measurement. Dielectric spectroscopy is very sensitive to intermolecular interactions. It is capable of monitoring cooperative processes via molecular spectroscopy. This provides a link between the investigation of the properties of the individual constituents of complex material and the characterization of its bulk properties.

In dielectric spectroscopy, one measures complex dielectric permittivity:

$$\varepsilon^*(\omega) = \varepsilon'(\omega) - j\varepsilon''(\omega), \tag{4}$$

Where $\omega = 2\pi\nu$ is the angular frequency, and ν is the frequency of measuring the electric field. ε' and ε'' are the real and imaginary parts of the complex dielectric permittivity, respectively. There are two borderline values of the dielectric permittivity: $\varepsilon'(\omega \to 0) \equiv \varepsilon_o$, is the static dielectric permittivity and$\varepsilon'(\omega \to \infty) \equiv \varepsilon_\infty$, is the high-frequency limit of the dielectric permittivity. For non-polar materials, the dielectric permittivity is equal to $\varepsilon(\infty)$ in the frequency range from 100 kHz to the far-infrared region i.e., about 10^{13} Hz. For polar liquids, mesomorphic systems like liquid crystals and plastic crystals, and orientationally disordered crystals the reorientation of permanent dipole moments leads to the dielectric absorption in the microwave or radio frequency ranges. $\varepsilon''(\omega)$ is related to the energy (W) absorbed per unit volume in the dielectric sample under test (SUT) by the formula

$$W = \frac{1}{2}\omega E_o^2 \varepsilon_o \varepsilon''(\omega) \tag{5}$$

where ε_o= 8.854187817 x 10^{-12} F/m is the dielectric permittivity of free space and E_o is the amplitude of sinusoidal measuring electric field of angular frequency ω. In linear dielectric spectroscopy, the amplitude of the measuring electric field should be chosen so that it does not suppress the structure of SUT, for instance, the helicoidal structure of the SmC* phase. One of the best ways to study the dielectric behavior of material consists of measuring the complex dielectric permittivity versus frequency at constant temperature and ambient pressure. In this way, one obtains the dielectric spectrum in the frequency domain. That is why this method is termed frequency domain dielectric spectroscopy (FDDS). Dielectric studies at varying temperature can be easily performed. Dielectric studies under high pressure have also been performed for liquid crystals [45] as well as for FLCs [48]. Using FDDS, one obtains an absorption peak in a certain frequency range. In the case of polar liquids and plastic crystals, it usually appears in the microwave frequency range (100 MHz – 10 GHz) whereas for liquid crystals and ferroelectric liquid crystals, the dielectric absorption shows the absorption peak at radio frequency. Usually, by using impedance analyzers, one measures experimentally capacity $C(\omega)$ of the cell filled with a DUT and the loss tangent [10]:

$$C(\omega) = \varepsilon_o \varepsilon'(\omega) C_o \tag{6}$$

and

$$\tan\delta = \varepsilon''(\omega)/\varepsilon'(\omega) \tag{7}$$

where C_o is the capacity of the empty cell and $\varepsilon'(\omega)$, the real part of the complex dielectric permittivity of SUT. It is worth mentioning that the Debye equation written in terms of the loss tangent has the form:

$$\tan\delta = \frac{[\varepsilon(0)-\varepsilon(\infty)]\omega\tau}{\varepsilon(0)+\varepsilon(\infty)}. \tag{8}$$

The position of the maximum of $\tan\delta$ does not coincide with the position of the maximum of $\varepsilon''(\omega)$). The former is shifted to higher frequencies by a factor $\sqrt{[\varepsilon(0)/\varepsilon(\infty)]}$. So, the relation between the critical frequency f_c and the frequency of $\tan\delta_{max}$ at f_{max} is as follows:

$$f_{\max} = \sqrt{\varepsilon(0)/\varepsilon(\infty)} f_c. \tag{9}$$

As a consequence of Eq. (9), there is a pronounced shift of relaxation peak for FLCs and highly polar materials for this factor may be as large as 100 or even higher. It is a real problem to overcome in FDDS studies on FLCs or in general on materials showing very large dielectric permittivity in the sub-hertz frequency range. In many cases, one observes a nice peak for $\tan\delta$ and after transformation of the dielectric spectrum into the dielectric loss $\varepsilon''(\omega)$) only the right-hand side part of the spectrum shows up due to the shift of the absorption curve to low frequencies. So, for systems displaying very high dielectric permittivity at low frequencies due to collective processes like Goldstone mode, one should measure the dielectric spectrum in the possible widest frequency range. Sometimes, it is even better to work with $\tan\delta$ dielectric spectrum only.

Dielectric spectroscopy detects stochastic reorientation of molecular dipole moments and collective fluctuations of spontaneous polarization in paraelectric, ferroelectric, and anti-ferroelectric liquid crystals. For liquids and solid rotator phases of organic polar compounds, the dielectric spectrum is usually of Debye type. However, some systems composed of flexible molecules [42] and some disordered solid phases [46] exhibit broad dielectric spectra and can be described by the Cole-Cole function [47].

$$\varepsilon*(\omega) = \varepsilon'(\omega) - j\varepsilon(\omega) = \varepsilon(\infty) + \frac{\varepsilon(0)-\varepsilon(\infty)}{1+(\omega\tau)^{(1-\alpha)}} \tag{10}$$

Where α is a parameter responsible for the symmetric distribution of the relaxation times and τ_o in this case is the most probable relaxation time. σ and ε_o are the conductivity and static permittivity, respectively. The last term represents the conductivity effect at low frequencies.

Furthermore, dielectric spectroscopy is a tool to know the response time of the molecules by the applied electric field, activation energy of the liquid crystal molecules, the refractive index of the material, the dielectric strength, and much more. The memory effect in FLC device can also be analyzed by FDDS technique. The low value of permittivity at low frequency and 0 V bias after removal of 30 V bias showed that the dielectric spectroscopy could detect what electro-optical could not.

Consequently, one should point out that dielectric spectroscopy has greatly been developed during the last three decades. FDDS is often used to study dielectric properties of new materials for science and technology. Using FDDS one can evaluate on the one hand the static dielectric permittivity $\varepsilon(0)$, the high frequency limit of the dielectric permittivity $\varepsilon(\infty)$, and such molecular parameters as permanent dipole moment and polarizability. On the other hand, real dielectric spectroscopy consisting of measurements the complex dielectric permittivity $[\varepsilon^*(\omega) = \varepsilon'(\omega) - j\varepsilon''(\omega)]$ in a wide frequency range, say from 1 mHz to 10 GHz. It allows one to study the diffusion coefficients of ions present in real dielectric materials, collective, and molecular dielectric relaxation processes. The former comes from components of polarization vectors whereas the latter results from the molecular reorientation around different axes and/or from intra-molecular motion. New FLCs and AFLCs display a rich dielectric spectrum.

In the present study, for dielectric measurements, the cells were prepared with homogeneous alignment, as in the case of optical studies. Here, the sheet resistance of the ITO coated glass should be less whereas for optical studies higher tolerance can be applied in the selection of ITO plates. This is because, at high frequencies, the effect of ITO resistance is seen as an increase in the dielectric absorption and a decrease in dielectric permittivity, which mimics a relaxation process in the dielectric spectrum. To separate this effect from the relaxation process of the material under study, the resistance of the ITO plate should be chosen such that the relaxation frequency for this process is sufficiently shifted towards the high-frequency region. For this, the resistance of the ITO layer must be approximately 30 $\Omega/\square$. So cells were prepared with ITO plates of very low resistance for dielectric measurements.

Dielectric studies of FLC materials have been carried out using HP 4192 A, Solartron 1260, and Wayne Kerr 6540A impedance analyzers. Capacitance and dielectric loss are measured at different temperatures and varying frequencies. Dielectric permittivity is calculated using the value of air capacitance calculated before the filling of liquid crystal material using the formula given above Eqs. (9) and (10). These types of measurements provide the electrical energy storage capacity of the FLC materials which later furnish the information about the molecular contribution.

6.4.3. Electro-Optical Measurement

Optical tilt angle measurements and texture micrographs were taken with the cell mounted on a rotatable stage of the polarizing microscope (Ziess, Axioscope 40 Pol) and connected with the integrated computer-controlled temperature controller (Julabo-F25 HE. Germany). Tilt angle is the primary order parameter of SmC* phase. Thus, its accurate measurement is very much essential. Moreover, the contrast of an FLC display also depends on the value of its tilt angle. For homeotropic aligned liquid crystals, the conoscopic method is used for tilt angle measurement [49]. For homogeneous planar aligned sample, the electric field-induced switching is used [49]. In the FLC sample during the application of an external field, the molecular director is switched by an angle 2θ (θ is the tilt angle of the material) in the plane of the substrates. From this technique, the tilt angle of the material can be determined.

In our case of homogeneously planar aligned FLC cells, the sample was mounted on the polarizing microscope with crossed polarizers. A field high enough to unwind the helix is applied to the sample so that molecules in the whole sample are aligned uniformly with a tilt angle, θ, away from the layer normal. The sample is rotated on a microscope table to get minima in optical transmission. When the field is reversed the molecules rotate along the tilt cone so that the final tilt becomes –θ from the layer normal as shown in Figure 6. Once again the microscope table is rotated to achieve the minimum transmission. The angle by which the microscope table was rotated from one minimum position to another minimum will be twice the tilt angle (2θ), Figure 6.

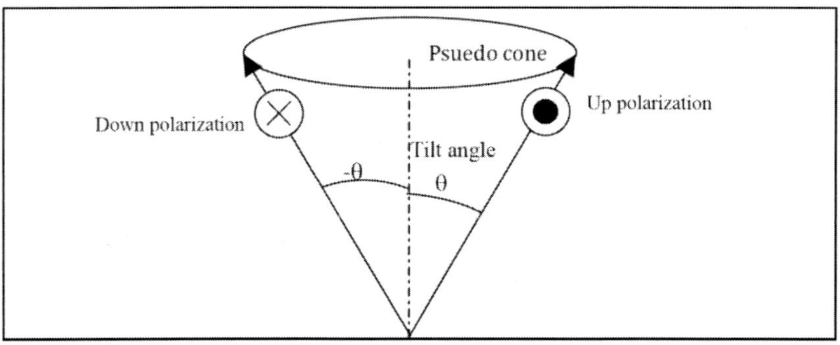

Figure 6. Mechanism of tilt angle measurement.

6.4.4. Spontaneous Polarization Measurement

Spontaneous polarization (P_s) is the intrinsic property of ferroelectric materials. In solid ferroelectrics, spontaneous polarization is the primary order parameter. But in FLCs, P_s is the secondary order parameter, while the primary order parameter is the tilt angle, θ. To achieve higher switching speed in SSFLC displays higher P_s materials are needed but at the same time it was found that bistability is affected in the case of higher P_s materials. Therefore, it is important to know the magnitude of the P_s to balance the switching speed and optical bistability.

This parameter can be measured in several ways. Sawyer-Tower technique is one of the standard methods to determine the P_s and hence the ferroelectric behavior, Figure 7, [50]. Pyroelectric methods can also be used to measure polarization provided that the polarization vanishes at a defined temperature [51]. Monitoring the amount of current through the circuit when the polarization is reversed from one state to another is another method [52, 53]. This particular method is useful in the case where ferroelectric switching speed is slow compared to the RC time constant of the circuit. However, for faster switching materials, the capacitive current and polarization reversal currents are often difficult to separate.

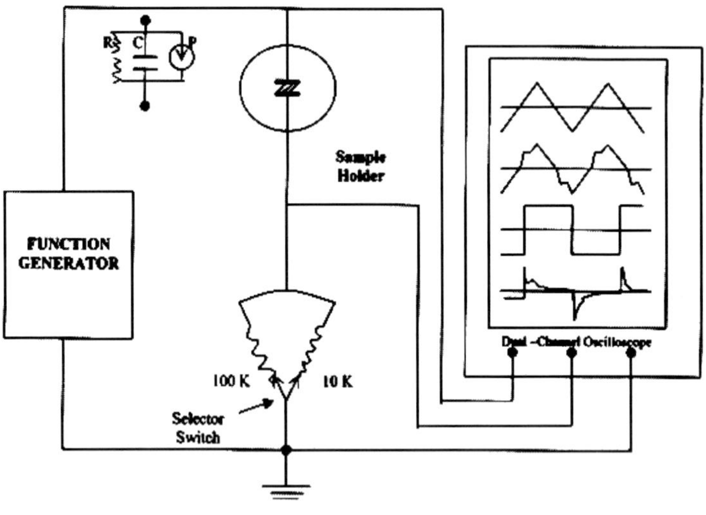

Figure 7. The experimental set-up used in the triangular wave method to estimate the value of P_s.

In the present chapter, the alternative switching method using a triangular wave has been used due to its unique advantages as the use of this wave form separates the capacitive contribution from interfering with current due to polarization reversal. The resistive component is easy to extract as this contribution varies linearly with applied voltage. To estimate the value of P_s [54], the triangular wave pulse method has been utilized. The experimental setup for polarization measurement used in the present studies is shown in Figure 8(A). The function generator delivers a triangular wave pulse in the frequency range of 0.1 Hz-100 kHz with $20V_{pp}$ voltage. The applied peak voltages are sufficient to cause a polarization reversal in the FLC. The FLC material sandwiched between two electrodes can be represented schematically by a parallel combination of a resistor (R) [55], a capacitor (C), and a polarization realignment device (P) as shown in Figure 8(A). The current I(t) induced in a ferroelectric by applying a voltage V(t) can be written as a sum of the following three contributions:

a) ionic current, I_R
b) charge accumulation in the capacitor (displacement current), I_c
c) polarization realignment current, I_p

Thus one obtains,

$$I(t) = I_R + I_C + I_P \tag{11}$$

$$\frac{V(t)}{R} + C\frac{dV(t)}{dt} + \frac{dP}{dt} \tag{12}$$

Where *P* is the amount of the charge induced by the polarization reversal in the cell on the application of an electric field. Figure 8(A)(e) shows the contribution of three components to the overall current response.

By selecting a suitable value of resistance *R* (about two orders of magnitude smaller than that of FLC sample resistance) we can get a suitable overall current profile to subtract the ionic and capacitive currents by drawing a base line. The area under the polarization current peaks gives the direct estimate of P_s [56].

i.e., *2aP = (area under the curve) (current * time)*
*= A (i * t)*

$$\text{Hence, } P_s = A\,(i * t)\,/\,2a, \tag{13}$$

where *a* is the area of the sample through which the electric field is applied. A is the area under the curve.

We have the voltage-time graph, so modifying the above formula

$$P_S = \frac{A(V*t)}{2*R*a} \tag{14}$$

where *R* is the resistance.

Area under the peak is calculated using integration in software or by using graph paper.

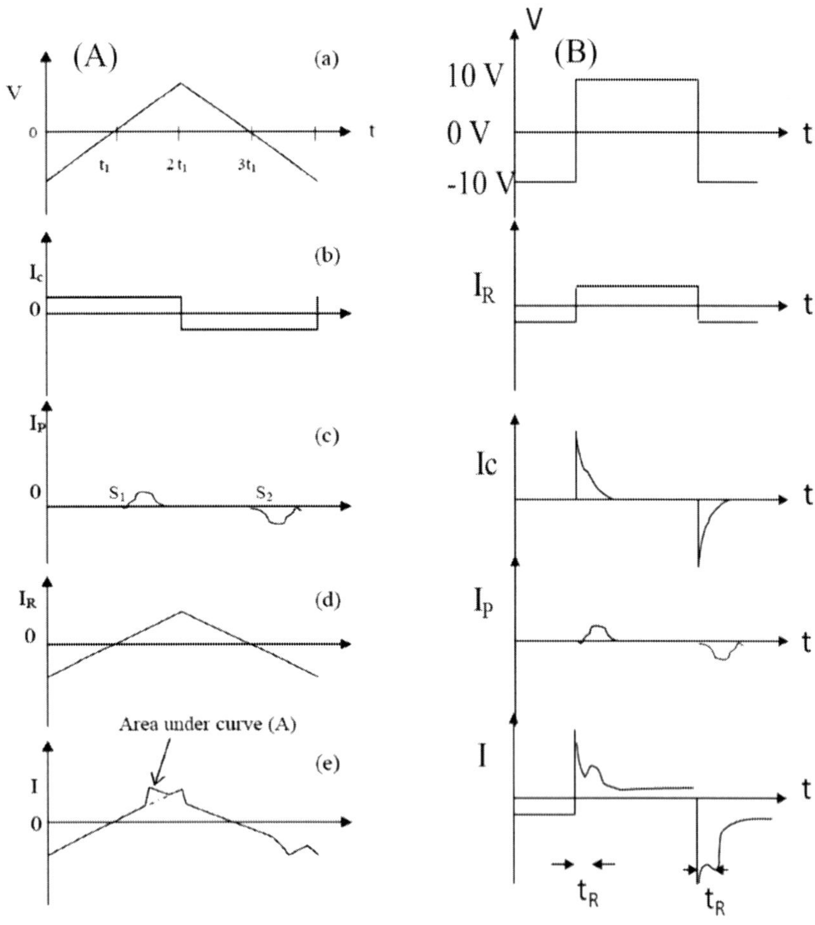

Figure 8. (A) Schematic illustration of the current induced on the application of a triangular pulse to FLC, (a) applied triangular wave (b) contribution due to charge accumulation I_c *(c)* polarization realignment current I_p *(d)* ionic current I_R *(e)* overall current profile. (B) Schematic representation of the current induced on the application of a square wave to FLC, (a) applied Square wave, (b) the contribution due to charge accumulation I_c, (c) the ionic current I_R., (d) polarization realignment current I_P, (e) overall current profile.

6.4.5. Response Time Measurement

Response time is considered one of the most important parameters of a liquid crystal device. Response time in SSFLC depends on the cell thickness, field strength, surface anchoring etc. as well as on the materials parameters like polarization, tilt angle, and rotational viscosity [57]. Response time can be measured either by monitoring the electrical response to an applied square/triangular wave [58, 59] or using the optical method [60].

6.4.5.1. Electrical Method

In electrical methods, the square wave method is much simpler than the triangular wave method. These square wave signals are applied to the FLC sample and the current response associated with the switching process is monitored as the voltage across a series resistance in the storage oscilloscope [59]. The schematic representation of different contributions to the current response is given in Figure 8(B).

The polarization bump occurs far away on the time scale from the square pulse edge of the applied voltage. The delay in time is caused by the response time of the FLC sample. Then, the time delay on the occurrence of the polarization bump from the applied square pulse edge directly gives the response time of FLC sample.

In triangular wave method, the frequency dependence of the polarization reversal current is utilized [58]. Here, the polarization reversal current peak is recorded at a particular frequency (say '*f*' Hz) in a CRO, and then a similar response at a frequency *2f* Hz is recorded after halving the time scale of CRO to obtain the same abscissa length and slope of the applied wavelength. The time difference between the occurrences of two polarization peaks gives the response time. However, the response time can be measured by using ALCT-P, INSTEC, Boulder, USA which is specifically made for three purposes polarization, response time, rotational viscosity measurements. However, other parameters like dielectric anisotropy etc. can also be measured by this device.

6.4.5.2. Optical Method

Photo diode helps to measure the optical response of FLC samples on a CRO. The time taken for a change in transmission from 10% to 90% of maximum transmission gives the rise time of the sample (τ_R) and the time taken from 90% to 10% decrease in transmission gives the decay time (τ_D) of the sample. Thus, we can directly measure the optical response time on the time axis of the cathode ray oscilloscope (CRO). Figure 9 shows how to measure the response time of an FLC device.

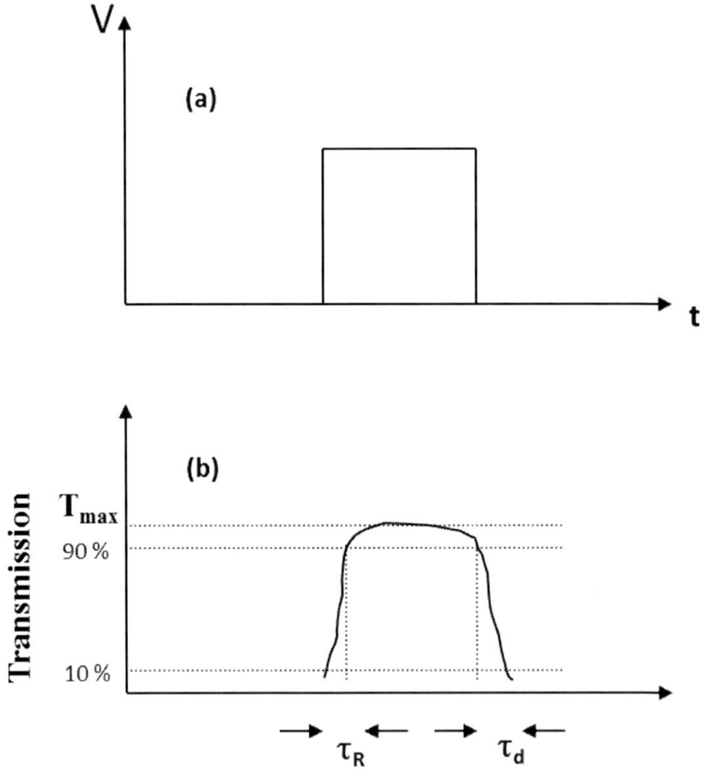

Figure 9. Schematic diagram of optical transmission of FLC to (a) an applied square pulse and showing (b) the optical time *(τ_R)* and optical decay time *(τ_d)*.

6.5. Results and Discussion

6.5.1. Gold Nanoparticles in Ferroelectric Liquid Crystals

Gold nanoparticles have been added to the complex matrix of FLC. GNPs have shown striking variations in the leading properties of commercially available FLC materials (Felix 17/100, Hoechst, Germany and FLC 6304, Rolic, Switzerland). The phase sequence of these materials is as follows [61]:

Felix 17/100: $Crystalline \xleftrightarrow{-28\,^{o}C} Sm\,C^{*} \xleftrightarrow{73\,^{o}C} SmA \xleftrightarrow{77\,^{o}C} Nematic \xleftrightarrow{84-87\,^{o}C} Isotropic$, and
FLC 6304: $Crystalline \xleftrightarrow{-14\,^{o}C} Sm\,C^{*} \xleftrightarrow{60.5\,^{o}C} SmA \xleftrightarrow{64\,^{o}C} Isotropic.$

Figure 10 shows the comparative study of tilt angle measurement of pure and gold nanoparticles doped Felix 17/100. The effective operating voltage of the bare FLC Felix 17/100 at which the tilt 2θ angle 25° is 1.6 V whereas after the addition of GNP in Felix 17/100 is 0.1V. The mere addition of AuNPs to Felix 17/100 is found to exhibit a strong interaction with the FLC molecules, inducing a long-range tilting of the dipoles resulting in a large tilt of FLC molecules. Unless there is an interaction between the nanomaterial and the FLC molecule, such a sudden change in the optical behavior cannot be accounted for. The simulation studies showed that the incident radiation of electric field 1 V/m at about 800 nm wavelength, which is greater than the size of GNP (10-15 nm), so radiation field can be assumed as the static field for sometimes, is enough to polarize the GNP [62]. The conversion of applied electric field strength into the near field can be estimated by calculating the near field efficiency as

$$Q_{\mathrm{nf}} = 2\sum_{n=1}^{\infty}|a_n|^2\left[(n+1)\left|h_{n-1}^{(2)}(x)\right|^2 + n\left|h_{n+1}^{(2)}(x)\right|^2\right] + (2n+1)|b_n|^2\left|h_n^{(2)}(x)\right|^2 \qquad (15)$$

Where $h_n^{(2)}(x)$is the Hankel function of the second kind and a_n and b_n are the scattering coefficients. x is the size parameter and defined as by $k \times R$ where $k=2\pi n/\lambda$. R and n are the radius of particles and refractive index of the surrounding medium, respectively.

The memory effect is one of the significant properties of FLC. The GNP doped Felix 17/100 has been studied [4, 5]. The electro-optical studies have been carried out on both the pure and GNP doped Felix 17/100. The square wave has been applied to the sample and the transmission has been recorded under crossed polarized microscope. At the time of +10V, the molecules occupy one of the states as per $k \times n$, due to the coupling of dipole moment with the applied electric field and the resulting transmission changes, as shown in Figure 11(a). If the material is bistable then the molecules will remain in the same state and transmission would be as it is in the absence of electric field and will not be as it was in the presence of electric field. Figure 11(a) shows that the transmission of bare Felix 17/100 material degrades after electric field goes off. In case of GNP mixed Felix 17/100, Figure 11(b), the transmission remains the same even in the absence of electric field. The studies have been further extended on the deformed helix FLC 6304 (deformed helix FLC,

DHFLC) material [5], which is a very short pitch material (visible range of wavelength), to frequency dependence. The observation shows that the transmission is unaffected after the removal of bias and remains in the same state for up to several days (four days), Figure 12(a)-(c). The previous state could be achieved by the application of the 35 Hz sine wave of 0.5 V amplitude, Figure 12(d).

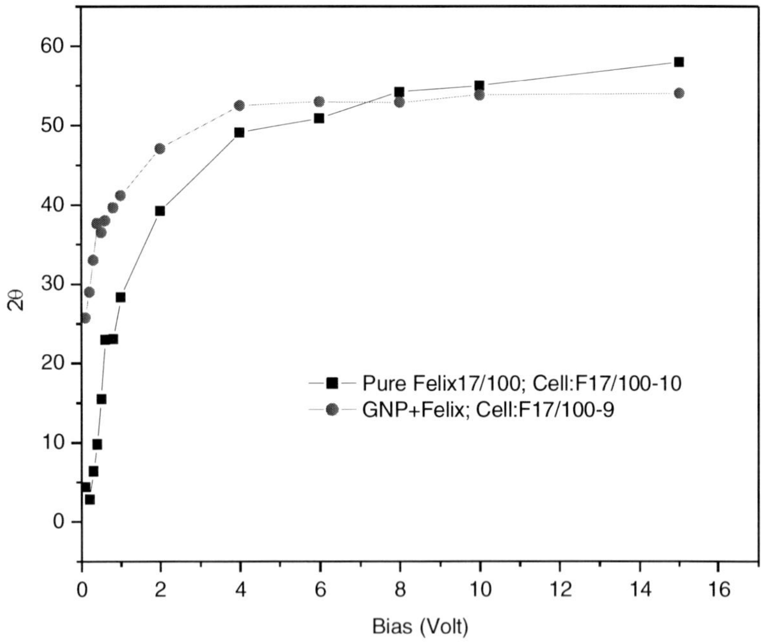

Figure 10. Behavior of optical tilt (x2) as a function of bias voltage in case of pure (square) FLC Felix 17/100 and Au NP doped FLC (star). With permission from AIP publishing [4].

The dielectric spectroscopy of this composition has been carried out for the determination of the memory effect [5]. In the absence of a bias electric field, the dielectric permittivity remains high at low frequency in the order of 100 and decreases as a function of frequency. The decreasing part of the curve shows the appearance of Goldstone mode which is the common appearance in the FLC materials. No new dielectric mode has been observed at this stage.

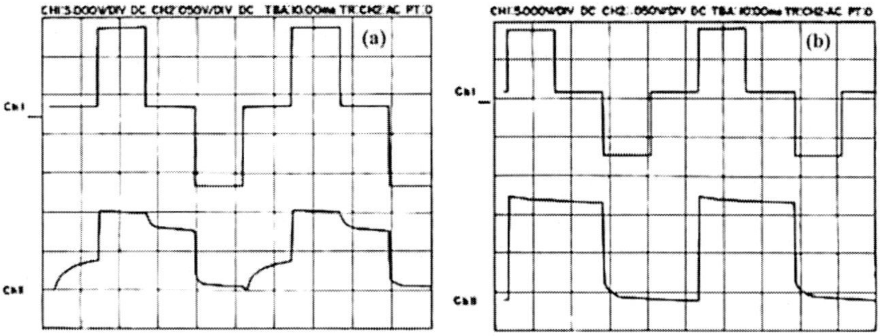

Figure 11. Optical response of (a) pure Felix 17/100 and (b) Au NP doped Felix 17/100, in 3 μm cell at 20 V and 25 Hz. CH1:[5.000 V/DIV] shows driving square voltage and CH2:[0.050 V/DIV] shows its optical response; outside mark on *y* axis (voltage axis) shows zero voltage level for respective channels. TBA shows the time scale per division (10.00 ms) on *x*- axis, which is the same for both channels [4].

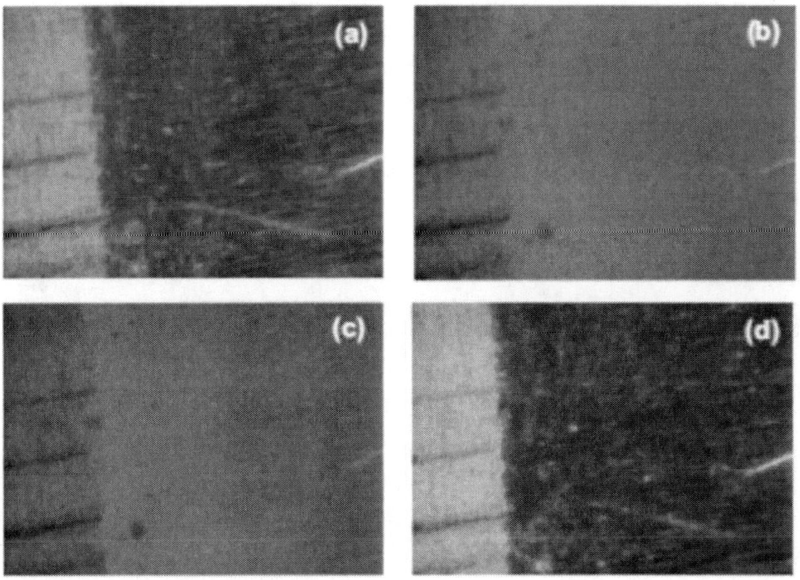

Figure 12. Optical micrographs of 3 μm GNPs doped DHFLC (FLC 6304) memory cell at room temperature at (a) 0 V, (b) 15 V bias, (c) 6 h after removal of bias, and (d) again 0 V [5].

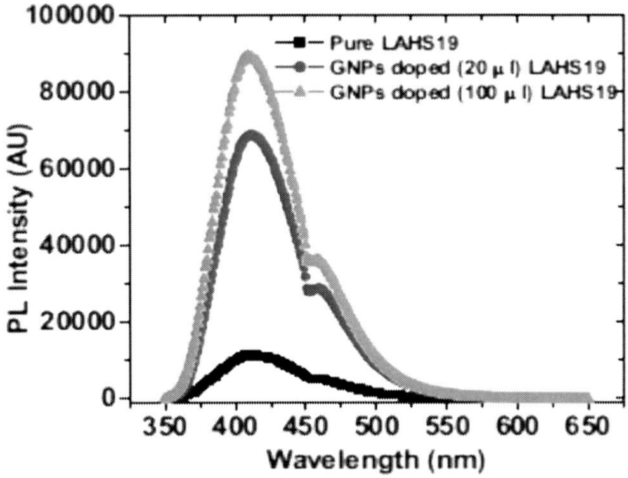

Figure 13. Comparison of PL intensity of pure DHFLC material LAHS19 and GNPs doped DHFLC material, excited at 340 nm, the emission wavelength is 411 nm [6].

The photoluminescence (PL) measurement has been performed on pure and GNP mixed DHFLC (LAHS19, newly synthesized material by Haase et al. at Darmstadt, Germany) [6]. The intensity of PL peak is stronger for well-aligned sample cell in comparison to unaligned at room temperature, Figure 13. Therefore, the experiment of PL measurement has been performed on the well-aligned pure and GNP added sample cells. The measurement shows the many times (about nine times) enhancement in the intensity of the peak. However, no shift in peak position has been observed with a concentration of GNP. The result has been attributed due to the coupling of surface plasmons with the dipole moment of FLC molecules. The alignment of the molecules contributes to the bulk enhancement. The PL phenomenon is the electronic transition delays the stability of electrons in the excited state for a longer duration and returns back after some time. In the case of present material, the electronic transition process is already available. But, the GNPs have paved the way to align the FLC molecules with higher ordering close to the surface due to either the surface chemistry or the induced polarization of the GNP, provided that sufficient energy is supplied to it. The emission of delayed radiation by the material is guided by the ordering of FLC molecules in the preferred direction rather than scattered one. The highly light scattering materials have been found suitable to emit the PL intensity. Therefore, the GNPs might be providing the scattering source due to the nature of creating the point defect in the FLC. However, further simulation and theoretical studies can insight into the material for PL enhancement and can open the ways for the synthesis of desirable materials suitable for higher PL intensities.

6.5.2. Isotropic Fluids in Ferroelectric Liquid Crystals

The dispersing mediums of most of the materials are water or ethanol. While mixing certain NPs, the dispersing medium is also added and tried to evaporate it from the system. But sometimes the content of dispersing medium may remain in the FLC matrix and may alter the result of the composites. It becomes significant to identify the impact of solvent, if not evaporated properly. Therefore, it becomes worth studying the impact of these two liquid materials on the dielectric and electro-optical properties of FLC materials.

Water is an inorganic medium so it becomes difficult to insert it into the matrix of FLC [9, 63]. One of the solutions to this problem could be that the water should be diffused into the SmA phase or nematic phase. In these phases, the molecular packing would be soft and water could be allowed to enter the matrix of FLC system. After diffusion, the slow cooling of the composite system would allow the system to retain the properties of pure SmC* phase but the existence of water molecules would occupy the equilibrium in the FLC matrix, most probably in between two smectic layers and would affect the natural phenomenon of the FLC material. The water molecules have shown their potential on the surface of nematic LC as a nano motor [64]. The interface studies of water with nematic LC have been studied [65, 66]. The diffusion studies of water in polycrystalline LC have been studied by using NMR technique and found anisotropic diffusion of water along and perpendicular to the smectic layers [67].

6.5.2.1. Electro-Optical Studies

6.5.2.1.1. Water/FLC System

The electro-optical studies of mere 1-2 wt % addition of water in FLC (commercial name, CS1026) show the generation of abnormal switching behavior. Generally, the switching is found to exist only under the region of the effective electrode but the addition of water has made the switching possible outside the electrode region. To analyze the behavior, the circular ring-like pattern was so chosen to demonstrate the in-plane isotropic nature of tilt gradient. Figure 14 shows the schematic geometry of the sample cell under study. Now, we have two portions in the same sample cell: one is conducting (circular ring electrode of diameter 2.32 mm) and another is non-conducting (inside the circular ring). The tilt angle measurement has been performed in two ways: one is in the conducting region at a different bias voltage and another is inside the ring electrode in the non-conducting region. Figure 15 shows the spatial measurement of tilt angle at constant voltage starting from the ring edge towards the center of the ring in the non-conducting region. The decreasing amplitude of tilt angle has been observed and confirmed the spatial variation of switching in the non-conducting region.

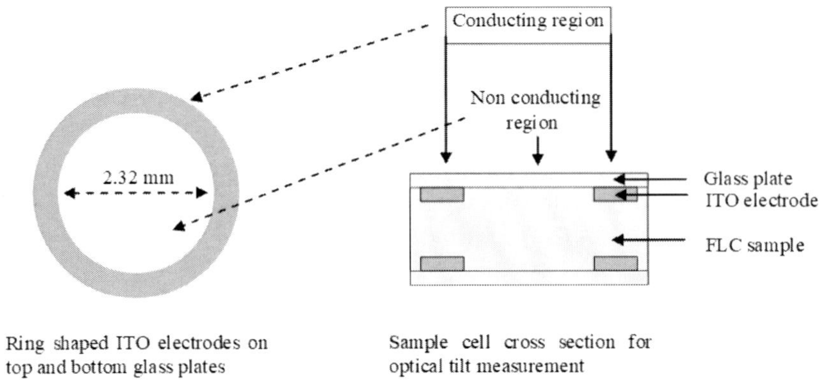

Figure 14. The top and cross-section view of the circular patterned ring ITO electrodes on the glass substrate.

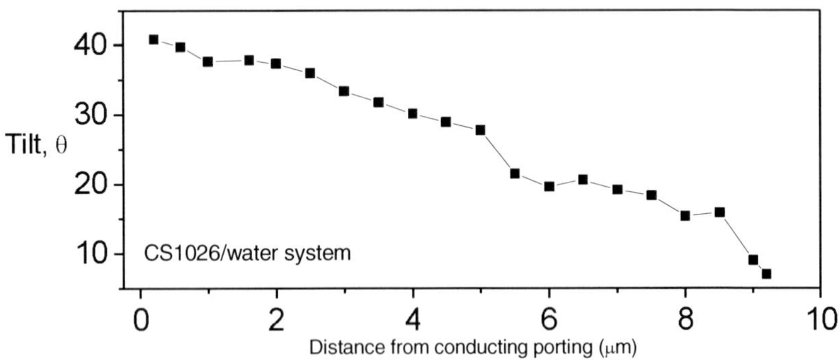

Figure 15. Behavior of tilt angle (2θ) as a function of distance (from the conducting edge) [9].

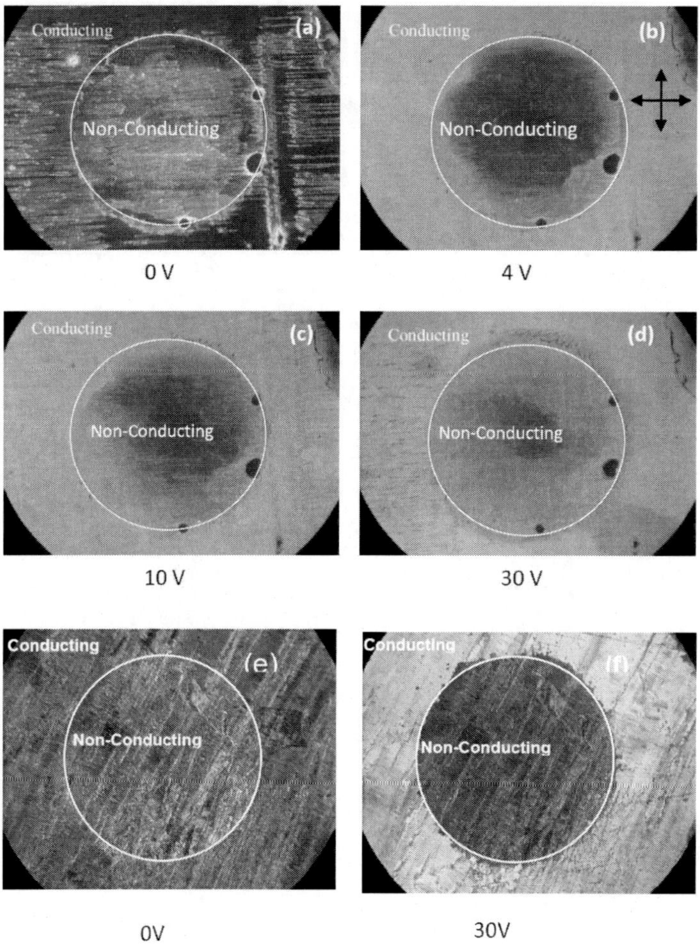

Figure 16. Optical micrographs of LC/water cell under crossed polarizers at room temperature at (a) 0 V, (b) 4 V, (c) 10 V, and (d) 30 V bias. (e) and (f) show the response in pure FLC sample at 0 V and 30 V respectively [9].

The DC voltage dependence of the optical tilt angle of FLC/water system is confirmed by the circular ring textures Figure 16. The optical textures of the circular ring electrode under crossed polarized microscope have also shown the spatial gradient in tilt angle from conducting edge to towards center in the variation. The contrast starts varying from 4 V and increases with the voltage. At 30 V the almost whole area inside the ring is covered from all sides. On the other hand, there is no variation in the texture of the pure FLC material as shown in Figure 16(e) and (f).

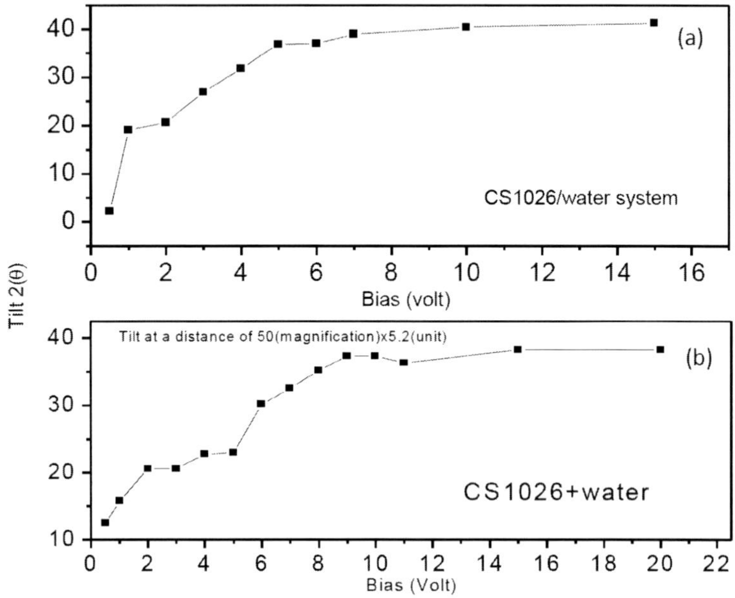

Figure 17. Behavior of tilt angle (2θ) as a function of (a) bias at a fixed distance in conducting region, and (b) bias at a fixed distance in non-conducting region [9].

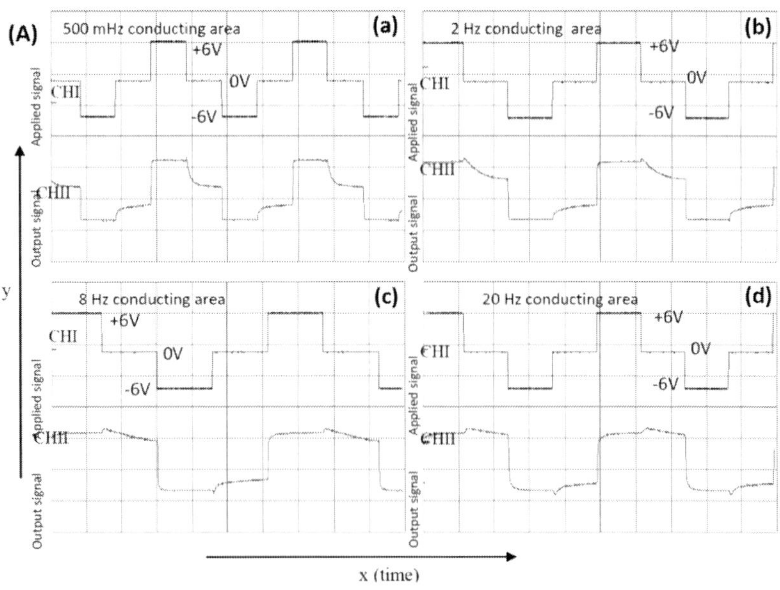

Figure 18. (Continued)

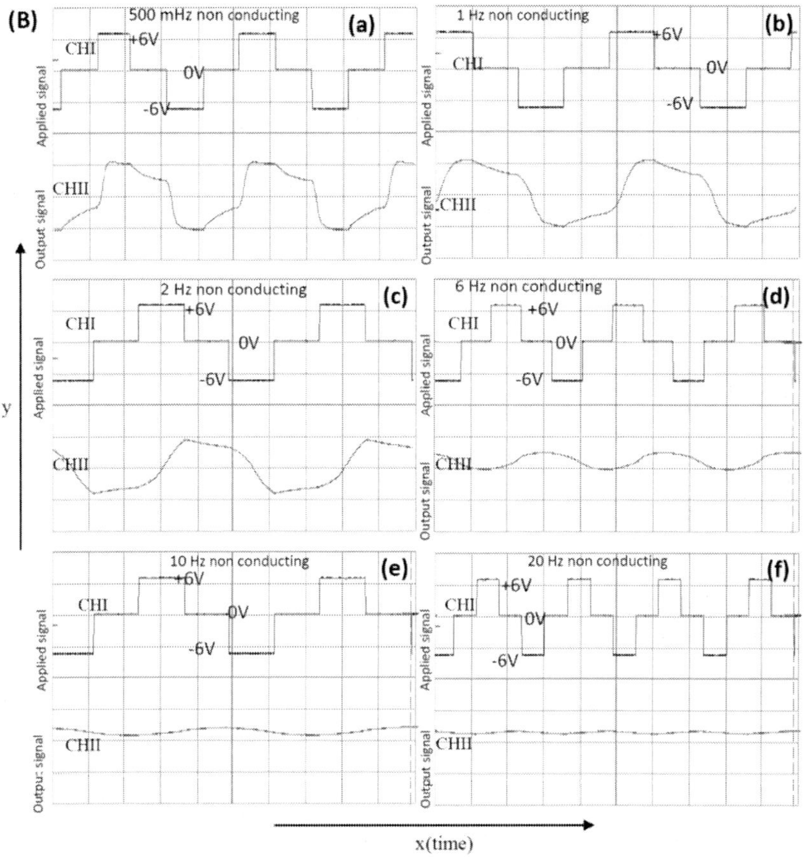

Figure 18. (A) Optical memory effect as a function frequency in conducting region of the sample cell of water doped CS 1026 (a) 500 mHz, (b) 2 Hz, (c) 8 Hz, and (d) 20 Hz. CHI:[5.000 V/DIV] shows driving square voltage and CHII: [0.020 V/DIV] shows its optical response; outside mark on the y-axis (voltage axis) shows zero voltage level for respective channels. TBA shows the time scale per division (10.00 ms) on the axis, which is the same for both channels. (B) Optical memory effect as a function of frequency in the non-conducting portion of the sample cell of water doped CS 1026 (a) 500 mHz, (b) 1 Hz, (c) 2 Hz, (d) 6 Hz, (e) 10 Hz and 20 Hz. CHI:[5.000 V/DIV] shows driving square voltage and CHII: [0.020 V/DIV] shows its optical response; outside mark on the y-axis (voltage axis) shows zero voltage level for respective channels. TBA shows the time scale per division (10.00 ms) on the axis, which is the same for both channels.

The tilt angle with bias in conducting region shows the saturation voltage at about 5 V whereas the same in the non-conducting region shows at about 9 V at room temperature, Figure 17. There is a difference in the saturation voltage which could be the difference due to the variation in the application of the external field and the effect of self-diffused water among smectic layers.

The application of frequency-dependent square delayed pulse applied to conducting electrodes shows the different behavior of optical switching in conducting and a non-conducting portion of the cell. Figure 18(A) shows the switching in the same phase with the input signal i.e., the FLC molecules respond almost at the same time as the input signal at a different frequency. Figure 18(B) shows the response of FLC molecules in the non-conducting portion at 6 μm away from the edge of the electrode. This Figure shows the switching of molecules out of phase with the input signal applied on the circular electrodes i.e., the molecular switching is lagging behind the switching under the electrode portion and the amplitude decreases with increasing in input frequency.

6.5.2.1.2. Ethanol/FLC System

The addition of ethanol in electroclinic type FLC shows the dramatic change in the electro-optical properties of the electroclinic FLC materials. Electroclinic FLC is a class of FLC materials in which the induced optical tilt angle can be observed in the SmA* phase by the application of an external electric field. Generally, the molecules in SmA phase are parallel to the layer normal but in ELC the molecules are aligned along the pseudo cone, as shown in Figure 2.

The optical tilt measurement shows the increment from 14o to 23° angle at 20°C temperature and varies as per the rule $\theta = \theta_o(T - T_C)^{\beta}$, where Tc and β are the transition temperature and exponent, Figure 19(a). Generally, the value of β is 0.5 for FLCs, but if we assume the transition at a temperature where the tilt angle becomes negligible, the value of β is found to be 0.54 in the case of pure ELC materials. After adding ethanol the phase transition increases from 28° to 52°C. The value of β is decreased to 0.47. The polarization value of SmC* phase has been enhanced from 43 to 67 nC/cm^2, Figure 19(b).

The order parameter does not vanish at the transition temperature of SmC* to SmA phase. On the other hand, the value of β for ELC is more than 0.5 which seems unsuitable for this material to be a proper ferroelectric order parameter. Therefore, this leaves the open platform for the search for new order parameters of SmC* to SmA phase transition for ELC materials.

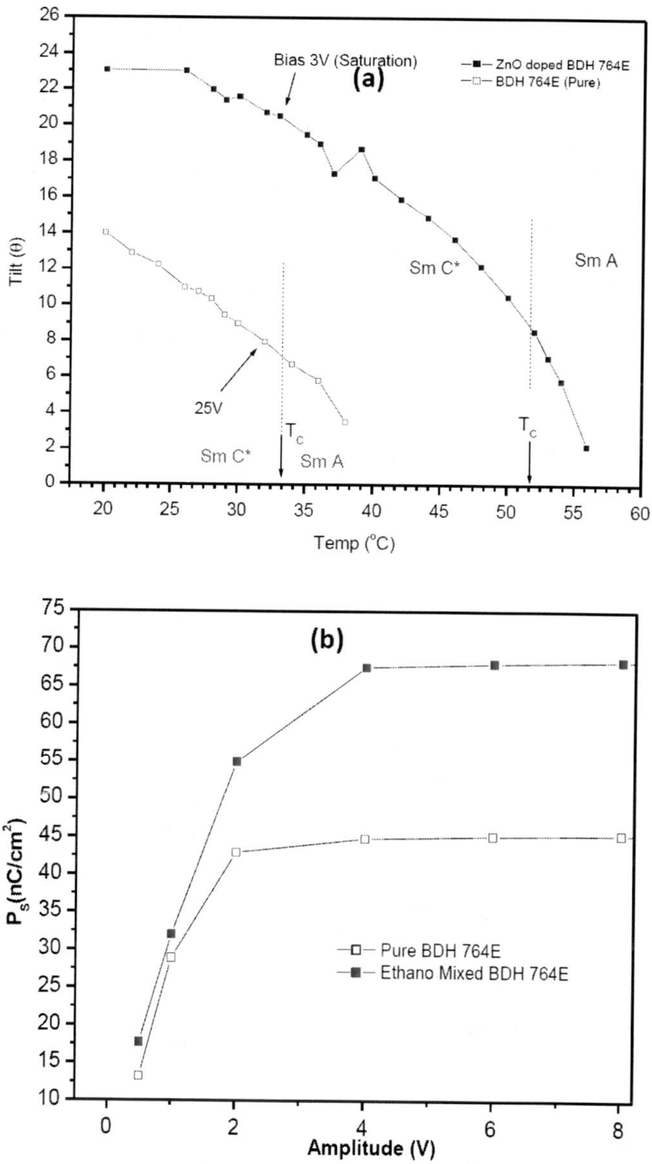

Figure 19. (a) Behavior of optical tilt (θ), as a function of temperature in case of pure BDH 764E (open square) and ethanol doped BDH 764E (closed square), (b) behavior of Spontaneous Polarization (P_s) as a function of bias in case of pure BDH 764E (open square) and ethanol doped BDH 764E (closed square). With permission from IOP publishing [7].

6.5.2.2. Dielectric Studies

6.5.2.2.1. Water/FLC System

Dielectric spectroscopy is useful to know the collective behavior of dielectric processes at various frequencies. The main observations of this study are the observation of a new relaxation process and the anisotropic behavior of the flow of water in SmC* matrix [63]. The new relaxation processes have been observed at a lower frequency than in Goldstone mode. The dielectric constant increases after the addition of water into FLC. It has also been observed that this process is visible only in tanδ graph and not in the imaginary part of the dielectric constant, $\varepsilon''(\omega)$. Its main reason is that after the conversion of $\tan\delta = \varepsilon''/\varepsilon'$into $\varepsilon''(\omega)$, the process shifts towards lower frequency, due to increment in $\varepsilon'(\omega)$ value, which is out of the window of the present experimental setup. Therefore, it has been dealt with tan□ throughout the study. The new relaxation peak has been observed just at a lower value of Goldstone mode frequency. The new peak in tanδ is very sensitive to DC biasing of the electric field. As can be seen in Figure 20(a), the DC bias-dependent dielectric constant as a function of frequency shows the decreasing nature as usual in our FLC sample. In Figure 20(d), a small DC bias voltage of 1 V suppresses the new process completely and only Goldstone mode is visible in the measurement. This shows that such a sensitive process may not be entirely due to FLC material. Now, it is required to check if it is associated with Goldstone mode or an independent process. If it is an independent process then it should be observable in homogeneous and homeotropic alignment. But, if, it is a conditional one, then it would be observable in one alignment only.

For the measurement of dielectric anisotropy, the room temperature study of the homogeneous and homeotropic alignment of water added FLC is shown in Figure 21. The data of homogeneous alignment, which is shown by filled squares, was taken just after filling and cooling the sample cell. The two peaks in tanδ versus frequency curve are visible at low frequency region where Goldstone mode usually appears. The peak at a higher frequency is due to the Goldstone mode and the peak at low frequency than the Goldstone mode relaxation frequency is due to the presence of isotropic material (filled square curve) i.e., water in the Felix 17/100 as explained in Figure 20. Then, the cell was converted into homeotropic alignment by the magnetic field and the same study was carried out. No relaxation peak due to water other than the peak due to molecular process at higher frequencies has been observed in tanδ versus frequency plot. This has been shown by the unfilled square curve. This indicates the effect of water attachment in the FLC matrix. The study reveals that the peak due to water along is associated with Goldstone mode only and not with other modes related to homeotropic alignment. This indicates the anisotropic nature of water distribution in FLC matrix.

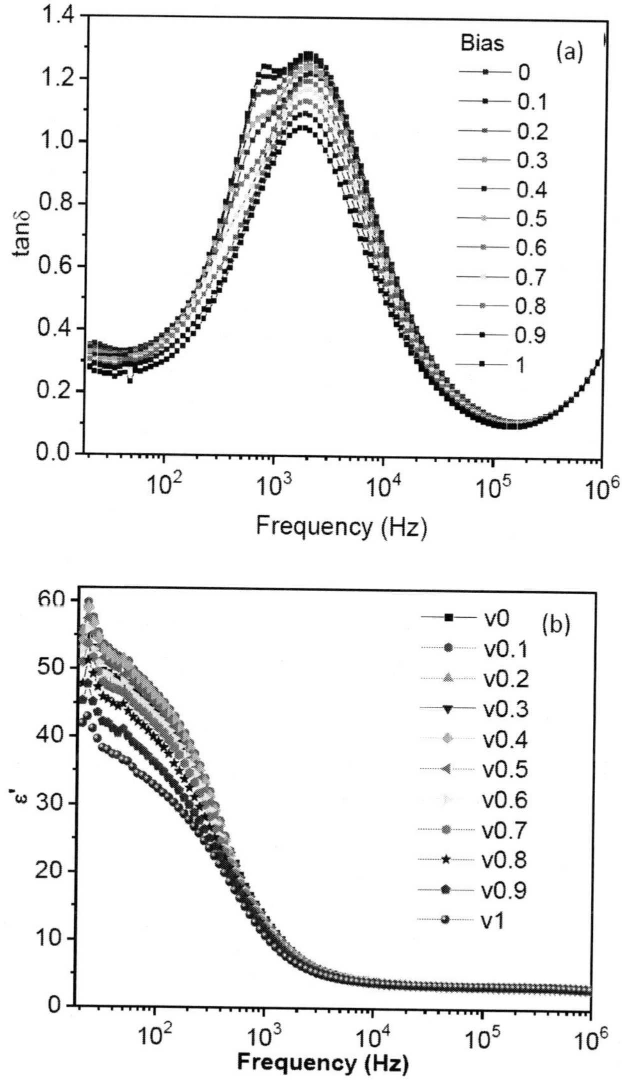

Figure 20. (a) Bias dependence of dielectric permittivity of water/FLC system, (b) suppression of second peak by bias application [63].

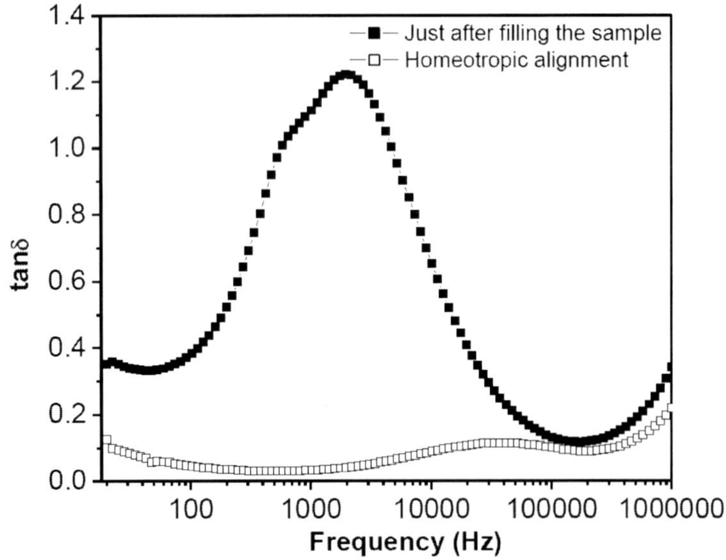

Figure 21. Tanδ study as a function of the frequency of water added FLC CS1026 in homogeneous (filled square) and homeotropic (unfilled squares) aligned states at room temperature.

To perform the water movement in the smectic layered structure, the pure FLC and water/FLC system have been filled up in the same sample cell having two isolated electrodes. This has been performed to avoid the variation in the result due to variation in a sampling of cells. Then the dielectric measurement has been performed concerning time as shown in Figure 22. Figure exhibits the appearance of peak due to water in tanδ versus frequency curve concerning time. There are two samples of different directions of alignment concerning the flow of water. In one sample, the flow of water from water mixed FLC to pure FLC is along the long molecular axis, Figure 22(a), whereas, in another, it is along the short molecular axis, Figure 22(b). After filling both the samples, the measurement was taken through a pattern having pure FLC material only to check the appearance of water which enters into the pure FLC side to water added FLC side. It exhibits the relaxation peak due to water near the Goldstone mode peak in the water added FLC sample while on the side of the pure FLC sample only the Goldstone mode relaxation process is visible, Figure 22(a). Here, since the observation is concentrated only on the side of the pure FLC sample, therefore, all the measurements for the water diffusion study have been carried out over the pure FLC sample side at room temperature concerning time. Water starts diffusing from one side to another and is analyzed through tanδ versus frequency graph with time. The appearance of the peak due to water has been measured in the region of the pure FLC sample. The peak is clearly visible after 200 hrs. Goldstone mode peak is unaffected in the pure sample due to water addition even in the presence of water.

The overall diffusion of water is very slow, Figure 22. Water flows from the water added FLC region to the pure FLC region of a sample cell and gets attached to FLC molecules, which is visible by the relaxation peak in tanδ curve of Figure 22(a). Figure 22(a) shows the diffusion of water along the long molecular axis i.e., perpendicular to the layers. The peak due to water starts appearing at about 72 hrs, Figure 22(a), while along a short molecular axis i.e., parallel to the smectic layer plane, it starts appearing after 132 hrs, Figure 22(b). This non-uniform flow of water in the FLC matrix might be due to the anisotropic nature of FLC material and shows the non-uniform distribution of water.

Thus, the unique behavior of FLC tailored by water has been studied by various techniques. The electro-optical behavior of switching in conducting and the non-conducting region has shown the evidence of a slower switching phenomenon in the non-conducting region than in conducting one of water added FLC sample. Water influences the dielectric constant of FLC sample also and shows the second relaxation peak near to Goldstone mode process. This peak is highly bias dependent and completely disappears at about 0.8 V. Goldstone mode is only visible which is the characteristic property of FLC materials. The anisotropic distribution of water has been observed, as the relaxation peak of water is only visible in the homogeneous alignment and not in the homeotropic one. This shows that water is associated with a unique process as can be seen in Figure 21. This is unique in the sense that the characteristic peak of water lies in the microwave region of the solar spectrum. This unique association leads to the enhancement in the anisotropy of the materials. The anisotropic approach makes the system transit towards lower temperatures from ferroelectric to the paraelectric phase transition. Thus, the higher diffusion is along long molecular axis i.e., perpendicular to the layers, which is due to the anisotropic distribution of water in the FLC matrix, as shown in Figure 22. Since, the diffusion along the molecular axis is large and short along the short molecular axis, through the $\varepsilon_{\parallel}$ component, which becomes stronger, and anisotropy increases as shown in Figure 23.

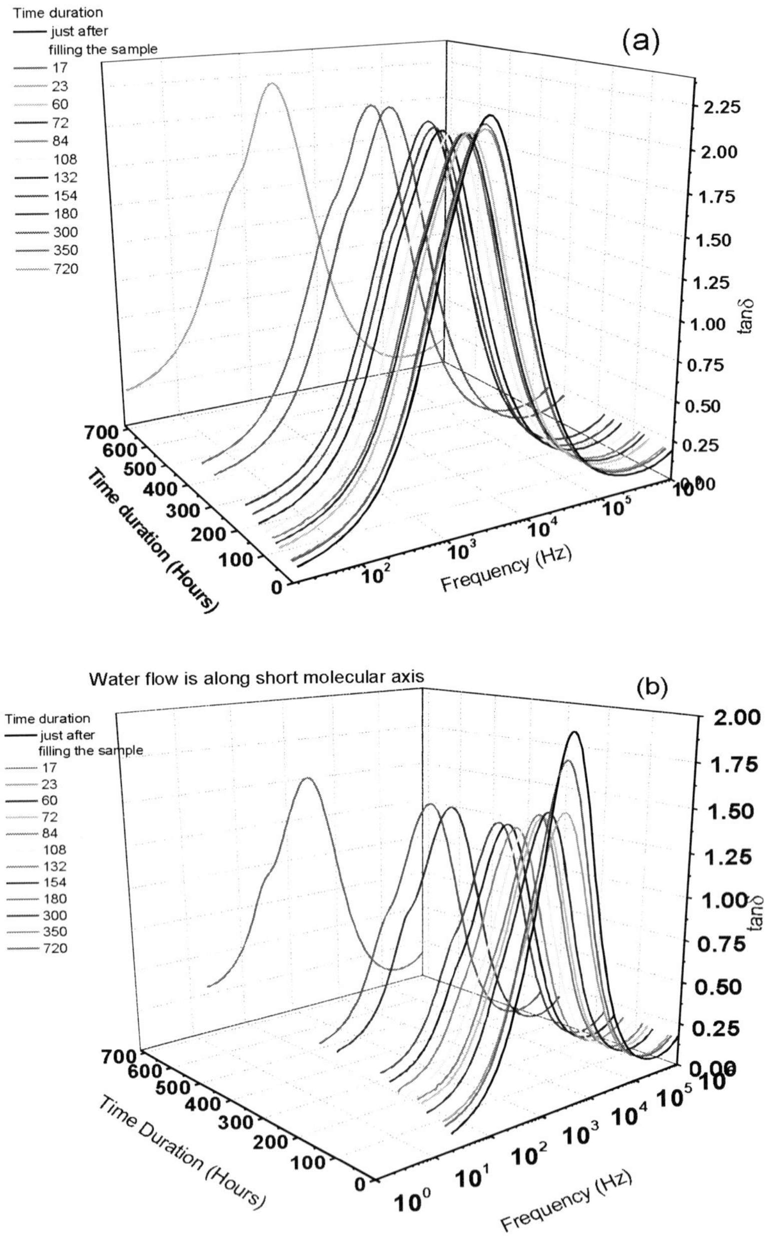

Figure 22. Flow of water from water added FLC to pure FLC in a sample (a) perpendicular, and parallel to the long molecular axis.

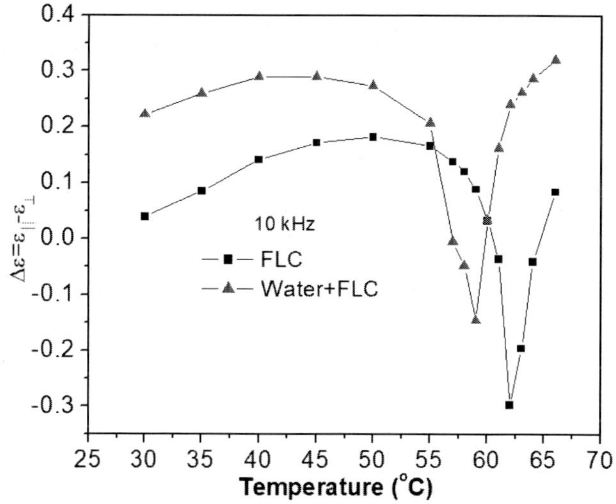

Figure 23. Dielectric anisotropy measurement of pure and water added FLC [63].

6.5.2.2.2. Ethanol/FLC System

The effect of ethanol on the dielectric properties of ELC material has been studied in two cells one with ethanol mixed in electroclinic liquid crystal and another of pure ELC for comparative study. Figure 24 shows the dielectric permittivity of pure FLC as a function of frequency at different temperatures in pure ELC material. As seen in the figure, the permittivity ($\varepsilon'(\omega)$) continuously decreases in SmC* phase near the transition temperature (28°C) of SmC* to SmA phase. Dielectric absorption is shown in the inset of Figure 24. If an ELC material is doped with 2-5% of ethanol, there is a drastic change in the dielectric relaxation process has been observed as shown in Figure 25(a). The high dielectric permittivity, $\varepsilon'(\omega)$, which is due to phason mode (Goldstone mode) in SmC* phase appears up to 23 °C above the critical temperature for pure FLC material as shown in Figure 25(b). This means that the transition temperature of ferroelectric SmC* to paraelectric, SmA phase is shifted towards a higher temperature by 23°C by doping ethanol in the ELC material. Even the relaxation frequency of Goldstone mode is shifted to 51 °C temperature as seen in Figure 25(b). Above 51°C temperature, the permittivity is due to soft mode which is strongly temperature dependent as in the case of conventional SmA phase. The texture observation also suggests that the dechiralization lines disappear at 51°C temperature, suggesting that the transition temperature has shifted from 28°C to 51°C temperature. This shift in the transition temperature from ferroelectric to paraelectric phase can be varied depending upon the concentration of the ethanol dopant in the ELC material, but in the present study, the highest temperature increment is 23°C.

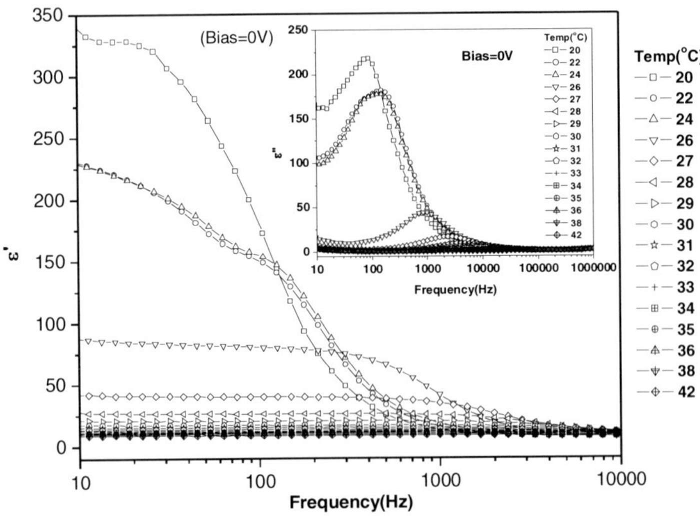

Figure 24. Dispersion curve of dielectric permittivity (ε′ vs freq.) and in the inset, dielectric absorption curves (ε″vs freq.) at different temperatures in pure BDH 764E [8].

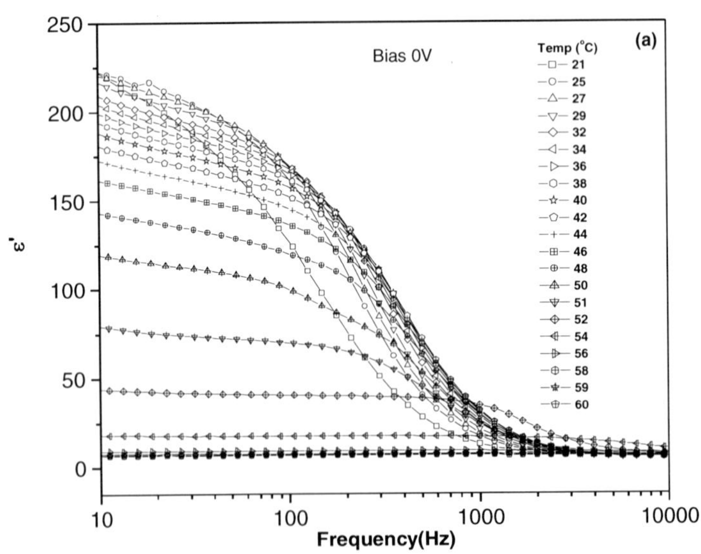

Figure 25. (Continued)

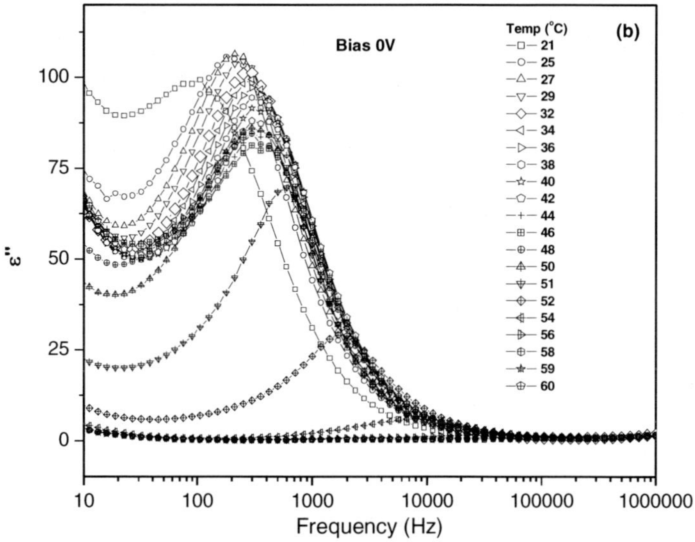

Figure 25. Dispersion curve of (a) dielectric permittivity (ε′ vs frequency) and (b) dielectric absorption curve (ε″ vs freq.) at different temperatures in ethanol doped BDH 764E [8].

The bias dependent behavior of $\varepsilon'(\omega)$ in ethanol added ELC material at lower frequencies is shown in Figure 26(a). As seen in the figure the permittivity due to Goldstone mode gets completely suppressed at 4V bias and no further change appears in the permittivity, $\varepsilon'(\omega)$, whereas the $\varepsilon'(\omega)$ in pure BDH 764E electroclinic liquid crystal material, does not get suppressed at a high bias field of 25 V as shown in Figure 26(b). The permittivity at low frequency in ethanol doped is around 5 and in pure ELC material 12. This means that one can still suppress the permittivity in pure ELC material by applying a higher bias field. This clearly shows that ethanol molecules are responsible for rearranging the liquid crystal molecules uniformly in the layer structure which results in to the suppression of $\varepsilon'(\omega)$ due to Goldstone mode (phason mode) at a very low field whereas in pure ELC material the static permittivity could not be suppressed even at a bias field of 25 V due to the randomization of the molecules in the layers [68]. Analysis of Cole-Cole plots of the ethanol doped BDH 764E (Figure 27(a)-(c)) at 25°, 49°and 53°C temperatures reveal that single collective processes are observed in deep SmC* and SmA phase. The low frequency values are due to the conductivity effect. However, two processes, Goldstone mode and soft mode are observed at 49°C near ferro to para phase transition (SmC* to SmA*) without suppressing the

Goldstone mode unlike in case of conventional ferroelectric liquid crystals. This means that after doping ethanol, these processes are visible without any bias application, which indicates the reduction of randomization of molecules.

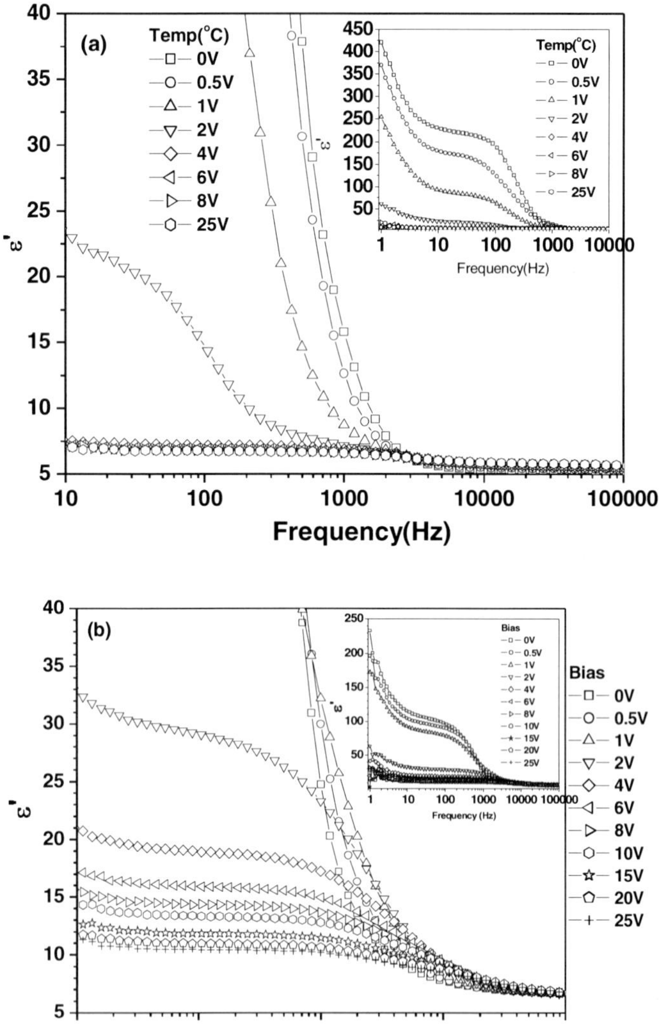

Figure 26. Effect of bias voltage on the suppression of dielectric permittivity (ε′) at room temperature in (a) ethanol doped BDH 764E and (b) pure BDH 764E. The inset shows the overall behavior of ε′ from 0V to 25V [8].

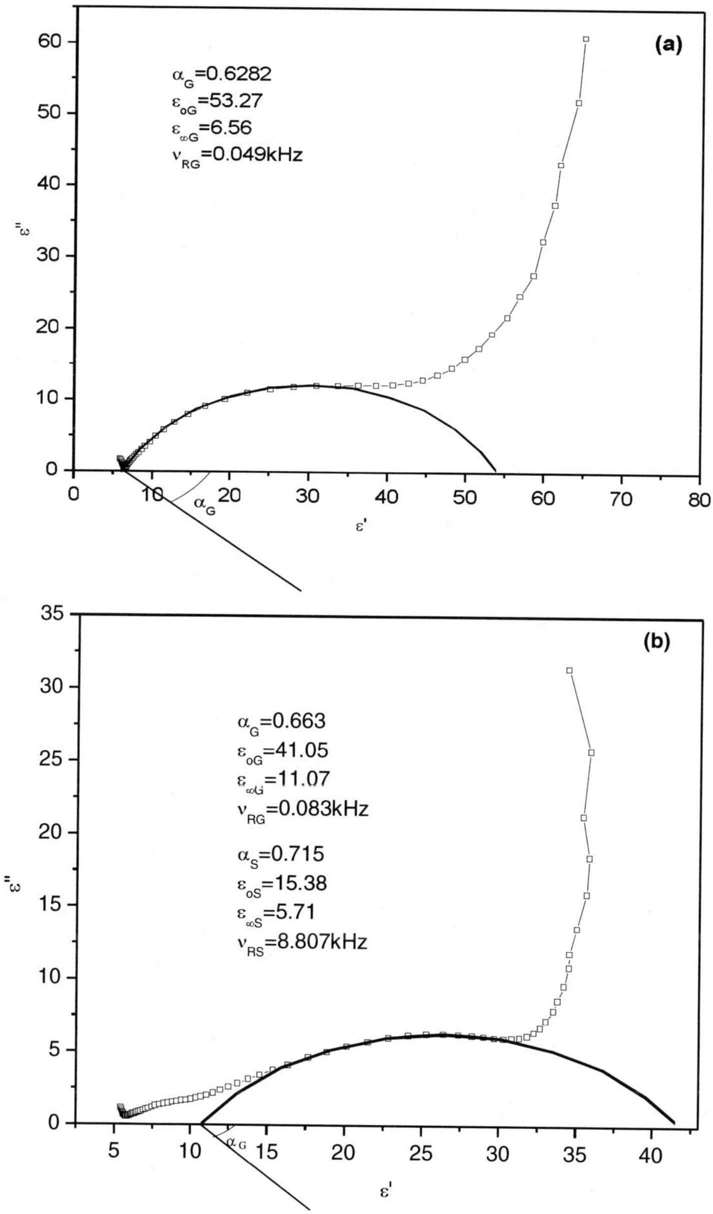

Figure 27. (Continued)

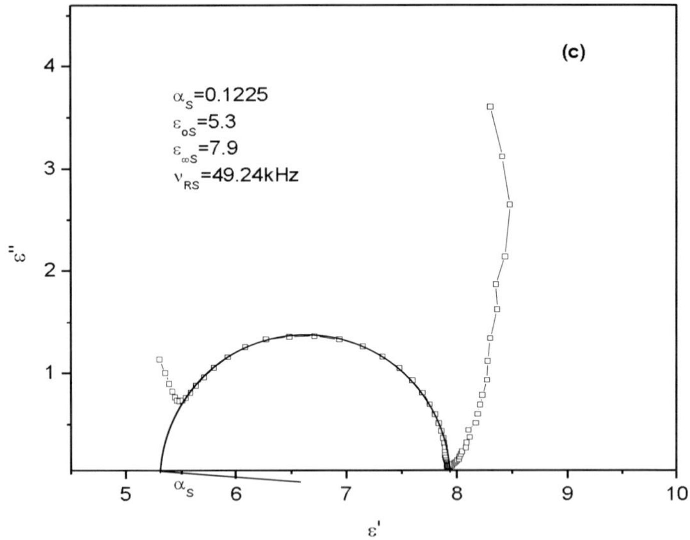

Figure 27. Cole-Cole plots of ethanol doped BDH 764E at (a) 25°C, (b) 49°C, and (c) 53°C respectively [8].

In Figure 28, the variation of the ratio of polarization to tilt angle at concerning temperature has been revealed. The relation between P/θ and the pitch value of the material is given as follows [69]

$$P/\theta = \chi(f \times q + z)(16)$$

where $q = 2\pi/L$; L is pitch, χ is dielectric susceptibility, and f and z are the flexoelectric and piezoelectric coupling coefficients between P and θ. In the case of pure ELC, the slope of P/θ is lower than ethanol doped ELC. The behavior of the helicoidal pitch of the material may depend on the other parameters also, such as f, χ, and z. The appearance of soft mode without any bias application in SmC* phase near T_c as explained above can be correlated with this gradual slope in the doped ELC. This gradual slope in doped ELC suggests that molecules are more aligned because of a reduction in phason fluctuations.

Figure 29 shows the inverse of dielectric strength and relaxation behavior of soft mode at 10 V bias in ferroelectric and paraelectric phases in ethanol doped ELC material. The behavior shows that the material is obeying Curie-Weiss law, which was not being obeyed in the case of pure ELC as per the earlier studies [4,70]. This means that the nature of material has been changed and ferro (SmC*) to paraelectric (SmA) phase transition is extended from 28°C to 51°C just by doping ethanol in the ELC material. At 10V bias in SmC* phase, there is no absorption peak in the whole SmC* phase, suggesting that the Goldstone mode is suppressed completely and soft mode appears near transition temperature of SmC* and SmA phase, [71, 72] as shown in Figure 27. It is worth mentioning here that in pure ELC material such behavior of soft mode near transition temperature was not observed in SmC* and SmA phase due to the randomization of liquid crystal molecules. The increase in the transition temperature of SmC* is due to the realignment of randomized liquid crystal molecules which contribute to the phase fluctuation mode in ferroelectric and paraelectric phase of liquid crystal material. This reduction in randomization of ELC materials in SmC* and SmA phase is due to the doping of ethanol in the material. Since ethanol has a dipole moment of 1.86 Debye, it can promote the pursuit of dipole-dipole interaction with ELC materials. When this interaction comes into the picture it dominates over the molecular interaction or random fluctuations and hence aids in the realignment of the randomizations. Once the FLC materials have less amount of randomization then they behave as if they are in SmC* phase and obey the Curie-Weiss law near the transition temperature of ferroelectric to paraelectric phase.

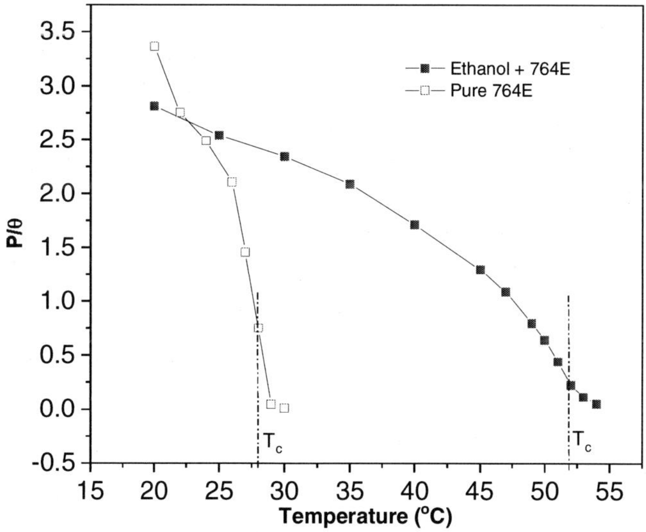

Figure 28. Behaviour of P/θ as a function of temperature in case of ethanol doped (closed square) and in case of pure BDH 764E (open square) [8].

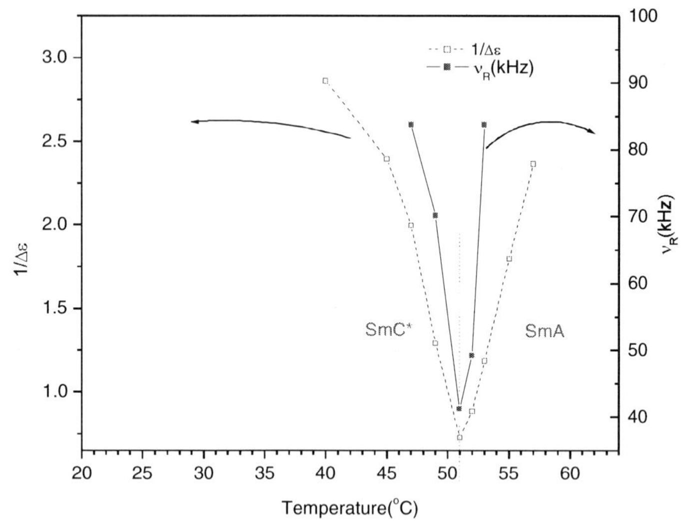

Figure 29. Behaviour of inverse of dielectric strength (open square) and relaxation frequency (closed square) at a bias of 10 V in SmC* and SmA phase near T_c which confirms the validity of Curie-Weiss law [8].

Conclusion

In conclusion, the gold nanoparticles and isotropic materials (water and ethanol) have been studied in the FLC matrix. The addition of GNPs has left some challenges such as the defect formation due to it in the layered structure of liquid crystals and the possibilities of coupling surface plasmons with the FLC dielectric, which could explain the reasons behind the existing experimental observations. On the other hand, the isotropic fluids, water, have created the new interesting phenomenon of side switching the molecules out of the electrodes. Besides this, a new relaxation peak just below the Goldstone mode, has shown the possibilities of the attachment of H_2O molecules with FLC in a unique way. However, the isolated peak of H_2O lies in the microwave region of the spectrum. But, it seems to be associated with Goldstone mode, i.e., the fluctuations of H_2O dipole at lower relaxation than the dipoles of Goldstone mode. Recently, we have explored the origin of this peak due to the interface of FLC with a substrate which is called partially unwound helical mode. In ethanol addition in FLC, the shifting of SmC* to SmA phase transition is observed from 28 to 53°C. This means the external entities have clearly changed the properties of FLC which could be explored for device applications also.

References

[1] Meyer, R. B., L. Liebert, L. Strzelecki, and P. Keller, Ferroelectric Liquid Crystals, *J. Phys. Lettres* 36, 69 (1975).

[2] Lagerwall, J. P. F., and F. Giesselmann, Current Topics in Smectic Liquid Crystal Research, *Chem Phys Chem.*

[3] Liu, J., M. G. Robinson, K. M. Johnson, D. M. Walba, M. B. Ros, N. A. Clark, R. Shao, and D. Doroski, The Measurement of Second-harmonic Generation in Novel Ferroelectric Liquid Crystal Materials, *J. Appl. Phys.* 70, 3426 (1991).

[4] Kaur, S., S. P. Singh, A. M. Biradar, A. Choudhary, and K. Sreenivas, Enhanced Electro-Optical Properties in Gold Nanoparticles Doped Ferroelectric Liquid Crystals, *Appl. Phys. Lett.* 91, 023120 (2007).

[5] Prakash, J., A. Choudhary, A. Kumar, D. S. Mehta, and A. M. Biradar, Nonvolatile Memory Effect Based on Gold Nanoparticles Doped Ferroelectric Liquid Crystal, *Appl. Phys. Lett.* 93, 112904 (2008).

[6] Kumar, A., J. Prakash, D. S. Mehta, A. M. Biradar, and W. Haase, Enhanced Photoluminescence in Gold Nanoparticles Doped Ferroelectric Liquid Crystals, *Appl. Phys. Lett.* 95, 023117 (2009).

[7] Choudhary, A., A. Kumar, S. Kaur, J. Prakash, and A. M. Biradar, Enormous Change in Tilt Angle and Transition Temperature of Smectic C * –Smectic A Phase in Electroclinic Liquid Crystal Material, *Jpn. J. Appl. Phys.* 48, 050201 (2009).

[8] Malik, A., A. Choudhary, J. Prakash, I. Coondoo, and A. M. Biradar, Enhancement of Ferro-Para Transition in Ethanol Doped Ferroelectric Liquid Crystals, *J. Appl. Phys.* 105, 034105 (2009).

[9] Choudhary, A., I. Coondoo, J. Prakash, K. Sreenivas, and A. M. Biradar, Electrically Tunable Spatially Variable Switching in Ferroelectric Liquid Crystal/Water System, *Appl. Phys. Lett.* 94, 174101 (2009).

[10] Haase, W., and S. Wróbel, editors, *Relaxation Phenomena* (Springer Berlin Heidelberg, Berlin, Heidelberg, 2003).

[11] Bahadur, B., *Liquid Crystal - Applications and Uses* (Volume 1) (World Scientific Publishing Company, 1990).

[12] Wulf, A., Steric Model for the Smectic- C Phase, *Phys. Rev. A* 11, 365 (1975).

[13] McMillan, W. L., Simple Molecular Theory of the Smectic C Phase, *Phys. Rev. A* 8, 1921 (1973).

[14] Cabib, D., and L. Benguigui, The Smectic C Phase of Liquid Crystals, *J. Phys.* 38, 419 (1977).

[15] Van Der Meer, B. W., and G. Vertogen, An Induced Dipole Model of the Smectic C Phase, *Le J. Phys. Colloq.* 40, C3 (1979).

[16] Priest, R. G., Simple Model for the Smectic A-Smectic C Phase Transition, *J. Phys.* 36, 437 (1975).

[17] Priest, R. G., Molecular Statistical Model of the Smectic A and Smectic C Phases, *J. Chem. Phys.* 65, 408 (1976).

[18] Matsushita, M., Mean-Field Theory of Smectic A-Smectic C Mesophase Transition, *J. Phys. Soc. Japan* 50, 1351 (1981).

[19] Goossens, W. J. A., The Smectic A - Smectic C Phase Transition: Sense and Nonsense, *J. Phys.* 46, 1411 (1985).

[20] Goossens, W. J. A., The Smectic A -Smectic C Phase Transition: A Molecular Statistical Theory, *Europhys. Lett.* 3, 341 (1987).

[21] Goossens, W. J. A., The Smectic A and the Smectic C Phase: A Coherent Molecular Picture, *Mol. Cryst. Liq. Cryst. Inc. Nonlinear Opt.* 150, 419 (1987).

[22] Osipov, M. A., Molecular-Statistical Theory of Ferroelectricity in Smectic C Liquid Crystals, *Ferroelectrics* 58, 305 (1984).

[23] Clark, N. A., and S. T. Lagerwall, Submicrosecond Bistable Electro-Optic Switching in Liquid Crystals, *Appl. Phys. Lett.* 36, 899 (1980).

[24] Goodby, J. W., and R. Blinc, *Ferroelectric Liquid Crystals: Principles, Properties and Applications* (Gorden and Breach Science Publishers, Philadelphia, 1992).

[25] Cahn, R. S., C. K. Ingold J., in *Chem. Soc.,* London (1951), p. 612.

[26] Cahn, R. S., C. K. Ingold, and V. Prelog, The Specification of Asymmetric Configuration in Organic Chemistry, *Experientia* 12, 81 (1956).

[27] Lagerwall, S. T., and I. Dahl, Ferroelectric Liquid Crystals, *Mol. Cryst. Liq. Cryst.* 114, 151 (1984).

[28] Saupe, A., On Molecular Structure and Physical Properties of Thermotropic Liquid Crystals, *Mol. Cryst.* 7, 59 (1969).

[29] Zeks, B., C. Filipic, and T. Carlsson, Theory of Ferroelectric Liquid Crystals, *Phys. Scr.* T25, 362 (1989).

[30] Antonii, F., Panacea Aurea-Auro Potabile. Bibliopolio Frobeniano, Hamburg, 1618, Cerca Con Google (n.d.).

[31] Faraday, M., X. The Bakerian Lecture. —Experimental Relations of Gold (and Other Metals) to Light, *Philos. Trans. R. Soc. London* 147, 145 (1857).

[32] Frens, G., Controlled Nucleation for the Regulation of the Particle Size in Monodisperse Gold Suspensions, *Nat. Phys. Sci.* 241, 20 (1973).

[33] Schmid, G., Large Clusters and Colloids. Metals in the Embryonic State. *Chem. Rev.* 1992. Vol. 92, (1709).

[34] Brust, M., and C. J. Kiely, Some Recent Advances in Nanostructure Preparation from Gold and Silver Particles: A Short Topical Review, Colloids Surfaces A Physicochem. *Eng. Asp.* 202, 175 (2002).

[35] Watson, K. J., J. Zhu, S. T. Nguyen, and C. A. Mirkin, Hybrid Nanoparticles with Block Copolymer Shell Structures, *J. Am. Chem. Soc.* 121, 462 (1999).

[36] Turkevich, J., P. C. Stevenson, and J. Hillier, *Disc. Faraday Soc.* 11, 1951, 55–75.(b) J. Turkevich, G. Kim, *Science* (80-.). 169, 873 (1970).

[37] Polte, J., T. T. Ahner, F. Delissen, S. Sokolov, F. Emmerling, A. F. Thünemann, and R. Kraehnert, Mechanism of Gold Nanoparticle Formation in the Classical Citrate Synthesis Method Derived from Coupled In Situ XANES and SAXS Evaluation, *J. Am. Chem. Soc.* 132, 1296 (2010).

[38] Doyen, M., K. Bartik, and G. Bruylants, UV–Vis and NMR Study of the Formation of Gold Nanoparticles by Citrate Reduction: Observation of Gold–Citrate Aggregates, *J. Colloid Interface Sci.* 399, 1 (2013).

[39] Maeda, M., M. Miyamori, and I. Suzuki, Optical Trtansmission And Polarization Switching Properties Of A Ferroelectric Liquid Crystal Cell With A Stable Twisted State, *Mol. Cryst. Liq. Cryst. Sci. Technol. Sect. A. Mol. Cryst. Liq. Cryst.* 366, 703 (2001).

[40] Beresnev, L. A., V. G. Chigrinovs, D. I. Dergachevt, E. P. Poshidaev, J. Fünfschilling, and M. Schadt, Deformed Helix Ferroelectric Liquid Crystal Display: A New Electrooptic Mode in Ferroelectric Chiral Smectic C Liquid Crystals, *Liq. Cryst.* 5, 1171 (1989).

[41] Fünfschilling, J., and M. Schadt, Fast Responding and Highly Multiplexible Distorted Helix Ferroelectric Liquid-crystal Displays, *J. Appl. Phys.* 66, 3877 (1989).

[42] Yoshino, K., S. Kishio, M. Ozaki, T. Sakurai, N. Mikami, R. Higuchi, and M. Honma, Low Threshold Field of Electro-Optic Effect in Ferroelectric Liquid Crystal with Extremely Large Spontaneous Polarization, *Jpn. J. Appl. Phys.* 25, L416 (1986).

[43] Kishio, S., M. Ozaki, K. Yoshino, T. Sakurai, N. Mikami, and R. Higuchi, Characteristics of Optical Switching and Memory Effects Utilizing Deformation of Helicoidal Structure of Ferroelectric Liquid Crystals with Large Spontaneous Polarization, *Jpn. J. Appl. Phys.* 26, 513 (1987).

[44] Fünfschilling, J., and M. Schadt, New Short-Pitch Bistable Ferroelectric (Sbf) Liquid Crystal Displays, *Jpn. J. Appl. Phys.* 30, 741 (1991).

[45] Jákli, A., S. Markscheffel, and A. Saupe, Helix Deformation and Bistable Switching of Ferroelectric Liquid Crystals, *J. Appl. Phys.* 79, 1891 (1996).

[46] Garoff, S., and R. B. Meyer, Electroclinic Effect at the A−C Phase Change in a Chiral Smectic Liquid Crystal, *Phys. Rev. A* 19, 338 (1979).

[47] Andersson, G., I. Dahl, W. Kuczynski, S. T. Lagerwall, K. Skarp, and B. Stebler, The Soft-Mode Ferroelectric Effect, *Ferroelectrics* 84, 285 (1988).

[48] Garoff, S., and R. B. Meyer, Electroclinic Effect at the A−C Phase Change in a Chiral Smectic Liquid Crystal, *Phys. Rev. Lett.* 38, 848 (1977).

[49] Martinot-lagarde, P., R. Duke, and G. Durand, Tempearture Dependence of Tilt, Pitch and Polarization in Ferroelectric Liquid Crystals, *Mol. Cryst. Liq. Cryst.* 75, 249 (1981).

[50] Sawyer, C. B., and C. H. Tower, Rochelle Salt as a Dielectric, *Phys. Rev.* 35, 269 (1930).

[51] Lines, M. E., and A. M. Glass, *Principles and Applications of Ferroelectrics and Related Materials* (Oxford university press, 2001).

[52] Merz, W. J., Domain Formation and Domain Wall Motions in Ferroelectric BaTiO3 Single Crystals, *Phys. Rev.* 95, 690 (1954).

[53] Merz, W. J., Switching Time in Ferroelectric BaTiO 3 and Its Dependence on Crystal Thickness, *J. Appl. Phys.* 27, 938 (1956).

[54] Bawa, S. S., A. M. Biradar, K. Saxena, and S. Chandra, Direct Pulse Technique for Spontaneous Polarization Dynamics and Molecular Reorientation Processes in Ferroelectric Liquid Crystals, *Rev. Sci. Instrum.* 59, 2023 (1988).

[55] Ouchi, Y., T. Uemura, H. Takezoe, and A. Fukuda, Correspondence between Stroboscopic Micrographs and Spontaneous Polarization Measurements in Surface Stabilized Ferroelectric Liquid Crystal Cells, *Jpn. J. Appl. Phys.* 24, L235 (1985).

[56] Miyasato, K., S. Abe, H. Takezoe, A. Fukuda, and E. Kuze, Direct Method with Triangular Waves for Measuring Spontaneous Polarization in Ferroelectric Liquid Crystals, *Jpn. J. Appl. Phys.* 22, L661 (1983).

[57] Kaur, S., A. K. Thakur, S. S. Bawa, and A. M. Biradar, Behaviour of Goldstone and Soft Mode at Tc in Ferroelectric Liquid Crystals, *Phys. B Condens. Matter* 344, 133 (2004).

[58] Bawa, S. S., A. M. Biradar, and S. Chandra, Frequency Dependent Polarization Reversal and the Response Time of Ferroelectric Liquid Crystal by Triangular Wave Method, *Jpn. J. Appl. Phys.* 25, L446 (1986).

[59] Xue, J.-Z., M. A. Handschy, and N. A. Clark, Electrooptic Response during Switching of a Ferroelectric Liquid Crystal Cell with Uniform Director Orientation, *Ferroelectrics* 73, 305 (1987).

[60] Skarp, K., I. Dahl, S. T. Lagerwal, and B. Stebler, Polarization and Viscosity Measurements in a Ferroelectric Liquid Crystal by the Field Reversal Method, *Mol. Cryst. Liq. Cryst.* 114, 283 (1984).

[61] Prakash, J., A. Choudhary, D. S. Mehta, and A. M. Biradar, Effect of Carbon Nanotubes on Response Time of Ferroelectric Liquid Crystals, P*hys. Rev. E* 80, 012701 (2009).

[62] Nedyalkov, N., T. Sakai, T. Miyanishi, and M. Obara, Near Field Properties in the Vicinity of Gold Nanoparticles Placed on Various Substrates for Precise Nanostructuring, *J. Phys. D. Appl. Phys.* 39, 5037 (2006).

[63] Singh, G., A. Choudhary, G. Vijaya Prakash, and A. M. Biradar, Anisotropic Behavior of Water in Ferroelectric Liquid Crystals, *Phys. Rev. E* 81, 051707 (2010).

[64] Tabe, Y., and H. Yokoyama, *Nature Mater.* 2, 806 (2003).

[65] Brake, J. M., M. K. Daschner, Y.-Y. Luk, and N. L. Abbott, Biomolecular Interactions at Phospholipid-Decorated Surfaces of Liquid Crystals, *Science* (80-.). 302, 2094 (2003).

[66] Kim, J.-W., H. Kim, M. Lee, and J. J. Magda, Interfacial Tension of a Nematic Liquid Crystal/Water Interface with Homeotropic Surface Alignment, *Langmuir* 20, 8110 (2004).

[67] Blum, F. D., A. S. Padmanabhan, and R. Mohebbi, Self-Diffusion of Water in Polycrystalline Smectic Liquid Crystals, *Langmuir* 1, 127 (1985).

[68] Huang, C. C., S. T. Wang, X. F. Han, A. Cady, R. Pindak, W. Caliebe, K. Ema, K. Takekoshi, and H. Yao, Experimental Investigations of One Liquid-Crystal Compound Exhibiting the No-Layer-Shrinkage Effect near the A–Sm−C*, *Phys. Rev. E* 69, 041702 (2004).

[69] Huang, C. C., and S. Dumrongrattana, Generalized Mean-Field Model for the Smectic- A– Chiral-Smectic- C Phase Transition, *Phys. Rev. A* 34, 5020 (1986).

[70] Thakur, A. K., A. Choudhary, S. Kaur, S. S. Bawa, and A. M. Biradar, A Dielectric Mode in Electroclinic Liquid Crystals, *J. Appl. Phys.* 100, 034104 (2006).

[71] Pavel, J., and M. Glogarová, The Effect of Biasing Electric Field on Relaxations in FLC Investigated by the Dielectric and Optical Methods, *Ferroelectrics* 121, 45 (1991).

[72] Gouda, F., G. Andersson, S. T. Lagerwall, K. Skarp, B. Stebler, T. Carlsson, B. Žekš, C. Filipič, and A. Levstik, Experimental and Theoretical Study of the Influence of a Bias Electric Field on the Dielectric Properties of the Chiral Smectic A* Phase, *Liq. Cryst.* 6, 219 (1989).

Chapter 7

Role of Dielectric Permittivity in Electrode Materials for Energy Storage Device Applications: A Fundamental and Brief Overview

Md. Yasir Bhat[1,*], Neakanshika Chadha[2] and Amrita Jain[3]

[1]Department of Chemistry, Khalifa University of Science and Technology, Abu Dhabi, UAE
[2]Kiet Group of Institutions, Delhi-NCR, Meerut Road, Ghaziabad, India
[3]Institute of Fundamental Technological Research, Polish Academy of Sciences, Adolfa Pawińskiego, Warsaw, Poland

Abstract

The rising demand and commercialization of power pulse systems led to the discovery of modern and advanced materials. In this context, exploring new dielectric materials is the first choice among researchers due to their low loss electromagnetic response, good thermal stability, and high energy density. On the basis of physical structure, the active materials with smaller remnant polarization, higher saturated polarization, and higher breakdown fields used in power electronics are extremely capable materials. Accordingly, there are a number of advanced materials used in next-generation power supercapacitors, here we highlight, four different kinds of materials, i.e., anti- ferroelectric, dielectric ceramic-glasses, relaxer ferroelectric, and polymer-derived ferroelectrics. This chapter covers the recent progressive efforts in the development of dielectric materials for better energy storage

* Corresponding Author's Email: mohammad.bhat@ku.ac.ae, yasirphysics89@gmail.com.

In: An Introduction to Permittivity
Editor: Parveen Saini
ISBN: 979-8-88697-042-5
© 2022 Nova Science Publishers, Inc.

performance, al, the future prospects is highlighted at the end to push their role in practical application.

Keywords: dielectric permittivity, electrodes, supercapacitors, polymer films, high-temperature dielectric materials, electrochemical performances

1. Introduction

Owing to the depletion of fossil fuels, the emergence of energy crises occurs that led to many conflicts worldwide. Apart from that, there are other issues such as energy efficiency, air pollution, and climate change, hence, to overcome these serious problems, different energy generation technology using wind, sun, and heat energy are generated that provides an alternative to fossil fuel energy resources with cleaner and safer renewable resources [1-3]. On the basis of storage ability, energy devices are basically divided into two categories, namely batteries and supercapacitors [2, 4-6]. Compared to batteries, capacitors usually show high power density i.e., 10^1-10^6 W kg^{-1} for ultracapacitors and ~10^8 W kg^{-1} for dielectric capacitors. However, this low energy density (typically below 30 Wh kg^{-1}) for capacitors can be enhanced, and nowadays commercially used conventional dielectric capacitors (which comprise dielectric ceramics or polymers) usually give an energy density values of 10^2 -10^1 Wh kg^{-1} [3, 7]. Particularly, electrochemical supercapacitors possess a complex physical structure with a maximum operating potential window (below 3.0 V) and high leakage current (low energy efficiency) that restrict their usage in advanced pulse power systems. Hence, there arises an urgent demand for high high-powered capacitors, the development of dielectric materials with improved electrical properties for energy-storage applications which has yet not much explored [7-9].

At present, many dielectric materials offer appreciable energy density and thus finds numerous application in defense, micro-electronics, power-pulse systems, power-grids, actuators, portable and wearable communication devices, etc. [8, 10]. Over the past few decades, ceramic and polymeric-based dielectrics with high capability to control and store electrical energy had a significant market share globally. The two important parameters i.e., static electric field strength and relative dielectric permittivity play decisive role in deciding the energy density for dielectric devices. The dielectrics store charges via polarization phenomenon in the presence of an external electric field, which migrates the charge (positive and negative) carriers on each electrode over time. Dielectrics that have high breakdown strength and also high permittivity are required to withstand the current demand for high energy density and to compete with other alternatives. Dielectric capacitors offer low-cost, good flexibility, and high electric field and are considered better than conventionally used electrolytic capacitors [11]. This chapter provides an overview of the different kinds of dielectric capacitors designed to date along with the role of permittivity in refining the overall performance of energy storage devices [12].

2. Theory of Dielectrics

2.1. Basic Fundamentals of Permittivity

The dielectric studies and their behavior in electric fields continue to mesmerize physicists and electrical engineers in a similar way. However, the fact, the dielectrics are poor conductors of electricity and play a fundamental part in the case of electric circuits that need a dielectric medium to build the circuit. The polarization of dielectric material takes place when the voltage is applied and the permittivity that a material showwhen it is exposed to an electric field depends on the frequency of the voltage source. A dielectric is an assembly of ions with positive and negative charges where charges cannot flow freely. The most important property of dielectrics is their permittivity along with the dielectric constant. The dielectric constant or relative permittivity is defined as the ability of a material to store electrical energy and is calculated as the ratio of the permittivity of the dielectric material and the permittivity of the vacuum depicted in Figure 1(a) [13]. All the materials have a dielectric constant greater than one. The electric field 'E' causes the bound charges to be separated. Permittivity is the measure of how the effects of the electric fields, a dielectric medium, each dielectric material has its own value of permittivity of material under a DC electric field is defined as in equations [6, 9, 14]:

$$D = \varepsilon_0 E + P \tag{1}$$

$$D = \varepsilon E = \varepsilon_0 \varepsilon_r E \tag{2}$$

where D denotes the electric displacement field, ε is the material's permittivity between the plates, ε_0 is the vacuum permittivity (8.8541878176 × 10 $^{-12}$ $C^2/J{\cdot}m$), ε_r denotes the material's relative permittivity. It is important that all dielectrics have the low-frequency limit of permittivity greater than the free space permittivity and their different measures the integrated loss over the entire logarithmic frequency range.

In general, dielectric polarizes by four mechanisms: one is due to the displacement of the outer electron cloud from the inner positive nucleus of an atom, the second happens through ionic response with the relative distance change in between them, third is orientation polarization due to the aligned dipoles along the electric field direction, and the fourth one is space charge polarization when the space charge appears and helps in the separation of both positive and negative charges The effect of the polarization of various mechanisms with changing frequency is depicted in Figure 1(b). The responded polarization P that reflects the induced dipole moment 'p' is defined as.

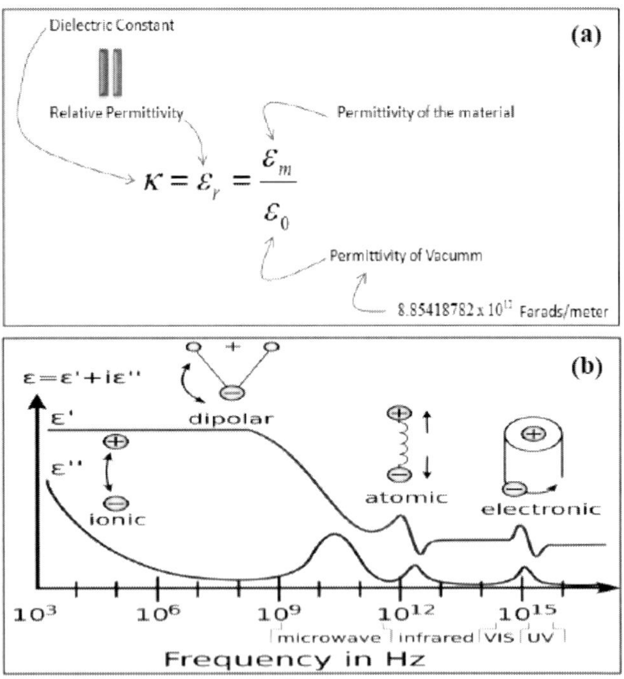

Figure 1. The frequency dependency of the real and the imaginary part of the permittivity spectrum for various mechanisms of polarization.

$$P=\frac{\sum_{i=1}^{N} p_i}{dV} \tag{3}$$

Using equations 1 and 2, the material polarization P under the electric field E can be shown as:

$$P= (\varepsilon_r-1)\, \varepsilon_0 E=\chi\varepsilon_0 E \tag{4}$$

where $\chi=\varepsilon r-1$ is the dielectric susceptibility of the material.

There is the displacement current passing through the dielectric when an AC electric field is applied.

Under the influence of AC electric current, the electric displacement D and polarization P became complex and is defined as:

$$D^* = \varepsilon_O E + P^* \tag{5}$$

$$P^* = (\varepsilon_r^* -1)\, \varepsilon_o E \tag{6}$$

where ε_r* represents the complex permittivity depicts the complex nature of electric displacement (D^*) and Polarization (P^*) and P. With the change temperature, discontinuities in the permittivity of a dielectric material have been observed and thus significant effect on the dielectric constant of a material has been observed. For example, with an increase in temperature it decreases the permittivity and the dielectric constant of material drops sharply when the temperature falls below the freezing point.

2.2. Dielectric Loss

When the voltage is variable the dielectric material dissipates some amount of energy which refereed as dielectric loss. The response to the external electric field generally depends on its frequency. When the AC electric field is applied to a dielectric material, the complex permittivity with the frequency can be expressed as:

$$\varepsilon_r^* = \varepsilon_r' - j\varepsilon_r'' \tag{7}$$

where j represents the imaginary unit, ε_r' is the real part of permittivity (related to the energy storage); ε_r'' depicts as the imaginary part of the permittivity (related to the dissipation or loss) of energy). The material's dielectric loss (δ) i.e., the intrinsic dissipation of electromagnetic energy in terms of heat can be expressed as:

$$\tan\delta = \frac{\varepsilon_r''}{\varepsilon_r'} \tag{8}$$

where tan δ is defined as the dielectric loss factor. The combination between ε_r' and ε_r'' can be expressed by Kramers-Krong relations [15]: The dielectric loss is one of the crucial parameters since it causes Joule heating of capacitors at higher frequencies, which may lead to the failure of the capacitors.

2.3. Relaxation Dielectric

Dielectric relaxation time describes the relaxation response of dielectric materials to an external electric field. It is the momentary delay in the dielectric constant caused by delay in polarization w.r.t varied electric field. Relaxation time 'τ' is determined by the obstruction of the molecular reorientation and ω is the angular frequency ($2\pi\nu$). Using a simple Debye equation expressed in Figure 2 and Figure 3 [15], dielectric relaxation is usually explained in various systems:

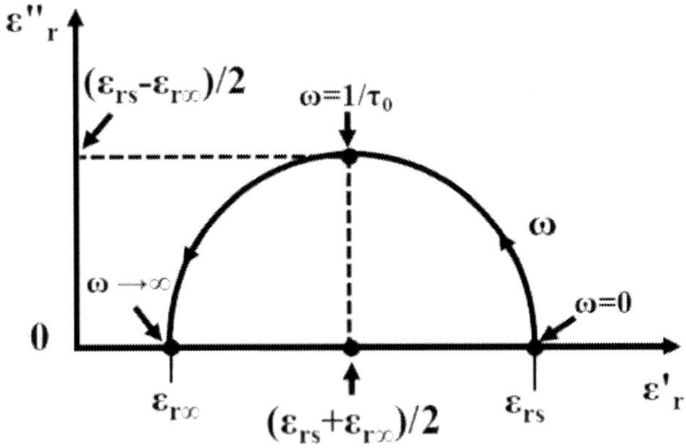

Figure 2. The schematic for dielectric material with only one relaxation time based on Debye equation.

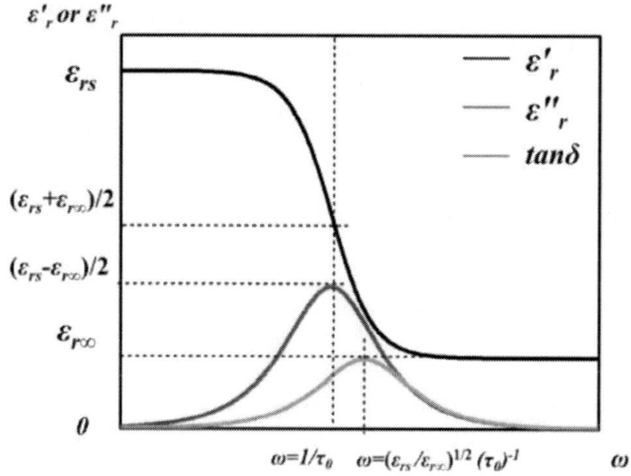

Figure 3. The schematic of ε'r, ε"r and tanδ as the functions of ω based on Debye.

$$\varepsilon_r^*(\omega) = \varepsilon_{r\infty} + \frac{\varepsilon_{rs}-\varepsilon_{r\infty}}{1+j\omega\tau_o} \tag{9}$$

$$tan\delta = \frac{\varepsilon_r''}{\varepsilon_r'} = \frac{(\varepsilon_{rs}-\varepsilon_{r\infty})\omega\tau_o}{\varepsilon_{rs}+\varepsilon_{r\infty}\,\omega^2\tau_o^2} \tag{10}$$

The $\varepsilon_{rs} - \varepsilon_{r\infty}$ is defined as the permittivity strength of the relaxation mechanism

3. Dielectric Materials

3.1. Types of Dielectric Materials and Their Application

Nowadays, dielectric materials were vastly utilized in a number of applications including energy storage, electronic packaging, satellite communication, dielectric resonators the basis of permittivity value, dielectric strength and dielectric loss, the materials are divided into three main categories.

Category I includes materials with a relatively low permittivity value (15 ~ 500) and a small dielectric loss ≤ 0.003;

Category II includes materials with high permittivity (<20,000); such as ferroelectric ceramics Category includes materials with a conductive phase having a low breakdown field. Based on the composition, dielectric materials are sub-divided into anti-ferroelectrics, dielectric glass-ceramic, relaxor ferroelectrics, and polymer-derived ferroelectrics. Basically, any dielectric material is either polar or non-polar. A non-polar inorganic material has usually low permittivity not have a permanent dipole moment whereas polar possesses permanent dipole moment. For a part of the polar molecules, there is a strong enough interaction between the dipoles to make them align in the same direction resulting to a polarization without applying the electric field to the material. These materials are known as pyroelectric materials and the polarization without the electric field is called spontaneous polarization. This spontaneous polarization can be switched in some of the pyroelectric materials which can be switched by the outside electric field and hence this material was named ferroelectrics [15, 16]. However, the spontaneous polarization charges with an applied electric field, ferroelectric materials usually display a very high permittivity and this makes these materials attractive a lot of attention in dielectric applications. The polarization- Effect Field relation of a ferroelectric material is depicted in Figure 3. This also displayed the basic characteristics of this material [17, 18]. These ferroelectric materials exhibit a high dielectric constant is strongly dependent on temperature, especially in the situations when it is around the phase transition temperature between the ferroelectricity and para-electricity. This temperature is a critical temperature for ferroelectric materials also named Curie temperature.

4. Application of Dielectrics

4.1. High/Low Permittivity

High permittivity composites are capable and have vast applications in microelectronics, electrical engineering, biomedical engineering, etc. Large group well designed. Low inductance and high voltage capacitors are used to provide the supply of high pulses of current for various pulsed power applications. These comprises of forming Marx generators, electromagnetic, pulse-forming networks, radar, pulsed lasers, fusion research and particle accelerators. In power supplies, reservoir capacitors are used where they put the smooth point of a full or half-wave rectifier and that can be used in charge pump circuits as energy storage elements in the generation of higher voltage as compared to the input voltage [19].

Due to the huge global demand of urgent demand for energy storage devices, the serious investigation of dielectric materials with enhances electrochemical performances has attracted an increase interest and excellent research has been published. However, the scenario on the research status of the dielectric materials for high energy storage performances is still in search. So in the current work, some of the basic principles are applied initially to respond to the fundamental ethics to study the potential materials. Further, the behavior of energy storage in four types of most investigated dielectric materials, enlisted as anti-ferroelectrics, dielectric glass-ceramics, relaxor ferroelectrics, and polymer-derived ferroelectrics. However, some general forecasts are established for future research. In order to improve the permittivity, efforts have been made to improve the breakdown strength by blending the ceramics or conducting materials into the matrix. However, the high loading of ceramic nanoparticles leads to high permittivity value but poor stability of the composites. The low dielectric constant substantially limits the energy density and on the basis of energy storage time, the commercial devices are termed as short time span and long time span.

5. Principles of Enhanced Energy Storage in Dielectric Capacitors

5.1. Introduction of Dielectric Capacitors

5.1.1. Classification of Supercapacitors

On the basis of the design, charge storage mechanism, and configuration, capacitors are broadly classified into four different categories, namely conventional capacitors, electrolytic capacitors, electrochemical capacitors and dielectric capacitors [20, 21]. Dielectric capacitors usually have a high power density of up to 10^5 kW kg^{-1} which is greater than electrochemical supercapacitors (0.01-10^3 kW kg^{-1}), while the energy density is small (~30 $Whkg^{-1}$). Figure 4 shows the ragone plot of commercially used energy storage devices.

A conventional capacitor comprises of two conducting metallic electrodes which are separated by insulating dielectric materials (such as paper, mica, glass, etc.), as shown in Figure 5. The equal and opposite charges are accumulated on the surfaces of each electrode and the voltage (V) is applied. The production of electric field result due to the insulating nature of the dielectric material keeps the charges separated, which allows the capacitor to store the energy as shown in Figure 5.

The capacitance (C) of such capacitors is given by:

$$C = \frac{dQ}{dV} = \frac{\varepsilon_r \varepsilon_o A}{d} \tag{11}$$

where C is the capacitance, d is the separation between two plates, A is the overlapped area of two plates, ε_r is the relative permittivity and ε_o is the permittivity of free space (~8.85 x 10^{-12} Fm^{-1}). It can be clearly seen that the capacitance directly depends on the area of the plates and the permittivity of the dielectric, whereas inversely depends on the separation between the plates. Depending on the external voltage V, the capacitance may vary due to different extent of polarization and can be given in terms of store charge ‘Q’ and voltage ‘V ‘. The energy stored ‘E’ by the movement of charges due to an external source can be obtained by:

$$E = \int_O^Q V dQ = \int_o^Q \frac{Q}{C} dQ = \frac{1Q^2}{2C} = \frac{CV^2}{2} \tag{12}$$

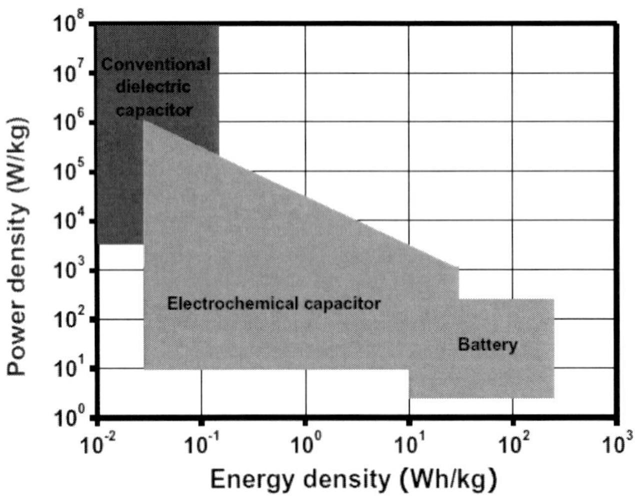

Figure 4. Ragon plot of commercially used energy storage devices.

The capacity to store electrical energy differs from one dielectric material to another material. The amount of electrical energy that a capacitor can store depends on the amount of polarization that occurs when voltage is applied between the plates. Clearly, the materials with high dielectric constant values can store more energy as compared to those with low dielectric constant values. The electrolytic capacitors used in DC power supply circuits provides high value of capacitance and the electrolyte (solid or liquid) itself acts as one of its plates in such type of capacitors The cathode is generally semi-liquid electrolyte solution in the form of jelly or paste. The insulating dielectric layer responsible for high capacitance is made up of oxides like Ta_2O_5 or Al_2O_3. Another class of energy-storage devices referred to as "Electrochemical Capacitors (ECs) or supercapacitors" explored a variety of active materials as electrodes (high surface-area carbons, transition metal oxides, nitrides, polymers, and their composites) and electrolytes (aqueous, organic, polymer-based electrolytes, etc), assembled in symmetric or asymmetric configurations. Diversity, design, and flexibility are crucial parameters for achieving effective performance in terms of power density and energy density. The overall optimized performance of supercapacitors depends on various components including the electrode materials used, electrolytes, dielectric properties, and the current collectors.

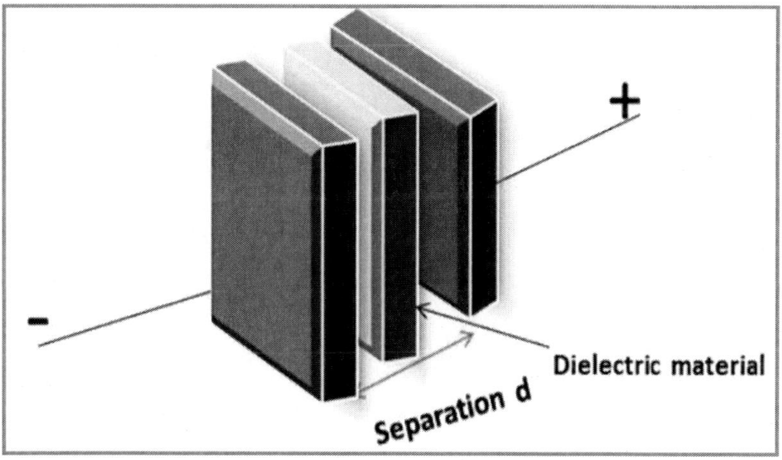

Figure 5. Schematic of a conventional capacitor.

6. Evaluation Techniques of Energy-Storage Density for Dielectric Capacitor

The energy stored per unit volume 'J' of a dielectric can be obtained in two different ways: static method and dynamic method. Figure 6 shows the circuit diagram for the energy storage density which is measured in a static way. In this scenario, firstly the capacitor is charged by an external bias, and hence the electrical energy is stored in the dielectric. After this step, the capacitors are connected to the load R which will complete the circuit by MOSFET switching. In this way, part of the energy storage is released and is governed by the transient current formed in the closed circuit.

By applying the *I(t)-t* curve, the curve related to the discharge energy could be expressed by the following expression:

$$W = \int I^2 (t)Rdt \tag{13}$$

where R denotes the load resistance, t; discharge time. The energy density; J is the work done per unit volume and can be expressed as:

$$\text{J}= \frac{W}{Ad} = \frac{\int_0^Q VdQ}{Ad} = \int_0^{E_{max}} DdE \tag{14}$$

where, E is the applied external field and D is electrical displacement ($\varepsilon_0\varepsilon_r E$) in dielectrics. For the high permittivity value, D becomes very close to electrical polarization P. In Figure. 7(a) the red area indicates the stored energy (J_s) and the green area indicates the recoverable electrical energy (J_r), their ratio is said to be energy-storage efficiency (η):

$$\eta = \frac{J_r}{J_s} \, x \, 100\% \tag{15}$$

Though, the permittivity is expressed as dP/dE, depicted in Figure 7(b), the formula can be given by the Equation. However in case of linear dielectric materials, whose permittivity does not depends on the external applied field, energy storage density is directly proportional to relative permittivity and external applied field, given by equation:

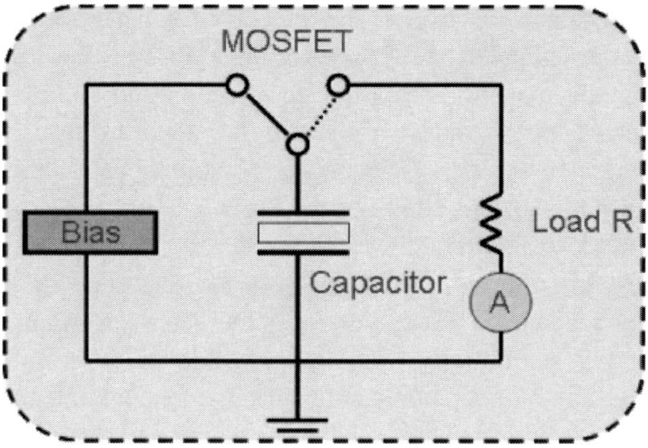

Figure 6. Scheme of the measurement circuit for the energy-storage density.

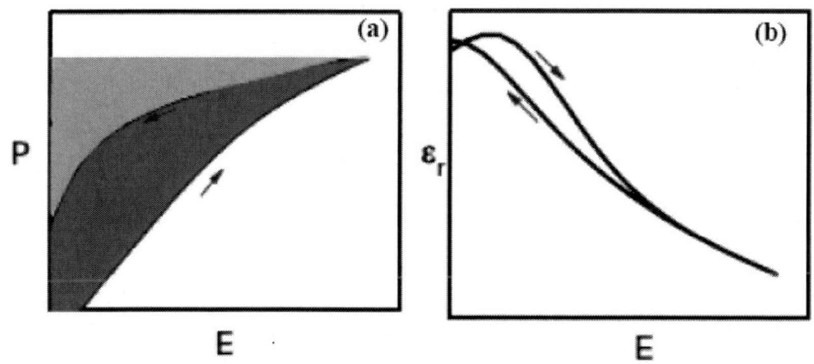

Figure 7. (Color online) typical dependence of (a) polarization and (b) permittivity on electric field of ferroelectrics in the first quarter.

$$J = \int_{o}^{E_{max}} \varepsilon_o \varepsilon_r \, E dE \tag{16}$$

$$J = \int_{o}^{E_{max}} P dE = \frac{1}{2} \varepsilon_o \varepsilon_r E^2 \tag{17}$$

7. Potential Dielectrics for High Energy-Storage Application

In order to achieve higher capacitance values, dielectric materials with high dielectric constants are needed, though, as discussed in above sections, other parameters that regulate the capacitance of a capacitor also includes the spacing between the electrodes and the effective plate area. The urgent demand for improving energy density for high pulsed power capacitors, and various dielectric materials has attracted attention and quite serious efforts have been made. So as to design dielectric capacitors with acceptably high energy storage and better efficiency for commercial applications, two parameters need to be controlled i.e., dielectric loss and value of permittivity. There are basically four kinds of dielectrics (i) linear dielectrics (ii) ferroelectrics (iii) relaxor ferroelectric (iv) anti-ferroelectrics. High energy density is not achieved in linear dielectrics (e.g., glass, aluminum oxide) due to their smaller permittivity and high breakdown field. In ferroelectric materials, the dipole orientation can be changed by applying an external DC electric field, keeping aligned with the adjacent dipoles in the same polarization orientation. Ferroelectric filler-based composite materials such as $BaTiO_3$ (Barium Titanium oxide, BT), $PbZrO_3$ (Lead Zirconium oxide, PZT), $BaSrTiO_3$ (Barium strontium titanium oxide, BST), etc. have been extensively studied by researchers. However, in ferroelectrics, small energy density and lower efficiency is observed due to higher saturated polarization and moderate electric field.

Because of the urgent demand of high energy density, relaxor ferroelectrics and anti-ferroelectrics attracted huge attention owing to their moderate breakdown field and larger saturated polarization.

In the meantime, due to the development of new techniques for the materials manufacturing process listed as glass-crystallization technique and composite technology, other types of polymer-based ferroelectrics and glass ceramics are found to be the potential electrode materials energy device applications. These material combines both the qualities like a higher breakdown field of linear dielectric and large polarization of ferroelectrics.

Within the category of anti-ferroelectric materials, there are many subcategories, including perovskite, pyrochlore, ceramics, liquid crystal, etc. Amid all of these anti-ferroelectrics, materials that have have perovskite structure are the most interesting ones and are usually expressed as ABO_3. Since 1961, anti-ferroelectrics $PbZrO_3$ having energy density lower than 1 J/cm^3, because of the lower breakdown field (below 60 kV/cm) was achieved. Further improvements were made and in 1971, a comparatively improved energy density of 2.1 J/cm^3 was achieved in 65-micron-thick $PbZrO_3$ thick films with a tiny amount of low-softening-point SiO_2_Bi_2O_3 glass by the screen-printing method. After the above work, another dielectric material of 30 m-thick of varying Pb and La ratio in anti-ferroelectric films with PbO, B_2O_3, SiO_2, ZnO glass, as a result, 3.1 Jcm^{-3} of energy density was obtained.

Up to now, it has been proved that $PbZrO_3$, $PbHfO_3$, $NaNbO_3$, and their possible combinations are the typical anti-ferroelectrics with perovskite-type structure at room temperature.

Recently, $BaTiO_3$-based composites also showed anti-ferroelectric-like behavior in certain specific conditions, such as, at higher temperatures and/or special sintering atmosphere. For example, Li et al., synthesized $(Na_{0.5}Bi_{0.5})$, TiO_3 $(Sr_{0.7}Bi_{0.2})$ TiO_3 multilayer ceramic capacitors by mixing anti-ferroelectrics and relaxor ferroelectrics, achieving an energy density of 9.5 J cm^{-3} and a very high energy storage efficiency of 92%.

However, generally, all above discussed four kinds of dielectrics listed as anti-ferroelectic, dielectric glass-ceramics, relaxor ferroelectric based polymer-derived ferroelectrics are the most attractive materials for the application of high energy devices and also small work has also been reported on other materials.

8. Novel Dielectric Polymers for Energy Storage

The inappropriate size of electronic devices and optimization of power system structure needs a strict and appropriate size requirements for capacitors as passive components [22, 23]. The experimental analysis of thickness reduction of thin polymer films boosts the enhanced electrochemical measurements and became the necessary aspects for the future technology development. This achievement of lower polymer thickness enhances the capacitance density, lightweight module and system, and less raw resource consumption. However, the thickness has an important role of the properties of dielectric, the surface measurements and respective comparison with synthesis measures of the dielectric films. Hence, there is a great effect on the thickness of the polymer film in the electrochemical measurements, role of fillers, high-temperature dielectric polymer technology, various process methods are hitting concerns of the polymer films. These dielectric polymer films needs proper process, scalability, commercial and industrial accessibility, and required component manufacturing. It needs future investigation to meet the concerned issues like feasible multilayer effects of polymeric films on an increment of the strength of dielectric materials.

9. Polymer Dielectrics as Capacitive Energy Storage for High-Temperatures

High temperature dielectric polymers are recent concern of the device technology and high energy concerns [24]. The growing demand of the electric vehicle transport and the increasing demand for the advanced development of electronics needs polymer dielectrics feasible of operating at high-temperature conditions. The most recent analysis and development in the dielectric polymers for the purpose of high-temperature electrochemical measurements are deeply discussed in the recent article [24]. The main concern of these devices is focused on the structure depending on high field's dialectic, electric properties and electrochemical performances such as capacitance, charge discharge efficiency, cyclic stability, and rate capability of dielectric polymers at high temperatures. There are advancements, advantages, challenges and concerns limitations of these high-temperature dielectric polymers of deep concern.

10. Future Prospects

The conclusion depicts the above-summarized examinations of the dielectric materials for the energy storage devices and is concluded in some interesting achievements for the same energy devices. But this study is not yet completed; still, there are various problematic issues that need to be solved properly for the application of the energy storage devices in practical applications. Here we will give the picture of the investigation regarding the dielectric materials for the energy device applications. As investigated the ultimate feasible energy storage density in anti-ferroelectric and relaxor ferroelectric bulk ceramics are usually very small and less than1 J/cm3 which is not enough to fulfill the practical demand. Due to some limitations in anti-ferroelectric materials that do not meet the demand in some cases, it's necessary to develop bulk ceramic or thick films (1100 nm) depicting the breakdown field better performer compared to the thin film form to enhance their electrochemical performances. The importance of the deep study of the polarization mechanism is important at the interface in a proper way to enhance the breakdown field and decrease the interface loss. Also, it should be taken into account there should be further optimization of the process of fabrication to control the permittivity and to make better the glass-ceramic and inorganic polymer interface systems.

Excellent electrochemical properties should be explored to fabricate new dielectric materials. Nowadays focus is on anti-ferroelectrics and relaxor ferroelectrics due to lead-containing issues but the number of unwanted issues are there and need to be modified and should be addressed in future studies.

Conclusion

In summary, the brief details are presented the current study on energy storage dielectrics. Research has been focused mainly on four types of dielectric materials mentions above briefly. Recently, modifying and improvement studies have been established between the polarization saturated and polarization remnant and higher dielectric breakdown field are the keen issues to generalize the electrochemical performance of energy storage devices in these dielectric materials. The dielectrics materials mentioned above have excellent electrochemical performance for energy storage devices. In comparison, energy storage properties have attracted more attention in polymer-based dielectric materials. A number of techniques such as growth orientation composite, stress phase control, and so on have been examined to utilize to optimize their electrochemical energy storage performances. Finally, due to the analysis of the current materials under research progress so modified research progress were further improved the energy storage behaviorthese dielectric materials. Also, a few unique examined parameters of the energy storage e dielectric materials were deeply addressed to meet the practical requirements. Finally, we believe that this fundamental and brief detail has captured an exciting overview and new developments in the same field and needs further deep studies about the improvement and development of new dielectric materials for the application of energy storage devices.

Acknowledgments

Author (Md. Y Bhat) is thankful to the *Khalifa University* of *Science and Technology- Abu Dhabi, (UAE)* for providing the necessary facilities and financial support as postdoctoral fellow. Author (Amrita Jain) has received financial support by the National Centre for Research and Development (NCBR, Poland); Project number: V4-Japan/2/17/AtomDeC/2022 under the

Visegrad Group-Japan 2021 Joint Call on Advanced Materials in cooperation with the International Visegrad Fund.

References

[1] V. Yu, Chabot, J. Zhang, *Electrochemical supercapacitors for energy storage and delivery: fundamentals and applications*, Taylor & Francis 2013.

[2] R. Carbone, *Energy storage in the emerging era of smart grids*, BoD-Books on Demand 2011.

[3] B. E. Conway, *Electrochemical supercapacitors: scientific fundamentals and technological applications*, Springer Science & Business Media 2013.

[4] M. Y. Bhat, N. Yadav, S. Hashmi, Gel Polymer Electrolyte Composition Incorporating Adiponitrile as a Solvent for High-Performance Electrical Double-Layer Capacitor, *ACS Applied Energy Materials* 3(11) (2020) 10642-10652.

[5] M. Y. Bhat, S. A. Hashmi, Solid-state pseudocapacitors based on MnO2-nanorod-electrodes and plastic crystal incorporated gel polymer electrolyte: synergistic effect of Li-salt addition in electrolyte and morphology of electrodes, *Journal of Solid State Electrochemistry* 23(2) (2019) 591-606.

[6] X. Wu, X. Chen, Q. Zhang, D. Q. Tan, Advanced dielectric polymers for energy storage, *Energy Storage Materials* 44 (2022) 29-47.

[7] G. Wang, L. Zhang, J. Zhang, A review of electrode materials for electrochemical supercapacitors, *Chemical Society Reviews* 41(2) (2012) 797-828.

[8] X. Hao, A review on the dielectric materials for high energy-storage application, *Journal of Advanced Dielectrics* 3(01) (2013) 1330001.

[9] K. Yao, S. Chen, M. Rahimabady, M. S. Mirshekarloo, S. Yu, F. E. H. Tay, T. Sritharan, L. Lu, Nonlinear dielectric thin films for high-power electric storage with energy density comparable with electrochemical supercapacitors, *IEEE transactions on ultrasonics, ferroelectrics, and frequency control* 58(9) (2011) 1968-1974.

[10] D. Malleo, J. Nevill, A. Van Ooyen, U. Schnakenberg, L. Lee, H. Morgan, Note: Characterization of electrode materials for dielectric spectroscopy, *Review of Scientific Instruments* 81(1) (2010) 016104.

[11] Y. Wang, X. Zhou, Q. Chen, B. Chu, Q. Zhang, Recent development of high energy density polymers for dielectric capacitors, *IEEE Transactions on Dielectrics and Electrical Insulation* 17(4) (2010) 1036-1042.

[12] L. Qi, L. Petersson, T. Liu, Review of recent activities on dielectric films for capacitor applications, *Journal of International Council on Electrical Engineering* 4(1) (2014) 1-6.

[13] Y. Tong, *Development of dielectric composites for dielectric and energy storage applications*, (2017).

[14] Z. M. Dang, J. K. Yuan, J. W. Zha, T. Zhou, S. T. Li, G. H. Hu, Fundamentals, processes and applications of high-permittivity polymer–matrix composites, *Progress in materials science* 57(4) (2012) 660-723.

[15] M. Barsoum, *Fundamentals of ceramics Institute of Physics Publishing,* Bristol and Philadelphia (2003).

[16] R. E. Newnham, *Properties of materials: anisotropy, symmetry, structure*, Oxford University Press on Demand2005.

[17] Ch. Kittel, *Introduction to solid state physics Eighth edition*, (2021).

[18] T. Mitsui, I. Tatsuzaki, E. Nakamura, *An introduction to the physics of ferroelectrics*, (1976).

[19] Z. Peng, P. Liang, B. Chen, Y. Xiang, H. Peng, D. Wu, X. Chao, Z. Yang, Low-temperature synthesis of CdCu3Ti4O12 powders with high dielectric permittivities, *Ceramics International* 45(9) (2019) 11899-11904.

[20] M. Y. Bhat, S. Hashmi, Mixture of non-ionic and organic ionic plastic crystals immobilized in poly (vinylidene fluoride-co-hexafluoropropylene): A flexible gel polymer electrolyte composition for high performance carbon supercapacitors, *Journal of Energy Storage* 51 (2022) 104514.

[21] M. Y. Bhat, N. Yadav, S. Hashmi, A high performance flexible gel polymer electrolyte incorporated with suberonitrile as additive for quasi-solid carbon supercapacitor, *Materials Science and Engineering*: B 262 (2020) 114721.

[22] D. Q. Tan, Review of polymer-based nanodielectric exploration and film scale-up for advanced capacitors, *Advanced Functional Materials* 30(18) (2020) 1808567.

[23] J. Ho, T. R. Jow, S. Boggs, Historical introduction to capacitor technology, *IEEE Electrical Insulation Magazine* 26(1) (2010) 20-25.

[24] H. Li, Y. Zhou, Y. Liu, L. Li, Y. Liu, Q. Wang, Dielectric polymers for high-temperature capacitive energy storage, *Chemical Society Reviews* 50(11) (2021) 6369-6400.

Chapter 8

Dielectrics in Electronic and Energy Storage Devices

Sonal Agrawal[1], Kumar Gaurav[1], Sushmita Dandeliya[2], Boddepalli SanthiBhushan[3], Parveen Saini[4] and Anurag Srivastava[1,*]

[1]Advanced Materials Research Group, CNT Lab,
ABV-Indian Institute of Information Technology and Management, Gwalior, India
[2]Department of Electronics and Communication Engineering, Indian Institute of Information Technology, Nagpur, India
[3]Department of Electronics and Communication Engineering, Indian Institute of Information Technology Allahabad, Prayagraj, India
[4]CSIR-National Physical Laboratory, Delhi, India

Abstract

Dielectrics are an important category of materials that are useful for a variety of industrial applications in electrical transmission, electronic devices, energy storage, etc. This chapter focuses on the basic aspects of permittivity, relative permittivity or dielectric constant, and the role of dielectrics in electronic devices (Field Effect Transistors (FETs) and Single Electron Transistors (SETs) and energy storage devices (capacitor, supercapacitor, and battery). In the realm of silicon electronics, the dielectrics are classified into 2 types viz. high-k (materials with a dielectric constant greater than SiO_2) and low-k (materials with a dielectric constant lesser than SiO_2). The high-k dielectrics come in handy during the process of device miniaturization. They also enhance certain performance parameters of the MOSFETs and

* Corresponding Author's Email: profanurag@gmail.com.

In: An Introduction to Permittivity
Editor: Parveen Saini
ISBN: 979-8-88697-042-5
© 2022 Nova Science Publishers, Inc.

SETs, which are discussed in detail in this chapter. Besides, the role of dielectrics in a capacitor and the choice of solvent for supercapacitors and batteries have been discussed in reference to the dielectric constant.

Keywords: permittivity, dielectric constant, single electron transistors, MOSFETs, supercapacitor, electrolyte, solvent, battery

8.1. Introduction

Based on the conductivity, the materials are categorized as conductors, insulators, and semiconductors. Knowing the fact that every material is made up of molecules which in turn made up of atoms and under the influence of an electric field these atoms undergo certain displacements and changes in their properties. In October 1745, Ewald Georg von Kleist, a German experimentalist connected a high-voltage electrostatic generator to a volume of water collected in a hand-held jar using a wire, reported first time the storing of charge. Just after one year, a Dutch physicist Pieter van Musschenbroek of the University of Leiden reported the first capacitor called "Leyden Jar". The new material property that backed this invention was "Dielectric". Dielectrics are the important materials for variety of industrial applications, including the electronic devices. In the electronic industry, the dielectric materials are used depending upon the permittivity of materials. Low k dielectric materials are used in digital circuits for isolation of metallic wires and as inter-level metal dielectrics, operating at switching frequencies in the gigahertz regime. High-k dielectrics are used for transistors, capacitors as well as memory applications. A wide variety of dielectric materials are used in capacitors as an insulator between the capacitor plates i.e., mica, ceramic, paper, marble, glass, etc. All of them contain a different value of permittivity or dielectric constant and the capacitance varies depending upon their permittivity. The advanced capacitors require storing a large amount of charge which is possible only with high dielectric constant materials. These dielectric materials not only work as a separator but also affect the electrical properties of the capacitor, which in turn influence the performance of a capacitor.

This chapter is mainly classified into four sections according to the role of permittivity in different applications. The first section is about different dielectric materials, their classes, types, and properties.The second section is the role of dielectric materials in the Field Effect Transistors (FETs). The third section is the role of dielectric materials in the Single Electron Transistors

(SET). The last section covers the role of permittivity in energy storage devices such as capacitors, supercapacitors, and batteries. The aim of the chapter is to present a review on permittivity and show how the dielectric materials are evolving for different applications.

8.2. Dielectric Constant and the Permittivity

The properties of dielectric materials, used in semiconductor fabrication and packaging play a very important role, to achieve the required performance of integrated circuits. Hence, most of the engineers working in the semiconductor industry require a basic understanding of dielectric properties. In general, the conductor starts conducting when electricity is applied, and an insulator opposes the flow of current due to the absence of free moving electrons. However, the dielectric is a special type of insulator that does not conduct current but gets polarized when the electric field is applied. The magnitude of the polarization is linearly proportional to the applied field and this proportionally constant is called the permittivity. Polarization of the material, shifts the electrons towards the positive voltage and upon removal of the applied voltage, the material either returns to its original state or may remain polarized if the molecular bonds of the material are weak.

Permittivity is the ability of a material to store electrical energy under the influence of an applied electric field. It is a measure of the electric *polarizability* of a *dielectric*, and is denoted by the *Greek letter* ε (epsilon). The relative permittivity (ε_r) is the ratio of the absolute permittivity of a material and the *vacuum permittivity* ε_0, Hence the relative permittivity of the vacuum is 1. The relative permittivity is a dimensionless quantity and is also called the dielectric constant of material. A high permittivity implies high polarizability of the material with the applied electric field.

8.3. Classification of Dielectric Materials

The dielectric materials can be classified on the basis of polarization and state, as depicted in Figure 1 below.

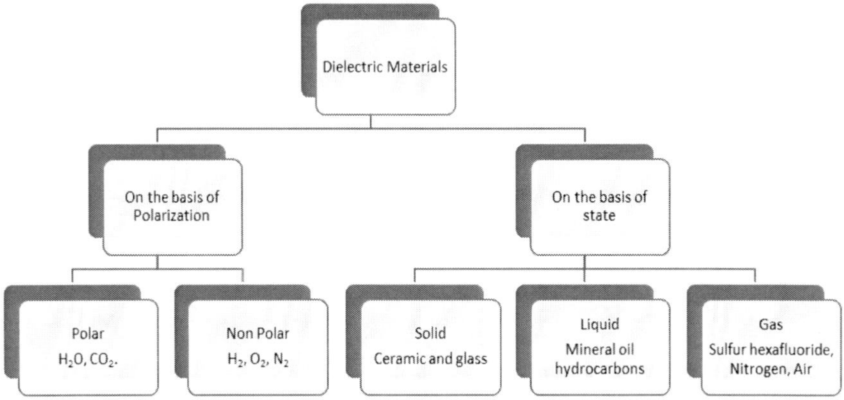

Figure 1. Types of dielectric materials.

Classification on the Basis of Polarization

On the basis of polarization, dielectrics can be classified into two categories i.e., polar dielectrics and non-polar dielectrics.

(i) Polar Dielectrics

Here, the molecules align with the applied electric filed and offers a dipole moment. The net dipole moment in the molecules becomes zero on the removal of applied electric filed. The center of mass of positive particles doesn't coincide with that of negative particles due to the asymmetric shape of molecules.

(ii) Non-Polar Dielectrics

Here, the center of mass of positive particles coincides with that of negative particles. These molecules maintain symmetry. The non-polar dielectrics show zero dipole moment in their normal state.

Example: H_2, O_2, N_2.

Classification on the Basis of State

Dielectric materials can be classified into three types based on their state i.e., solid, liquid, and gas. Each type has different dielectric properties and different applications.

(i) Solid Dielectrics

Most of the available dielectric materials are in solid form, and hence used for applications such as capacitors, transformers, switches, overhead lines, cabling, etc.

Examples: Ceramic, glass, plastic films, resins, varnishes, silicones, vulcanized adhesive tapes, mica, etc.

(ii) Liquid Dielectrics

These dielectrics are in the liquid form and used for applications such as insulation and cooling for transformers, reactors, capacitors, rheostats, etc.

Examples: Hydrocarbons of mineral oil, silicone fluids, synthetic esters, etc.

(iii) Gas Dielectrics

These dielectrics are in gaseous form and used for applications such as transformers, switchgears, circuit breakers, gas-insulated lines (GILs), gas-insulated transformers (GIT), etc.

Examples: Sulfur hexafluoride, nitrogen, air, carbon dioxide, hydrogen, etc.

8.4. Properties of Dielectric Materials

The following imperative properties need to be considered before using the dielectric material for a specific application.

(i) Electric Susceptibility

Denoted by χ, it represents the measure of polarization of a material with the applied electric field. Susceptibility also determines the material's electrical permeability.

(ii) Dielectric Polarization

Dielectric polarization indicates the separation of negative and positive charges within the dielectric system. This separation of charges leads to the formation of a dipole moment. The relationship between the dipole moment (M) and the electric field (E) provides the dielectric properties of a material. After the removal of applied electric field, the atoms/molecules revert to their original state, and the time taken for this reversion is called the relaxation time.

(iii) Dielectric Breakdown

If a strong electric field is exerted on a dielectric (insulator), the material may lose its insulating behavior and start to conduct at a certain value of an electric field. This irreversible process is called dielectric breakdown. The dielectric breakdown may cause permanent damage to the insulating ability of the material.

(iv) Dielectric Dispersion

Dielectric dispersion is defined as $P(t) = P\,[1\text{-}\exp(-t/t_r)]$

Where, P is the maximum polarization magnitude and t_r is the relaxation time of dielectric material for a particular polarization process.

8.5. Applications of Dielectric Materials

- In electronic devices (FETs, MOSFETs, SETs, etc.).
- In energy storage, such as capacitors, supercapacitors, batteries, etc.
- In electrets, which are electrostatic equivalents to the magnets.
- As insulating and cooling mediums for transistors, transformers, etc.
- In liquid crystal displays (LCDs).
- Dielectric resonator oscillator, and so on.

The dielectrics find applications in several devices. However, we focus here on two important applications i.e., electronic devices (FETs, MOSFETs, SETs) and energy storage devices (Capacitors, Supercapacitors, Batteries).

8.5.1. Dielectrics in Field Effect Transistors

The Metal Oxide Semiconductor Field Effect Transistors (MOSFETs) (as shown in Figure 2), which are the basic building blocks of the modern ICs, are being subjected to continuous scaling in order to increase the packing density and thereby the computing power. These MOSFETs are made up of silicon (Si) semiconductor owing to its abundance, lower leakage current, higher peak inverse voltage, lower exciton Bohr radius, and higher thermal range over its close counterparts such as germanium. However, the continuous scaling of silicon MOSFET size has brought it into the nano-regime, where the device is facing reliability and performance issues due to higher leakage and short channel effects [1]. To avoid these issues, the conventional gate dielectric (silicon dioxide - SiO_2) has been replaced with the high relative permittivity materials called high-k dielectrics [2]. The dielectrics used in the MOSFETs can be broadly divided into 2 categories i.e., Low-k and High-k dielectrics. Dielectrics with relative permittivity lesser than silicon ($\varepsilon_r \approx 3.8$) are called low-k dielectrics, whereas those with ε_r higher than silicon are called high-k dielectrics. The improvement in device performance has been reported with the use of high-*k* dielectric materials such as hafnium silicate ($HfSiO_4$), lanthanum oxide (La_2O_3), titanium dioxide (TiO_2), yttrium oxide (Y_2O_3), aluminium oxide (Al_2O_3), hafnium oxide (HfO_2) and zirconium oxide (ZrO_2) [2]. It has been reported elsewhere that TiO_2 is a good dielectric material for the replacement of SiO_2 as the gate dielectric in MOSFETs, owing to its non-toxic nature and higher chemical stability [3].

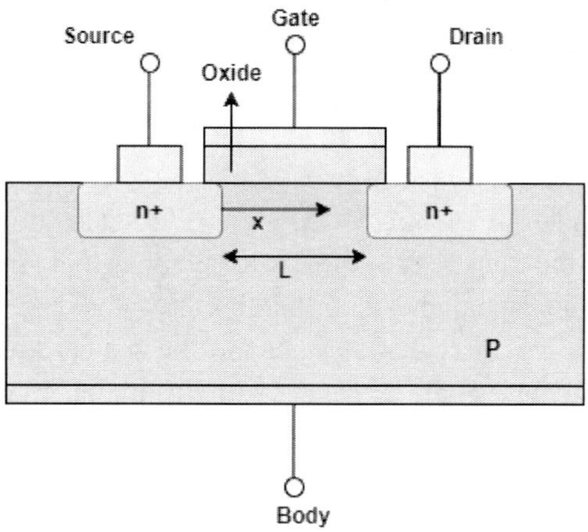

Figure 2. Schematic of MOSFET.

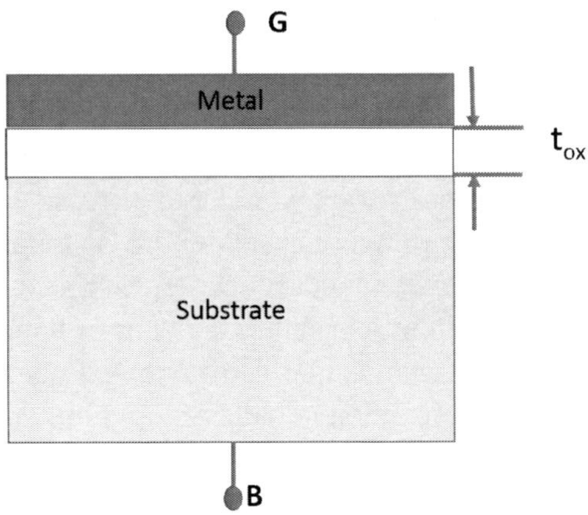

Figure 3. Schematic of MOS capacitor.

8.5.1.1. Dielectric Role in Gate Control of MOSFET Channel

In a MOSFET, the gate oxide capacitance controls the flow of current through the device. To understand this phenomenon, let us consider the MOS capacitor model shown in Figure 3, where the gate electrode represents one plate and silicon substrate represents the other electrode.

The oxide capacitance (C_{ox}) is given by,

$$C_{ox} = \frac{k\varepsilon_o}{t_{ox}} \tag{1}$$

Here, t_{ox} represents thickness of oxide layer, k is the dielectric constant (relative permittivity) of the oxide material and ε_o is the permittivity of free space ($8.854*10^{-12}$ F/m). From eq.(1), the oxide capacitance is inversely proportional to the oxide layer thickness and directly proportional to relative permittivity of material. As the MOSFET size is scaled down to increase the packing density, the thickness of the oxide layer has to be scaled down, which increases the oxide capacitance and thereby the gate control over the channel region. However, the SiO_2 thickness cannot be scaled down beyond certain limit due to the increased gate leakage arising from tunneling of electrons from channel to gate. It has been reported in Ref. [4] that the SiO2 thickness can not be reduced below 1.2 nm. Hence, in order to increase the oxide capacitance without reducing the oxide thickness, the high-k dielectrics are introduced in the MOSFETs [5]. Figure 4 shows various high-k dielectrics and their band gaps [6-8].

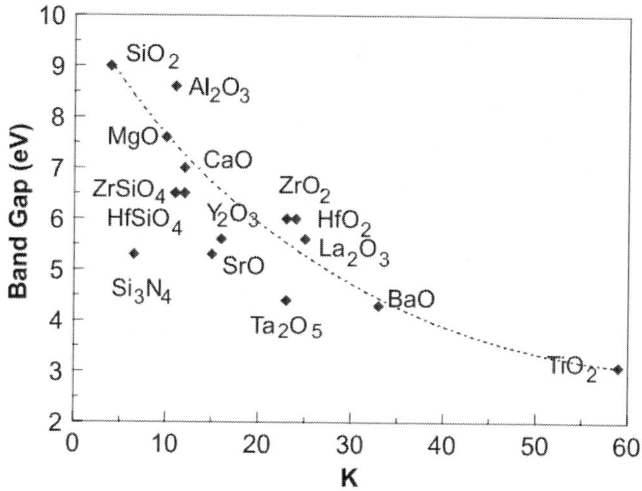

Figure 4. Dielectric constant of various oxides Vs their band gap [9].

Therefore, it is necessary to replace the silicon dioxide with high-k materials like Hafnium oxide, Aluminium oxide, Titanium oxide, etc. so that the oxide capacitance of the device can be enhanced without increasing the leakage current. The selection of dielectric depends on some desirable properties like

- High dielectric constant
- Stable in contact with Si substrate
- Compatible with CMOS processing and gate metal
- Lower power consumption
- High Barrier height
- Reduced leakage current
- Lower direct tunnelling effect

The performance of various FET and MOSFET devices like, Silicon double gate field-effect transistor, silicon gate all around (GAA) nanowire FET (NWFET), Double gate MOSFETs, GAA MOSFETs, carbon nanotube (CNT) FETs, etc. have been studied by the researchers with different gate dielectric materials i.e., Silicon dioxide (SiO_2), aluminium oxide (Al_2O_3), hafnium oxide (HfO_2), lanthanum oxide (La_2O_3), and titanium dioxide (TiO_2), etc. to understand the suitable gate dielectric for those devices and to reduce the short channel effects with scaling.

In a report by *Mech* et al. [2], the performance of Si, InAs, and CNT FETs with different dielectric materials have been studied (see Figure 5 and Figure 6). The results show TiO_2 as gate oxide offers better transfer characteristics, improved drain induced barrier lowering, and high conductance. Hence, TiO_2 may be preferred over SiO_2, Al_2O_3, HfO_2, and La_2O_3. The transfer characteristics of **(a)** Si double gate FET, **(b)** Si nanowire FET, **(c)** InAs nanowire FET, and **(d)** CNT FET are plotted at a drain bias of 0.5V for different gate oxides namely SiO_2 ($\varepsilon_r \approx 3.8$), Al_2O_3 ($\varepsilon_r \approx 10$), HfO_2 ($\varepsilon_r \approx 25$), La_2O_3 ($\varepsilon_r \approx 30$), and TiO_2 ($\varepsilon_r \approx 80$) are shown in Figure 5. The transconductance observed from the transfer characteristics follows the pattern $TiO_2 > La_2O_3 > HfO_2 > Al_2O_3 > SiO_2$, which is in the same pattern as the relative permittivity. Similar pattern has also been witnessed for the subthreshold slope (SS), drain induced barrier lowering (DIBL), leakage current (I_{off}). This indicates that the transconductance, SS, DIBL, and I_{off} of the device are highly dependent on the relative permittivity of the gate oxide, and a proportionality is witnessed in this case.

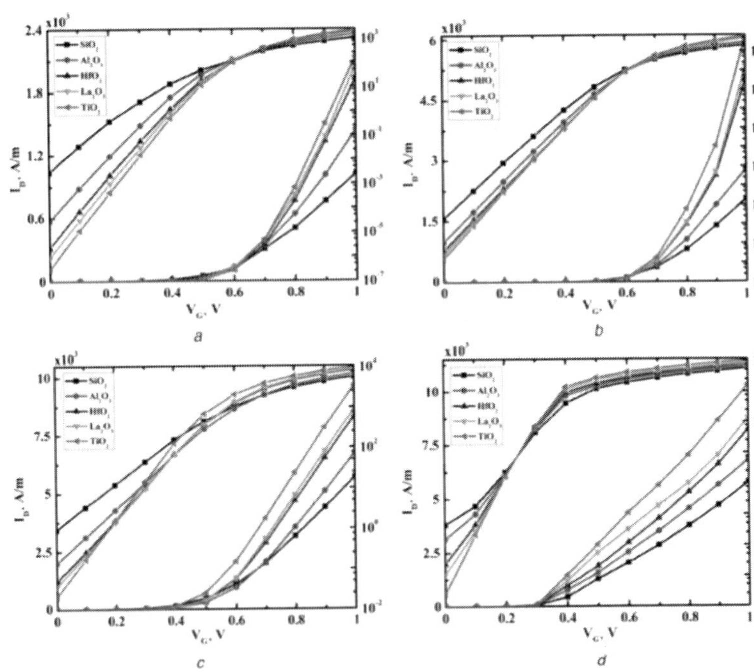

Figure 5. Transfer characteristics of (a) Si double gate FET, (b) Si nanowire FET, (c) InAs nanowire FET, and (d) CNT FET for different dielectrics in linear and logarithmic scales [2].

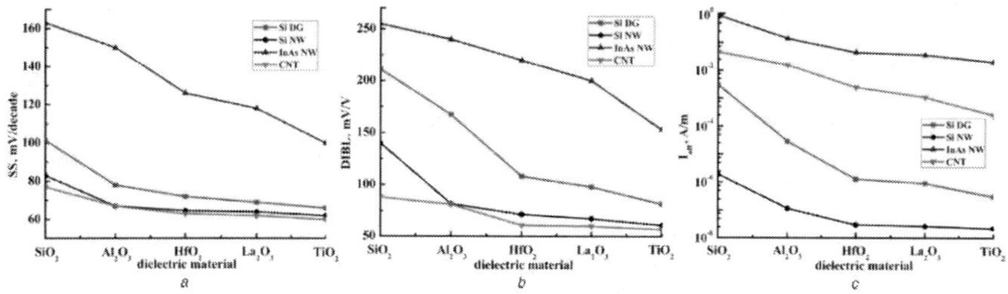

Figure 6. Variations of (a) SS, (b) DIBL and (c) Leakage (I_{off}) for various transistors with different dielectrics [2].

In another report [10], the effect of different dielectric materials (Silicon Nitride (Si_3N_4), Aluminium Oxide (Al_2O_3), Zirconium Oxide (ZrO_2), and Hafnium Oxide (HfO_2)) on the transfer characteristics, subthreshold slope, threshold voltage, and electrical performance of double Gate (DG) and Gate-All-Around nanowire (GAA) MOSFET have been explored. The results shows that out of the chosen dielectric materials HfO_2 as gate dielectric offers better performance in comparison to other dielectric materials, owing to its relatively higher permittivity. In another report, Santhibhushan et al. [11] studied the performance of junction less double gate MOSFETs with different oxide dielectrics. The analysis revealed that the transconductance of junction less MOSFET improves with the use of high-k dielectric materials.

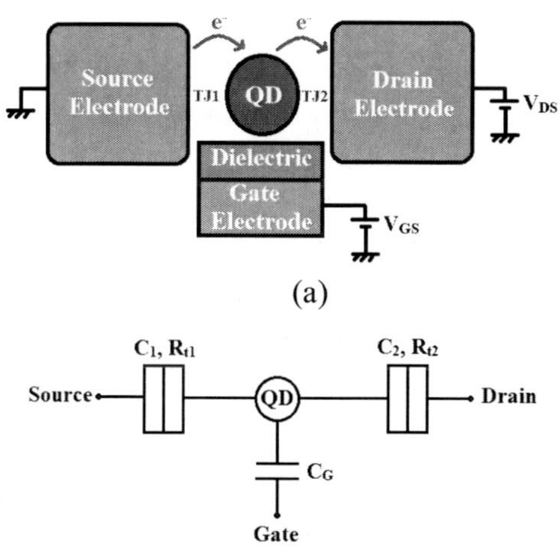

(b)

Figure 7. (a) Schematic of SET, and (b) Equivalent model of SET.

8.5.2. Dielectrics in Single Electron Transistors

Single electron transistor (SET) is a three terminal device that works based on intrinsically quantum phenomenon ‘tunneling’. The tunneling phenomenon follows basic the principle of quantum mechanics ‘all the microscopic material particles in motion can have wave characteristics with a wavelength inversely related to the size of particle, thus an electron in motion can penetrate through a potential barrier of sufficiently small size due to the wave-particle duality’. The structure of SET device contains an electrically conductive quantum dot (QD) or island, which is the heart of device. Two electrodes namely ‘source’ and ‘drain’ to provide the bias, and a third terminal called ‘gate’ to control the flow of carriers through the QD (see Figure 7). The source and drain electrodes are separated from the QD by thin insulating tunnel junctions (TJ1 and TJ2), while a thick dielectric layer separates the gate from QD to avoid unwanted electron tunneling into gate [12-14]. The dielectric material influences the gate capacitance, which in turn plays a major role in charging the quantum dot. Thus, the charging energy (i.e., the energy needed to add one electron to the QD from source) of the QD is not just dependent on the QD material and electrode material, but also on the gate dielectric material. This section will elaborate the importance of the relative permittivity of dielectric material on the performance of SET.

Lee et al. [15] fabricated SETs utilizing MoS_2 nanosheet as quantum dot and zinc metal contacts. The authors studied the device performance for two different dielectrics viz. the conventional SiO_2 dielectric and Al_2O_3 high-k dielectric. The authors reported a charging energy of 22 meV for SiO_2 dielectric SET, and 17.6 meV for the Al_2O_3 dielectric SET (see Figure 8, the charging energy can be measured as half the height of central coulomb blockage diamond). This indicates a significant reduction in the energy needed to add an electron to the QD. This reduction in charging energy for Al_2O_3 dielectric SET can be attributed to the enhanced gate coupling coefficient (α), back gate capacitance (C_{bg}) and total capacitance ($C_\Sigma=C_{bg}+C1+C2$) in comparison to the SiO_2 dielectric SET. For the Al_2O_3 (SiO_2) dielectric SET, the gate coupling coefficient α is estimated as 0.0285 (0.00147), C_{bg} is estimated as 0.267 aF (0.0107 aF), and C_Σ is estimated as 9.37 aF (7.27 aF). Thus, the high-k dielectric reduced the charging energy by 20%, and increased the gate coupling by about 20 times.

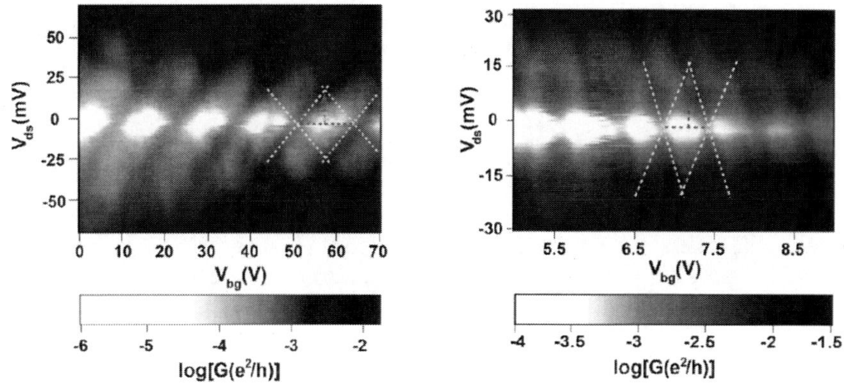

(a) SiO_2 **(b) Al_2O_3**

Figure 8. Conductance vs gate and source/drain voltages plot for (a) SiO_2, and (b) Al_2O_3 dielectric based SETs [15].

In another report, Anu et al. [16] designed SETs based on thiophene metal organic complex, and also studied the role of dielectric constant on the device performance. The authors have varied the relative permittivity of dielectric from 3.9 (SiO_2) to 7 (Si_3N_4), 15 (Y_2O_3), 25 (HfO_2, ZrO_2), 30 (La_2O_3), 50, and 80 (TiO_2). For each value of relative permittivity they have calculated the first and second ionization and affinity energies of the QD (see Table1). First affinity E_A (ionization E_I) energy indicates the energy needed to add (remove) a single electron to (from) QD. The second affinity E_A^{-1} (ionization E_I^{+1}) energy indicates the energy needed to add (remove) two electrons to (from) QD. In general, we require less first ionization/affinity energies and higher second ionization/affinity energies, for proper enabling of single electron tunneling. As seen from Table1, the second affinity (ionization) energies are always higher than the first affinity (ionization) energies, and both the ionization as well as affinity energies decrease with increasing relative permittivity. Among the considered dielectrics, TiO_2 offers the lowest first ionization and affinity energy indicating the least amount of required energy for an electron tunneling through the QD. Thus, a high relative permittivity of dielectric materials are beneficial for the SET.

Table 1. Ionization and affinity energies of thiol-ended thiophene SET for various gate dielectrics [16]

Ionization/Affinity Energy	Dielectric Material (relative permittivity)						
	SiO_2 (3.9)	Si_3N_4 (7)	Y_2O_3 (15)	HfO_2/ZrO_2 (25)	La_2O_3 (30)	- (50)	TiO_2 (80)
E_I	6.29	6.15	6.02	5.97	5.97	5.94	5.92
E_I^{+1}	9.24	8.76	8.36	8.20	8.16	8.07	8.03
E_A	-2.84	-3.02	-3.17	-3.23	-3.24	-3.27	-3.29
E_A^{-1}	-0.47	-0.99	-1.45	-1.63	-1.68	-1.78	-1.83

8.5.3. Dielectrics in Energy Storage

The energy can be stored either in the form of electrical, electrochemical, mechanical, or thermal energy, as depicted in Figure 9. This section focuses on the role of dielectrics in the electrical energy storage devices capacitors and supercapacitors, and the electrochemical storage device battery. These three devices have both advantages and disadvantages of their own as shown in Figure 10. All these devices require dielectrics for their operation, though their operation is slightly different. The capacitors use dielectric material as a separator for parallel plates, and the energy is stored in the form of an electric field between the parallel plates. The supercapacitors use dielectric material as solvent and the energy is stored in the form of Helmholtz layers. The batteries also use dielectric material as solvent, but the energy storage is done electrochemically in the bonds.

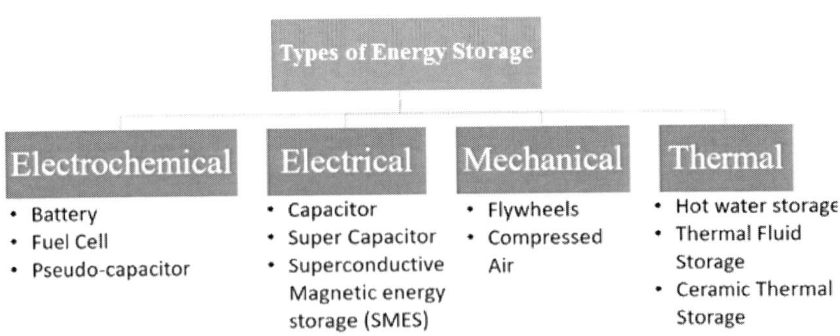

Figure 9. Different energy storage mechanisms.

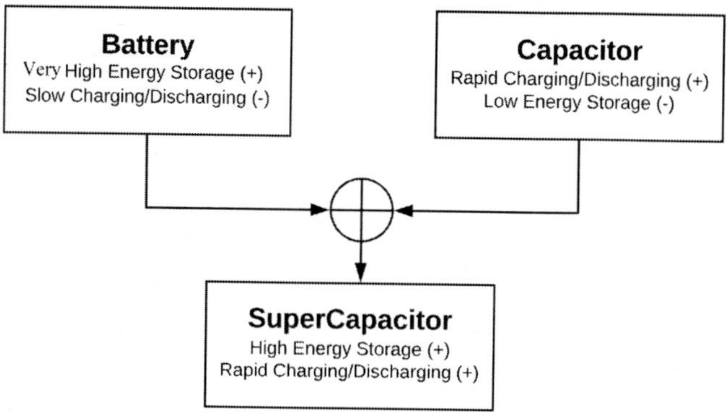

Figure 10. The attributes of capacitor, supercapacitor, and battery.

8.5.3.1. Capacitor

A parallel plate capacitor is made by separating two conducting plates with the help of a dielectric (see Figure 11). The capacitance of such a device depends on three parameters (i) permittivity of the dielectric (ε), (ii) area of the cross section of conducting plate (A), and (iii) distance of separation between the conducting plates (d), so, $C = \varepsilon\frac{A}{d}$. The permittivity (ε) can be further written as $\varepsilon = \varepsilon_r \varepsilon_0$. Where, ε_0 is the permittivity of free space ($8.854 \times 10^{-12} F/m$), and ε_r is relative permittivity or dielectric constant of the dielectric material. Table 2 shows the widely used dielectric materials of a parallel pate capacitor and their relative permittivity. A dielectric with high relative permittivity yields higher capacitance in a parallel plate capacitor. It is worth noting that the relative permittivity of a material is constant at low frequencies and varies over high frequencies and temperature [17].

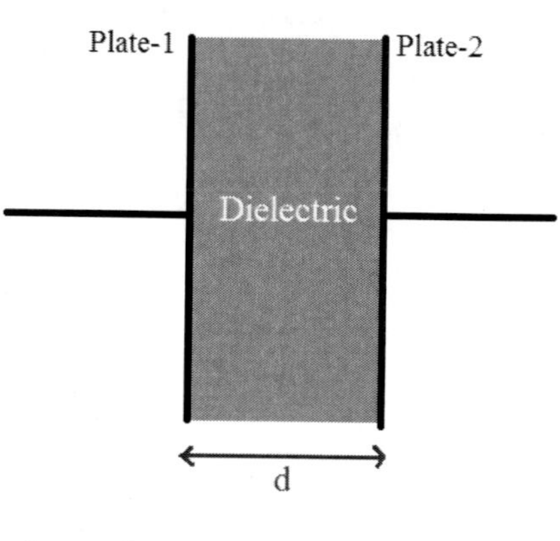

Figure 11. Schematic of a parallel plate capacitor.

Table 2. Commonly used capacitor dielectric materials and their relative permittivity (ε_r)

Material	Relative Permittivity (ε_R)
Air	1
Hydrogen	1
Rubber	2.1
Polypropylene	2.2-2.36
Parafinn wax	2-2.4
Polyimide	2.25
Paper	3.85
Silicon dioxide	3.9
Ceramic	4.5-6.7
Mica	5.6-8.0
Marble	8.3
Glass	5-10
Aluminium oxide	8.6
Ebonite	2.7-2.9
Silicon	11.68
Acetonitrile	38.8
Calcium titanate	150
Strontium titanate	200
Barium titanate	200-12000

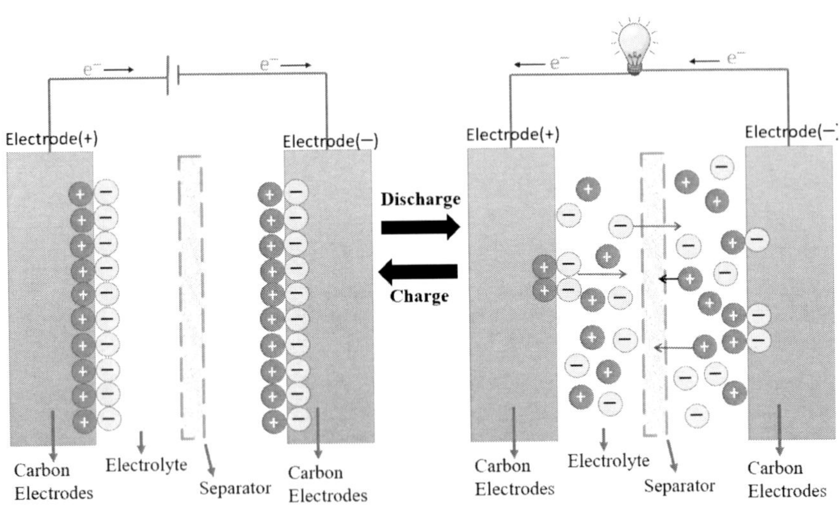

Figure 12. Schematic of supercapacitor/battery after full charge (left), and during discharge (right).

8.5.3.2. Supercapacitor and Battery

The structure of both the supercapacitors and battery looks the same (see Figure12), except that they use different kinds of electrodes and ions. Electrostatic Double Layer (EDL) supercapacitors use the physical movement of ions and the energy is stored in the Helmholtz layers, whereas the batteries use chemical bonding to store the energy. A typical EDL supercapacitor uses graphite electrodes, whereas a typical Li-ion battery uses a Lithium metal oxide cathode and a graphite anode. Both the supercapacitors and batteries use a dielectric solvent (inorganic solvent: water; Organic solvents: dimethyl carbonate (DMC), N,N dimethylformamide (DMF), and propylene carbonate (PC), tetraethylammonium tetrafluoroborate ($N(Et)_4BF_4$), etc. [18]) to solvate the solute ions. The solvent does not have free carriers hence it is electrically neutral. However, the application of potential difference causes the solute ions to move towards the opposite polarity electrodes, creating current flow through the device. The choice of solvent has to be made very carefully as it can dictate the energy density, electrochemical window, and stability/life of the device. As reported in [19, 20], the dielectric constant and viscosity are the two important properties of a solvent that can affect its ionic conductivity. The dissociation of salts in the solvent is decided by the dielectric constant, whereas the salt ion mobility is decided by the viscosity. The dissociation of salts is proportional to the dielectric constant, whereas the ionic mobility is inversely proportional to the viscosity. Hence, to get an appropriate solvent for the energy storage devices, a low viscosity solvent (ex: dimethyl carbonate) is usually mixed with high dielectric constant solvent (ex: ethylene carbonate). It has been reported elsewhere that the binary solvent based electrolytes offer better ionic conductivity than the single solvent based electrolytes [19]. In general, a low viscosity solvents also possess low dielectric constants [21]. Though, a lower viscosity improves the ionic mobility, the low dielectric constant associated with the same diminishes the ionic dissociation and encourages ion pairs, which degrades the ionic conductivity. Hence, binary solvents made by mixing the low viscosity solvent and high dielectric constant solvents are preferred.

8.5.4. Dielectrics in Solar Cells

A solar cell is an energy conversion device that converts sunlight into electric energy using the photovoltaic effect. It is also known as a photovoltaic cell (PVCell). After 50 years of the discovery of the photovoltaic effect in 1883, an American scientist Charles Fritz created the first working solar cell, with selenium. A solar cell is basically a p-n junction diode, whose electrical characteristics such as current, voltage, and resistance vary when exposed to light. Alike other devices, discussed in earlier section, in solar cells also, dielectrics play an important role in controlling their performance. The working mechanism of a solar cell is as follows.

In the solar cell, a p-type material is connected to n-type material, and they form a PN junction. When sunlight is incident on the cell, some of the photons of the light are absorbed by the cell, those absorbed photons which have energy greater than the energy gap, give energy to the electron. After getting the energy electron excites, and jumps out from the bond and creates one electron-hole pair. The light-generated electrons near the p-n junction move towards the n-side of the junction due to the electric field across the junction. Similarly, the holes move towards p-side of the junction. Hence, a potential difference is created between the two sides of junction. If an external circuit connects these two sides, the current starts to flow from the positive to negative terminal of the solar cell.

8.5.4.1. Performance Parameters of Solar Cell

(a) Short Circuit Current

Maximum current density or short circuit current density rating is the ratio of short circuit current to exposed surface area of the cell.

$$J_{SC} = \frac{I_{SC}}{A} \tag{2}$$

Where, I_{sc} is short circuit current, J_{sc} maximum current density and A is the area of a solar cell.

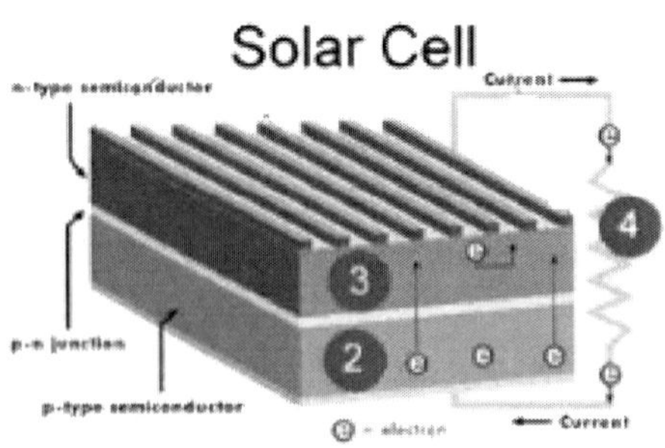

Figure 13. Solar cell model.

(b) Open Circuit Voltage

It is the voltage across the terminals of the cell when no load is connected to the cell. It is normally denoted by V_{oc}.

(c) Fill Factor

The ratio between the product of current and voltage at maximum power point to the product of short circuit current and open circuit voltage of the solar cell.

$$Fill\ Factor = \frac{P_m}{I_{sc} \times V_{oc}} \tag{3}$$

(d) Efficiency

It is defined as the ratio of maximum electrical power output to the radiation power input to the cell and it is expressed in percentages.

$$Efficiency\ (\eta) = \frac{P_m}{P_{in}} \tag{4}$$

8.5.4.2. Role of the Dielectric to Realize Highly Efficient Solar Cell

Nanostructures are being widely used in solar cells due to their extraordinary photon management properties. However, nanostructured solar cells are not as efficient as commercial solar cells due to the poor pn junction quality and high surface recombination velocity.

In recent years, low-cost dielectric materials with a large bandgap (above 3 eV) are widely used in c-Si solar cells, such as silicon nitride (SiN_x), aluminum oxide (Al_2O_3), and silicon dioxide (SiO_2) [22]. In addition to these, dielectric materials have been reported to provide excellent surface passivation for c-Si solar cells [23]. Therefore, by nanostructuring the dielectric materials, antireflection and light trapping could be achieved without sacrificing the junction quality and surface passivation. Hence, in a report [24], a new approach is adopted to design the whole-wafer nanostructures of dielectric layer on thin c-Si for solar cell light trapping and results show that the periodic nanostructure arrays on dielectric materials could suppress the reflection loss over a wide spectral range. Moreover, by applying the nanostructured dielectric layer on 40 μm thin c-Si, the reflection loss is suppressed to below 5% over a wide spectra and angular range. These results suggest that the dielectric layer has significantly improved solar cell performance and efficiency. Hence, the dielectric materials are required for high-efficiency and low-cost c-Si solar cells.

Dielectric materials also affect the open circuit voltage of organic super cells, in an experimental report [25], the studies on the effects of dielectric constant on open circuit voltage (V_{oc}) have been performed on P3HT: PCBM based solar cells, and the results suggest that the dielectric constant of the bulk heterojunction can control the V_{oc} in the non-thermal annealing treatment solar cells. Hence, the controlling of the medium dielectric constant forms a new mechanism to reduce the V_{oc} loss toward device efficiency improvement in organic solar cells. On the other hand, the exciton energy of solar cell also depends on the dielectric constant. As we know, the dielectric constant of inorganic semiconductors are low. The low dielectric constant of organic semiconductors (OSCs) can lead to the formation of polarons and Frenkelexcitons, resulting in high exciton binding energies [23], in comparison to the high-k dielectric materials. The low dielectric constants in organic semiconductors can enhance the Coulomb interaction between holes and electrons, thereby increasing the attraction and repulsion effects. In the operation of a solar cell, the Coulomb interactions are determined by the dielectric constant negatively affect charge separation and prompts charge recombination. Therefore, an increase of the dielectric constant of solar cell is generally preferred for OSCs due to several merits like, it reduces bimolecular recombination as well as reorganization energy, it also reduces the exciton binding energy, and Coulomb attraction between charge transfer (CT) excitons.

In the OSCs, the exciton binding energy $E_b$is sensitive to the dielectric constant. The exciton binding energy of follows the relation [23],

$$E_b = \frac{q^2}{4\pi\varepsilon_0\varepsilon_r} \tag{5}$$

Here, q is the elementary charge, r is the separation distance between the electron-hole pair. From equation (5), it is quite clear that increasing ε_r will reduce the binding energy of CT excitons, if r is constant.

In another report [26], a generic absorber/buffer/window solarcell architecture of solar cell has been used to analyse the effect of dielectric material in different scenarios, which affects the power conversionefficiency significantly. First is effect of dielectric on the depletion width, the depletion width of P-N heterojunction is given by

$$W = \left[\frac{2V_{bi}\varepsilon_n\varepsilon_p(N_d+N_a)^2}{eN_dN_a(\varepsilon_nN_d+\varepsilon_pN_a)}\right]^{1/2} \tag{6}$$

Here, W is the total width of the depletion region, n and p are the static permittivity's of the n- and p-type semiconductors, V_{bi} is the built-in voltage in equilibrium, e is the elementary charge and N_a and N_d are the acceptor and donor densities of the p- and n-type materials, respectively. If both light absorption and collection efficiency are low in the n-type heterojunction, then the depletion region lies in the p-type material, which has a width

$$W_p = \left[\frac{2V_{bi}\varepsilon_n\varepsilon_pN_d}{eN_a(\varepsilon_nN_d+\varepsilon_pN_a)}\right]^{1/2} \tag{7}$$

Which predicts that the width of the depletion region increases asthe permittivity of the two heterojunction materials increases. If one of the two sides of the junction is heavily doped with respect to the other, then only the permittivity of the lightly-doped side of the junction responsible for depletion region widening. In another scenario permittivity also affects the built-in voltage, by the relation [26], describing the effect of the materials permittivity on device performance is the following:

$$\frac{V_{bi,n}}{V_{bi,p}} = \frac{\varepsilon_pN_a}{\varepsilon_nN_d} \tag{8}$$

Here, $V_{\mathrm{bi,n}}$ and $V_{\mathrm{bi,p}}$ are the built-in voltage drops in the n-type and p-type junction materials.

Conclusion

This chapter discusses the fundamental aspects related to permittivity, relative permittivity or dielectric constant, and the role of dielectrics in electronic devices (Field Effect Transistors (FETs) and Single Electron Transistors (SETs)), energy storage devices (capacitor, supercapacitor, and battery) and solar cells. The dielectrics can be broadly classified into high-k (materials with a dielectric constant greater than SiO_2) and low-k (materials with a dielectric constant lesser than SiO_2), in the vicinity of silicon electronics. As the technology is scaled into nano-regime, the conventional SiO2 dielectric is no longer relevant due to the increasing gate leakage. On the other hand, the usage of high-k dielectrics are observed to enhance the transconductance and short channel effects of the FETs. Likewise, the usage of high-k dielectrics in SETs reduces the charging energy, and increases the gate coupling coefficient, gate capacitance, and total capacitance of the device. Similarly, the high -k dielectric have low exciton energy as well as can make highly efficient devices.Moreover, the role of dielectrics in a capacitor and the choice of solvent for supercapacitors and batteries have been discussed in reference to the dielectric constant. To attain a high ionic conductivity and high performance of supercapacitors/batteries, it is suggested to use binary solvents by mixing low viscosity solvents and high dielectric constant solvents.

References

[1] Taur, Y., and Ning, T. H. (2021).*Fundamentals of modern VLSI devices. Cambridge university press.*

[2] Mech, B.C., and Kumar, J. (2017). Effect of high-k dielectric on the performance of Si, InAs and CNT FET. *Micro & Nano Letters*, 12(9), pp.624-629.

[3] Mondal, S., and Kumar, A. (2016). Tunable dielectric properties of TiO2 thin film based MOS systems for application in microelectronics. *Superlattices and Microstructures*, 100, pp.876-885.

[4] Jing, K.H., Arshad, M.M., Huda, A.R.N., Ruslinda, A.R., Gopinath, S.C., Ayub, R.M., Fathil, M.F.M., Othman, N., and Hashim, U. (2016), July. Gate dielectric scaling in MOSFETs device. *In AIP Conference Proceedings* (Vol. 1733, No. 1, p. 020073). AIP Publishing LLC.

[5] Oswal, R. (2014). Investigation of different dielectric materials as gate insulator for MOSFETs.

[6] Robertson, J. (2005). Interfaces and defects of high-K oxides on silicon. *Solid-State Electronics*, 49(3), pp.283-293.

[7] Robertson, J. (2011). New high-k materials for CMOS applications.

[8] Robertson, J. (2002). Band offsets of high dielectric constant gate oxides on silicon. *Journal of non-crystalline Solids*, 303(1), pp.94-100.

[9] Robertson, J. (2004). High dielectric constant oxides. *The European Physical Journal-Applied Physics*, 28(3), pp.265-291.

[10] Kosmani, N.F., Hamid, F.A., and Razali, M.A. (2020). Effects of high-k dielectric materials on electrical performance of double gate and gate-all-around mosfet. *International Journal of Integrated Engineering*, 12(2), pp.81-88.

[11] Santhibhushan, B., Dobwal, P., and Srivastava, A. (2015). TCAD Assessment of Dielectrics and Channel Doping Impact in Junctionless Double Gate MOSFET. *Journal of Computational and Theoretical Nanoscience*, 12(8), pp.1924-1930.

[12] Srivastava, A., Santhibhushan, B., and Dobwal, P. (2014). Performance analysis of impurity added benzene based single-electron transistor. *Applied Nanoscience*, 4(3), pp.263-269.

[13] SanthiBhushan, B., Khan, M.S., Srivastava, A., and Khan, M.S. (2016). First principle analysis of (10-Boranylanthracene-9-yl) borane-based molecular single-electron transistor for high-speed low-power electronics. *IEEE Transactions on Electron Devices*, 63(3), pp.1232-1238.

[14] Gaurav, K., SanthiBhushan, B., Ray, S.J., and Srivastava, A. (2019). Acridinium Based Organic Molecular Single Electron Transistor for High Performance Switching Applications. *IEEE Transactions on Nanotechnology*, 18, pp.1148-1155.

[15] Lee, K., Kulkarni, G., and Zhong, Z. (2016). Coulomb blockade in monolayer MoS 2 single electron transistor. *Nanoscale*, 8(14), pp.7755-7760.

[16] Anu, A. Sharma, Khan, M.S., Srivastava, A., Husain, M., and Khan, M.S. (2017). High-performance single-electron transistor based on metal–organic complex of thiophene, First principle study. *IEEE Transactions on Electron Devices*, 64(11), pp.4628-4635.

[17] Giusti, G., Tian, L., Jones, I.P., Abell, J.S., and Bowen, J.(2009). Microstructure–Property relationships in thin film ITO. *Thin Solid Films*, 518(4), pp.1140-1144.

[18] Xiao, D., Dou, Q., Zhang, L., Ma, Y., Shi, S., Lei, S., Yu, H., and Yan, X. (2019). Optimization of organic/water hybrid electrolytes for high-rate carbon-based supercapacitor. *Advanced Functional Materials*, 29(42), p.1904136.

[19] Kamath, G., Cutler, R.W., Deshmukh, S.A., Shakourian-Fard, M., Parrish, R., Huether, J., Butt, D.P., Xiong, H., and Sankaranarayanan, S.K.(2014). In silico based rank-order determination and experiments on nonaqueous electrolytes for sodium ion battery applications. *The Journal of Physical Chemistry C*, 118(25), pp.13406-13416.

[20] Pal, B., Yang, S., Ramesh, S., Thangadurai, V., and Jose, R.(2019). Electrolyte selection for supercapacitive devices: A critical review. *Nanoscale Advances*, 1(10), pp.3807-3835.

[21] Che, H., Chen, S., Xie, Y., Wang, H., Amine, K., Liao, X.Z., and Ma, Z.F.(2017). Electrolyte design strategies and research progress for room-temperature sodium-ion batteries. *Energy & Environmental Science*, 10(5), pp.1075-1101.

[22] Fraas, L.M., and Partain, L.D. (2010). *Solar cells and their applications* (Vol. 236). John Wiley & Sons.

[23] Fu, Z., Zhang, X., Zhang, H., Li, Y., Zhou, H., and Zhang, Y. (2021). On the understandings of dielectric constant and its impacts on the photovoltaic efficiency in organic solar cells. *Chinese Journal of Chemistry*, 39(2), 381-390.

[24] Chen, Y., Kang, Y., Jia, J., Huo, Y., Xue, M., Lyu, Z., Liang, D., Zhao, L., and Harris, J.S. (2017). Nanostructured dielectric layer for ultrathin crystalline silicon solar cells. *International Journal of Photoenergy*, 2017.

[25] Zang, H. (2012). Magnetic, Optical and Dielectric Effects on Photovoltaic Processes in Organic Solar Cells.

[26] Crovetto, A., Huss-Hansen, M.K., and Hansen, O. (2017). How the relative permittivity of solar cell materials influences solar cell performance. *Solar Energy, 149*, pp.145-150.

Chapter 9

Dielectric Properties and Electromagnetic Interference (EMI) Shielding Behavior of Conducting Polymer-Based Hybrids and Composites

Parveen Saini*
Conducting Polymers, Graphene Technology and Waste Management Group, Photovoltaic Metrology Section, Advance Materials and Devices Metrology Division, CSIR-National Physical Laboratory, Dr. K. S. Krishnan Marg, New Delhi, India

Abstract

The explosive growth of portable electronics and telecommunications has resulted in another kind of pollution popularly known as electromagnetic interference (EMI) as a byproduct. The cross talk and EMI among the electronic gadgets tends to degrade their performance and exert adverse effects on organisms. In the past, conducting polymer based hybrids and composites have emerged as potential candidate as EMI countering material due combination of electrical, dielectric/magnetic and physico-mechanical properties. In such systems permittivity and related dielectric properties plays very important role in governing the EMI shielding performance. This chapter enlightens the fundamentals of EMI shielding, measurement techniques, governing equations and potential material candidates with special reference to polymer based hybrids and composites and correlation between their dielectric properties and EMI shielding performance.

* Corresponding Author's Email: pksaini@nplindia.org.

In: An Introduction to Permittivity
Editor: Parveen Saini
ISBN: 979-8-88697-042-5
© 2022 Nova Science Publishers, Inc.

Keywords: electromagnetic interference (EMI) shielding, microwave absorption, electrical conductivity, hybrids, composites, conducting polymers, polyaniline, carbon nanotubes, graphene, shielding effectiveness, reflection loss, absorption loss, complex permittivity, dielectric constant, dielectric loss

9.1. Introduction

EMI shielding is recognized as the offshoot of proliferation of portable electronics, telecommunications and instrumentation [1, 2]. It's a novel yet silent & invisible kind of pollution which not only tries to degrade the performance of appliances but may also adversely affect the health of human beings [3, 4]. This requires suitable countermeasures to protect the devices and humans form EMI effects by developing efficient and viable EMI shielding technologies & materials. Consequently, over the past three decades, a number of strategies and materialistic candidates including conducting materials, dielectric/magnetic systems and combination thereof in terms of hybrids and composites have been developed [5–19]. Particularly, dedicated efforts have been made to utilize the good properties of above materials by using them in various combinations especially the polymer based hybrids or composites wherein the polymers are used in combination with metallic, dielectric or magnetic materials [7]. As, the selection of materials, the design of shield structure and realization of intended performance, require basic understanding of shielding phenomenon and influence of dielectric properties on shielding performance, these are comprehensively discussed in the next section.

9.2. EMI Shielding: Process and Governing Equations

The EMI shielding is the process of achieving a certain level of radiation attenuation (or reduction in magnitude/strength) using a suitably designed shield (Figure 1) placed in the path of incident electromagnetic (EM) radiation. In electromagnetics, the term shield specifically refers to an enclosure that completely encloses an electronic product or a portion of that product, and acts as a barrier to the transmission of the EM radiations across its boundaries. The EM radiation blocking efficiency or shielding effectiveness (SE) of a shield is measured in terms of reduction of strength of incident radiation upon transmission through the shield, expressed on a logarithmic scale as [1, 3, 9, 20]:

$$SE_T\,(dB) = SE_R + SE_A + SE_M = 10\log_{10}\left(\frac{P_T}{P_I}\right) = 20\log_{10}\left(\frac{E_T}{E_I}\right) = 20\log_{10}\left(\frac{H_T}{H_I}\right) \quad (1)$$

where P_I (E_I or H_I) and P_T (E_T or H_T) are the power (electric or magnetic field intensity) of incident and transmitted EM waves respectively.

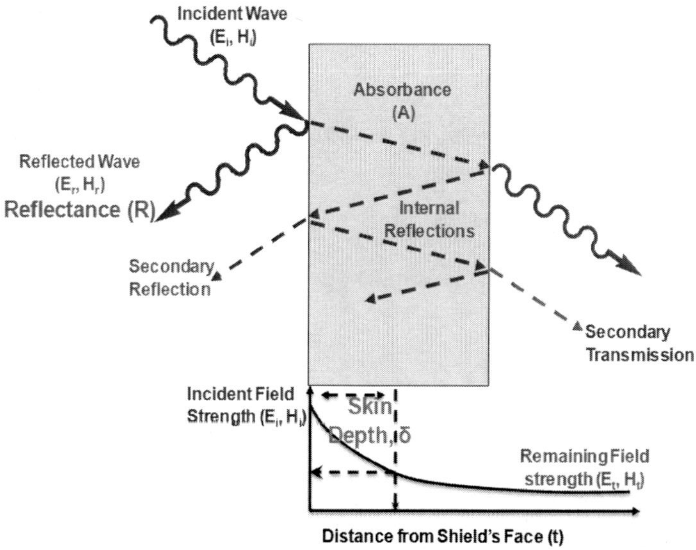

Figure 1. Schematic representation of EMI shielding phenomenon, shield/radiation interaction, and related reflection, absorption and multiple reflection processes.

It can be seen that three different radiation blocking processes viz. reflection (SE_R), absorption (SE_A) and multiple reflections (SE_M) contribute towards overall SE_T, through the interaction of radiation with the interfaces and bulk of shield material. The intensity of radiation falls exponentially as it passes through the thickness (t) of the shield, such that characteristic thickness (δ) known as skin depth can be identified where the incident radiation falls to 37% of its original value.

9.2.1. Theoretical Shielding Effectiveness

According to the distance (r) between the radiating source and the observation point, an electromagnetic radiative region can be divided into three parts relative to the wavelength λ of the electromagnetic wave [7, 9]. The region within the distance $r < \lambda/2\pi$ is the near field while the distance $r > \lambda/2\pi$ is the far field. Most cases and situation involve the far field situation where the radiation may be treated as plane wave. Under this regime, the theoretical shielding sub-processes (reflection, absorption and multiple-reflection phenomenon) can be related with the electromagnetic attributes (permittivity/permeability) & thickness of shield and frequency of incident radiation, by specific empirical relations, as presented in next section.

9.2.1.1. Shielding Effectiveness Due to Reflection

The shielding effectiveness due to reflection (SE_R) is related to the relative mismatch between the incident wave and the surface impedance of the shield. The magnitude of reflection loss good conductor approximation can be given by following generalized expressions [1, 9]:

$$SE_R = C + 10\log_{10}\left(\frac{\sigma_r}{\mu_r}\cdot\frac{1}{f^n r^m}\right) \qquad (2)$$

where σ is the relative conductivity referred to copper; f is the frequency in Hz; μ is the relative permeability referred to free space; r is the distance from the source to the shielding in meter. The values of constants C, n and m are listed in Table 1 for calculating reflection losses (in dB) for plane waves, electric fields, and magnetic fields, respectively.

Table 1. Constants for predicting SE_R under plane wave, magnetic and electric fields

Type of field	C	n	m
Electric	322	3	2
Plane wave	168	1	0
Magnetic	14.6	-1	-2

Therefore, SE_R for a plane wave radiation it can be expressed as:

$$SE_R = -10\log_{10}\left(\frac{\sigma}{16\omega\varepsilon\mu_r}\right) \quad (3)$$

The above expression shows that SE_R is a function of the logarithmic ratio of conductivity (σ) and permeability (μ) of the shield material i.e., quantity (σ/μ).

9.2.1.2. Shielding Effectiveness Due to Absorption

Shielding effectiveness due to absorption (SE_A) is a function of the physical characteristics of the shield and unlike SE_R, it is independent of the type of source field. Therefore, the term SE_A is same for all three waves. As shown in Figure 1, when an electromagnetic wave pass through a medium its amplitude decreases exponentially. This decay or absorption loss occurs because currents induced in the medium produce ohmic losses and heating of the material, such that E_t and H_t can be expressed as $E_t = E_i e^{-t/\delta}$ and $H_t = H_i e^{-t/\delta}$ [1, 7]. The distance required by the wave to be attenuated to 1/e or 37% is defined as the skin depth (δ) which can be expressed as:

$$\delta = \frac{1}{\sqrt{\pi\mu f\sigma}} \quad (4)$$

Therefore, the magnitude of absorption term (SE_A) in decibel (dB) can be expressed by following generalized equation [1, 7]:

$$\mathrm{SE_A} = 3.34t\sqrt{f\sigma_r\mu_r} \quad (5)$$

Where t is shield thickness in inches and f is frequency in Hertz. The parameters 'μ_r' and σ_r represent conductivity and permeability respectively relative to copper. The rearrangement of above equations gives following expression for SE_A in decibel (dB):

$$\mathrm{SE_A} = -20\left(\frac{t}{\delta}\right)\log_{10} e = -8.68\left(\frac{t}{\delta}\right) = -8.68t\sqrt{\sigma f\mu} \quad (6)$$

The above expression revealed that SE_A is proportional to the square root of the product of the permeability (μ) times the conductivity (σ) of the shield material i.e., quantity $(\sigma.\mu)^{1/2}$. Therefore, a good absorbing material should possess high conductivity and high permeability, and sufficient thickness to achieve the required number of skin depths at the lowest frequency of concern.

9.2.1.3. Multiple Reflections

If the shield is thin, the reflected wave from the second boundary is re-reflected off the first boundary, and then it returns to the second boundary to be reflected again, as shown in Figure 1. This process can be neglected in the case of a thick lossy shield, because the absorption loss is high enough to completely attenuate the radiation by the time it crosses the shield thickness. The factor SE_M can be mathematically expressed as [7]:

$$SE_M = 20\log_{10}(1 - e^{-2t/\delta}) \tag{7}$$

Therefore, when SE_A is $\geq$ 10 dB [5], SE_M can be safely neglected. In practical calculation, SE_M can also be neglected for electric fields and plane waves. However, in case of thin less lossy samples or porous systems, the re-reflections can't be neglected and plays important role in governing the overall attenuation.

9.2.2. Experimental Shielding Effectiveness

Experimentally, SE can be measured using instrument called vector network analyzer (VNA), wherein the incident and transmitted waves can be mathematically represented by complex scattering parameters (or S-parameters) i.e., S_{11} (or S_{22}) and S_{12} (or S_{21}) respectively [7]. These parameters can be conveniently correlated with reflectance (R) and transmittance (T) i.e., $T = |E_T/E_I|^2 = |S_{12}|^2 = |S_{21}|^2$, $R = |E_R/E_I|^2 = |S_{11}|^2 = |S_{22}|^2$, giving absorbance (A) as: A = (1-R-T). When SE_A is greater than 10 dB, SE_M becomes negligible (~ -1.0 dB) and can be safely neglected [15] so that SE_T can be expressed as: $SE_T = SE_R + SE_A$. In addition, the intensity of the EM wave inside the shield after primary reflection is based on quantity (1-R), which can be used for normalization of absorbance (A) to yield effective absorbance as: $A_{eff} = [(1\text{-}R\text{-}T)/(1\text{-}R)]$. Therefore, experimental reflection and absorption losses can be expressed as [9, 20]:

$$SE_R = 10\log_{10}(1 - R) \tag{8}$$

$$SE_A = 10\log_{10}(1 - A_{eff}) = 10\log_{10}\left[\frac{T}{(1-R)}\right] \tag{9}$$

Therefore, from the knowledge of Reflected and transmitted signals, VNA can easily compute reflection and absorption loss components of total shielding. In addition to these parameters, the dielectric & magnetic properties i.e., complex permittivity & permeability can be computed using Nicolson-Ross-Weir algorithm, which correlates these complex quantities with the experimentally measured S-parameters.

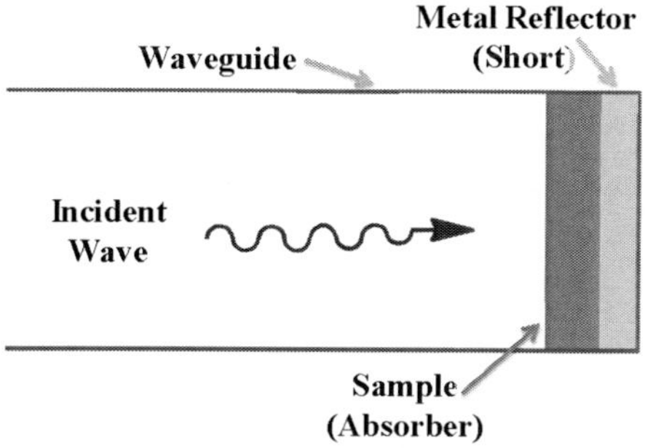

Figure 2. Schematic of testing configuration in single port measurement.

In addition, many times, single port technique is used to assess the EM radiation absorption capability of the material. In such measurement, a metal short is used to terminate the waveguide end such that the incident signal is directed towards the metal-short end (Figure 2) and the reflected signal is recorded in the absence and presence of sample (absorber). Finally, the difference in the strength of reflected signal (R') can be used to compute the return loss or reflection loss (R_L) as [7]:

$$R_L = 10\log_{10}R' = 10\log_{10}(S_{11}')^2 = 20\log_{10}(S_{11}') \tag{10}$$

where R' and S_{11}' are reflectance and corresponding scattering parameter in single port configuration.

9.2.3. Electromagnetic Attributes

The electromagnetic attributes of a shield like complex permittivity [ε^* = (ε'-jε'')] and permeability [μ^* = (μ'-jμ'')] can be retrieved from the experimental S-parameters [S_{11} (or S_{22}) and S_{21} (or S_{12})] using suitable algorithms and models like Nicolson-Ross-Weir (NRW) [both ε_r and μ_r], NIST iterative [ε_r and $\mu_r = 1$], New non-iterative [ε_r and $\mu_r = 1$] and Short circuit line (SCL) [only ε_r] techniques [7]. Each of these conversion technique has different advantages and limitations. The selection of the technique depends on several factors such as the measured S-parameters, sample length, desired output properties, speed of conversion and accuracies in the converted results. Among above-mentioned procedures, Nicholson-Ross-Weir (NRW) technique is the most widely used regressive/iterative analysis as it provides direct calculation of both the permittivity (ε^*) and permeability (μ^*) from the input S-parameters [7, 9, 21, 22]. Parameter ε' or ε_r (real permittivity) represents the charge storage (or dielectric constant) whereas ε'' (imaginary permittivity) is a measure of dielectric dissipation or losses. Similarly, μ' (or ε_r) and μ'' represents magnetic storage and losses respectively.

9.2.4. Shield Parameters and Material Considerations

The theoretical shielding expressions revealed that in order to meet design requirements and for extending efficient shielding action, shield may possess mobile charge carriers (electrons or holes) which may result in reflection losses as well as ohmic losses into the system (causing heat dissipation), thus resulting in attenuation. Shield may also possess electric and/or magnetic dipoles which can interact with the orthogonally pulsating electric (E) and magnetic (H) vectors of the incident EM radiation [4, 9], thus resulting in corresponding losses via absorption phenomenon. Another mechanism of shielding is multiple refection for which material should display porous/pseudo-porous architecture consisting of thin electrically conducting regions separated by low loss medium. In practical situations, these mechanisms may coexist with relative dominance of one process over other, thus deciding the preferred mechanism and attained attenuation. Among conducting materials, conventional metals are widely exploited for EMI shielding [1, 7] primarily due to high electrical conductivity in addition to high strength and good mouldability. However, their disadvantages like intrinsic high reflectivity (thus dominance of reflection i.e., >99% reflection), corrosion proneness, high density and energy intensive processing, limit their use for practical applications. The carbon based materials have shown some advantages above metals, but for real life utility they need to be exploited in the form of hybrid/composite system [7]. Similarly, the conducting polymers also need to be combined with other materials to realize practically viable EMI shielding attributes [7, 9, 10]. In general, the materials need to be combined in the form of alloys, blends, hybrids, composites, to realize the desired electromagnetic attributes, dominance of absorption, acceptable EMI shielding performance and requisite physico-mechanical properties. The next section deals with the details of polymer based hybrids and composites, their electromagnetic response and EMI shielding performance, using suitable examples form available literature.

9.2.4.1. Polymer Based Hybrids and Composites

Hybrids represent mixed phase systems wherein the one or more phases are mixed together (and may or may not be chemically linked) such that the properties are derived from the constituent phases. The hybrids are more complex system and different from composites wherein two non-chemically interacting material constituents viz. the matrix (or continuous phase) and reinforcement (or dispersed phase) can be identified [7]. Besides, a third phase so-called interfacial region is also present which is responsible for communication between the matrix and filler [4]. Interface possesses unique combination of properties that are not exhibited by filler or matrix alone and is responsible for the macroscopic properties of formed composites. In case of bulk composites, the concentration of interfacial regions is too low compared to matrix & reinforcements and often neglected. However, in case of nanocomposites, fillers possess nanometer (1 nm = 10^{-9} m) range dimensions (~10,000 times finer than a human hair) and extend ultra-large interfacial area per volume to host polymer matrix. Therefore, at nanoscale level, the concentration of bulk and interfacial regions become comparable. Consequently, significant changes in the properties of nanocomposites are observed compared to their bulk counterparts such as enhanced strength, better optical properties, good electrical properties, improved thermal/oxidative stability and enhanced EMI shielding performance [7]. In polymer based hybrids and composites two types of systems are identified, one wherein various fillers (metallic particles, carbon materials, conducting polymers or dielectric/magnetic particulates) are incorporated within the conventional insulating polymer skeleton [7, 10–12, 14, 23–26]. In other system, various fillers (metallic particles, carbon materials or dielectric/magnetic particulates) are incorporated inside conducting polymer matrix/framework [2, 6, 7, 11, 15–18, 27–32]. The properties of these systems are governed by the nature/properties of constituents, their relative proportion and interaction between them. The most insulating polymeric systems possess poor electrical, dielectric or magnetic properties and are transparent to electromagnetic radiations, such that the electrical and electromagnetic properties of the hybrids/composites are governed by the filled inclusions Therefore, the selection of a particular filler depends on the desired properties e.g., when electrical conduction is important electrically conductive fillers are used, whereas for dielectric/magnetic properties, fillers with magnetic/electric dipoles are preferred [7]. In case of conducting polymer based framework, the system itself is non-transparent towards incident EM radiation, and tends to attenuate the same by reflecting & absorption processes. However, a variety of conducting or dielectric/magnetic fillers may still be incorporated to manipulate the attenuation level and dominant shielding mechanism. Therefore, in the recent past a numerous effort have been made to combine various phases like insulating polymers, conducting polymers, carbon black, graphite, carbon nanotubes (CNTs), graphene, and dielectric or magnetic particulates [7, 9, 11, 15, 17, 19, 25, 28–30, 32–37] to form EMI shielding hybrids and composites. As already pointed, metals are considered most popular materials for EMI shielding, when there is no boundation of absorption dominance and only electromagnetic isolation is main aim. Due to high electrical conductivity and reflection dominated high attenuation for a very thin shields [7]. It is to be noted that for a highly conducting material (e.g., metals like Cu, Ag or Ni), only conductivity (σ) and magnetic permeability (μ) are important, such that the reflection loss (SE_R) is dependent upon their ratio (i.e., σ/μ) whereas the absorption loss (SE_A) is a function of their product (i.e., $\sigma.\mu$) [9]. Therefore, highly conducting and magnetic metals (ε'~1) are expected to display enhanced absorption and reduced reflection, which is indeed the case for ferromagnetic materials like Ni, permalloy, superpermally (compared to non-magnetic metals like Cu). However, in the case of moderately conducting materials (e.g., alloys, hybrids, composites, blends etc.) permittivity (ε') also plays a significant role (besides σ and μ) in deciding absolute values of SE_R and SE_A [7]. Such compounds are also capable of displaying dynamic dielectric and/or magnetic losses (i.e., ε'' and μ''), upon impingement by incident electromagnetic waves. Nevertheless, metal based compositions are suffered from such as high reflectivity, corrosion susceptibility, weight penalty and uneconomic processing. Among other alternatives, carbon based materials (graphite, expanded graphite, carbon black, carbon nanotubes and graphene) have also been widely explored for possible applications in EMI shielding [7]. However, graphite shows poor dispersion and high percolation threshold inside polymeric matrices. Similarly, graphene and CNTs are economically non-viable, difficult to produce at bulk scale, involve tedious purification procedures, require cumbersome functionalization steps and have toxicity issues. Besides, being a nanofiller, their dispersion issues and inherent agglomeration tendency, poses serious drawback limiting incorporation of not more than 1-2 weight % filler inside host matrices, such that achieved EMI shielding levels remained too low to satisfy attenuation requirements of >20dB [7]. In this context, highly doped and metallic forms of conducting polymers better known as intrinsically conducting polymers (ICPs), have emerged as promising candidate due to scaling of their conductivity with frequency (e.g., in the microwave region of 20 MHz – 100 GHz), which motivated to exploit them for microwave applications [4, 7, 38, 39]. They possess unique shielding mechanism of reflection plus absorption (due to doping induced inherent polarization) rather than dominated reflection for metals and carbons. Particularly, the unique properties like doping induced tunable conductivity (between insulating and metallic limits), adjustable permittivity/permeability via synthetic means (including through accomodability for fillers), low density, non-corrosiveness, facile processing (via melt or solution routes), and controllable electromagnetic attributes, further strengthen their candidature as futuristic shielding materials [7]. Although pure ICPs have been employed as EMI shielding materials, their mechanical and EM properties can be further improved by formation of blends (by incorporation inside various solution processable or thermoplastic host polymeric matrices) or advanced composites (via combining with suitable electrically conducting, dielectric/magnetic filler or high performance polymeric hosts like thermosetting polymers) [4, 7, 9]. It should be noted that ICPs can be used either as conducting filler (guest) for various insulating matrices (hosts) or as an electrically conducting matrix (host) for variety of filled inclusions (guests) viz. electrically conducting [e.g., metals, graphite, carbon back, carbon nanotubes, graphene], dielectric (e.g., $BaTiO_3$ or TiO_2) or magnetic (e.g., γ-Fe_2O_3, Fe_3O_4, $BaFe_{12}O_{19}$) inclusions. The, utilization of ICPs as host matrix is considered to be superior choice compared to conventional (insulating) polymer based matrices primarily due to microwave non-transparency and design flexibility. The next section is focused on the recent progresses in the dielectric properties and EMI shielding performance of ICPs based hybrids and composites.

9.2.4.1.1. EMI Shielding Performance of ICPs Loaded Hybrids/Composites

In many cases, highly doped ICP particles are used as conducting fillers (in place of metal or carbon based materials) for various insulating polymer host matrices to form electrically conducting blends [2, 4, 7, 10, 11, 12, 19, 23–26, 40]. It is speculated that, like metal/carbonaceous materials filled blends/composites, these conducting systems also provide EM attenuation due to electrical conductivity value as well as due to interfacial charge localization induced polarization & dielectric losses. The Poly(3-octyl thiophene) based blends with molten polystyrene (PS), polyvinyl chloride (PVC) and ethylene-co-vinyl acetate (EVA) matrix was shown to display good attenuation values [24]. Similarly, conducting thermoplastic PANI/PVC and PANI/Nylon blends were shown to display far field SE of -70 dB, with good agreement with the theoretical calculations [10, 23]. Again both polarization and conductivity plays the role in imparting high attenuation values. In addition to these works, many other studies revealed that for a given ICP–polymer matrix system, the electrical properties of the resultant composites critically depend on nature of ICP, its loading level and the processing technique. The increasing filler loading is shown to enhance the number of conducting links with proportionate improvement of inter-particle charge transport as well as proportion of interfaces (and related interfacial polarization & loses), ultimately resulting in improvement of electrical conductivity as well as permittivity and loses [4, 7, 9]. Studies have shown that combination of dielectric particulates such as BaTiO3, polyaniline and conducting carbon in polyurethane (PU) matrix [36] result in good complex permittivity and microwave absorption properties (Figure 3) in X-band (8.2–13.5 GHz).

An optimized composite sample with synthesized BaTiO3 and polyaniline exhibited a maximum reflection loss of -25 dB (>99% power absorption) at 11.2 GHz with a bandwidth of 2.7 GHz at sample thickness of 2.5 mm. The measured absorption values have been validated by theoretical calculations. In addition to combination of fillers, it has also been observed that the selection of proper processing technique facilitates the dispersion of ICPs particles, enabling percolation threshold realization at relatively lower loading level.

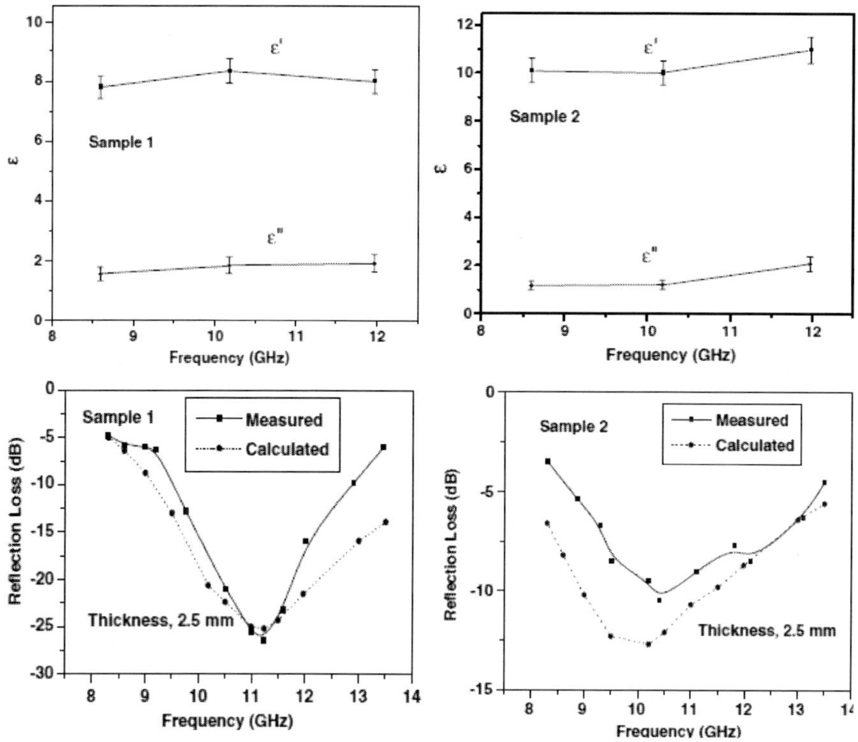

Figure 3. Complex permittivity (ε' & ε'') spectra and measured/calculated reflection loss spectra of composite sample-1 and sample-2 [36].

For example, at any given loading level, a well dispersed ICP by absorption transferring process [14] (Figure 4) resulting in higher conductivity compared to that less dispersed system formed by blending process. It is worth mentioning that incorporation of ICP in polymeric matrices not only leads to establishment and improvement of electrical conductivity but also contribute towards improvement of dielectric properties (e.g., real/imaginary permittivity, Figure 4a and 4b). Such polarization and related relaxation phenomenon contribute towards energy storage and losses respectively, which are expected to improve shielding performance via secondary shielding mechanism (i.e., absorption). It has also been shown that well-dispersed PANI/epoxy hybrids formed via absorption-transferring process, not only impart higher conductivity but also facilitate better charge delocalization leading to huge negative permittivity [hallmark feature of left handed materials (LHM)] [14, 19]. The observed EMI SE in an electric field at low frequency (100–1000 MHz) was found to be -30 to -60 dB (Figure 4c and 4d). Interestingly, though EMI shielding is closely related to electrical conductivity but the factors affecting EMI are more complex because they not only involve conduction (leading to ohmic losses) but also polarization phenomenon (due to presence of polarons/bipolarons, dipoles and filler-matrix interfacial polarization). For examples, when PANI-DBSA NPs (filler) were added in the epoxy (matrix), both the conductivity and dipole density increased due to formation of conductive networks. But any excess addition of PANI-DBSA (i.e., beyond the percolation threshold of 28 wt % PANI-DBSA) did not benefit the electrical conduction, because of NP aggregation. However, it did result in more dipolar and interfacial polarization in the hybrid material. Therefore, 38 wt % PANI-DBSA/epoxy improved the EMI shielding efficiency without much enhancement of conductivity. In another study, PANI/PS blends display interesting dielectric properties, such that both SE_R and SE_A increases with absolute value of complex dielectric permittivity and loss tangent. Accordingly, total shielding effectiveness (SE_T) also scales with polarization.

It is interesting to note that the absolute value of $\tan\delta$ also plays a critical role in determining the attained attenuation. For example, when $\tan\delta >> 1$, shielding is solely determined by σ. However, when $\tan\delta \sim 1$, both σ and ε_r must be considered for calculation of SE_T, SE_R and SE_A values [7]. As for most practical polymer composite/hybrid systems $\tan\delta$ is not much greater than unity, higher shielding efficiency is anticipated for systems possessing higher σ and ε_r values. For examples, the attenuation of PANI-MWCNT/PS based blends (Figure 5) with $\tan\delta$ ~1-2, is governed by polarization and conduction processes. Studies have shown that high dielectric constant fillers (e.g., dielectrics like $BaTiO_3$, TiO_2) with ICPs facilitates the improvement of absorbing properties. For example, $BaTiO_3$/PANI hybrid filler incorporated polyurethane (PU) system displayed maximum return loss (R_L) of -25 dB (>99% power absorption) at 10 GHz, with strong dependence on thickness and dielectric attributes [33].

Likewise, in case of PANI/MWCNT-PS blends [11], with increase in filler loading, real and imaginary parts of permittivity and permeability increases (Figure 5a & 5b), resulting in proportionate increase in losses due to reflection (SE_R) and absorption (SE_A). In another studies, efforts have also been made to incorporate variety of conducting, dielectric and magnetic fillers loaded ICPs within polymeric matrix to manipulate dielectric/magnetic properties and related attenuation levels. These allow formation of multiphase composites or hybrids with combined contributions from conduction, dielectric and magnetic losses. For example, PANI/MWCNT/epoxy hybrids[41] display R_L values of ~-14 dB (Figure 6a) and ~-33dB (Figure 6b) at ~12 GHz and ~26 GHz respectively, for 5mm thick sample. Similarly, PANI/$BaTiO_3$-epoxy composites [41] display R_L of ~13.5 dB (Figure 6c) and ~27 dB (Figure 6d) at ~12 GHz and ~33.5 GHz respectively. In these samples, both conductivity as well as enhanced dielectric properties (due to interfacial polarization) plays important role in regulating the achieved attenuation, matching frequency and absorption bandwidth.

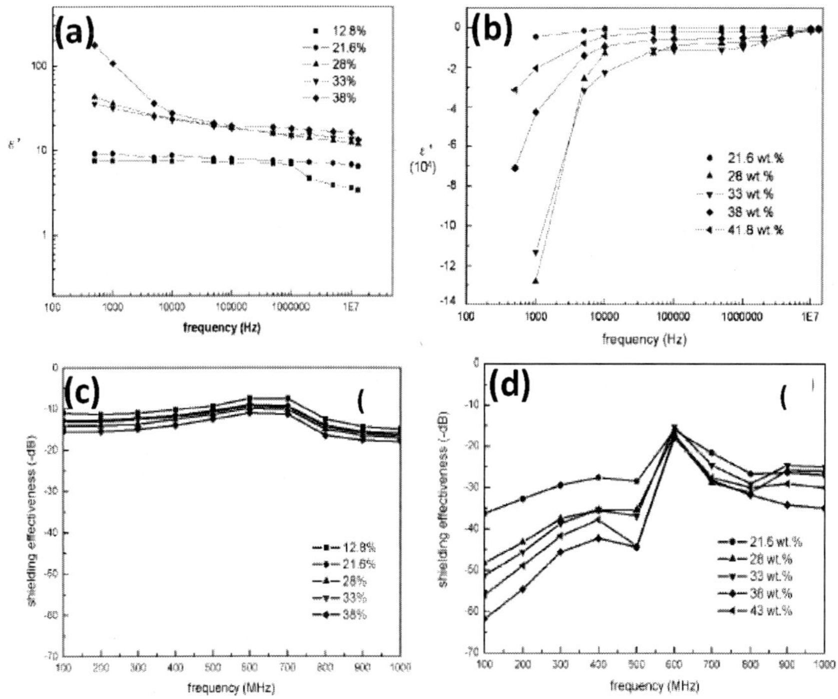

Figure 4. Frequency dependence of the dielectric constant (ε′) for PANI-DBSA/epoxy hybrids prepared using the (a) blending and (b) absorption transferring processes. EMI shielding effectiveness toward electric fields plotted with respect to frequency (100 MHz-1 GHz) for PANI-DBSA/epoxy hybrids prepared using the (c) blending and (d) absorption-transferring processes. Reprinted from [14] with permission from ACS.

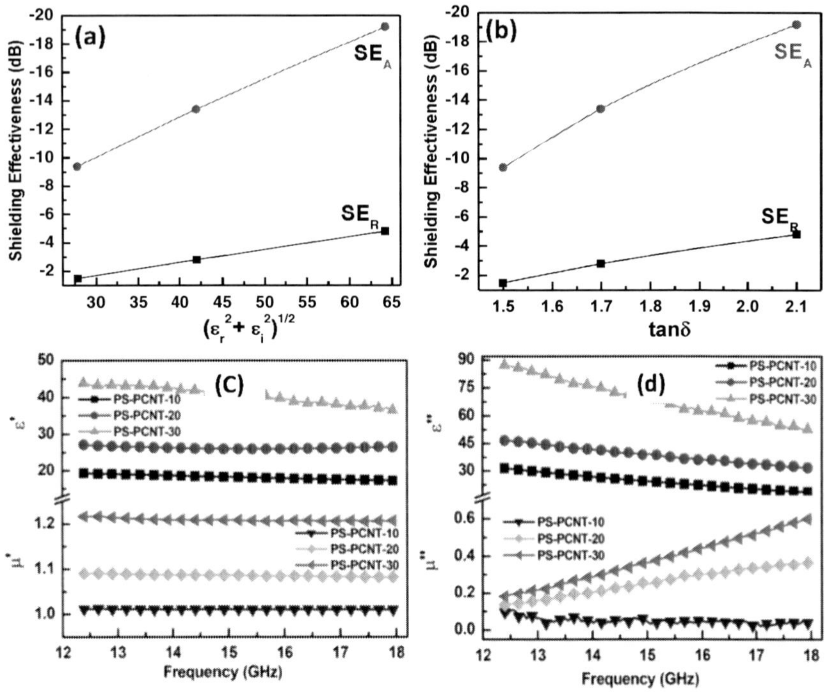

Figure 5. Dependences of losses due to absorption (SE_A) and reflection (SE_R) of the PANI-MWCNT/PS composites as a function of absolute value of (a) complex permittivity $\left(\varepsilon_r^2 + \varepsilon_i^2\right)^{\frac{1}{2}}$ and (b) loss tangent $(\tan\delta)$, and Frequency dependence of (c) dielectric constant (ε') & real permeability (μ') & (d) dielectric loss (ε'') & magnetic loss (μ'') of PANI-MWCNT/Polystyrene. Reprinted from [11] with permission from Elsevier.

The incorporation of additional magnetic phase is known to introduce additional magnetic losses in addition to dielectric losses. For example, the PANI/Fe_3O_4/epoxy hybrids [13, 43] display additional magnetic losses compared to PANI/epoxy composites which show only dielectric losses. It is worth mentioning that due to underlying dielectric and magnetic losses, these materials with high permittivity and/or permeability are expected to display good microwave absorption response. It was observed that epoxy composite [43] (Figure 9.34) containing 15% of PANI and 10% of Fe_3O_4 ($\varepsilon' = 10$) gives return loss (R_L) of -42 dB (at 16.3 GHz) whereas composite of 15% of PANI and 25% of Fe_3O_4 ($\varepsilon' = 17$) gives R_L of -37.4 dB (at 14.85 GHz). In comparison, a purely dielectric composite with 20% of PANI ($\varepsilon' = 8.5$) show a R_L of only -11 dB (at 18 GHz).

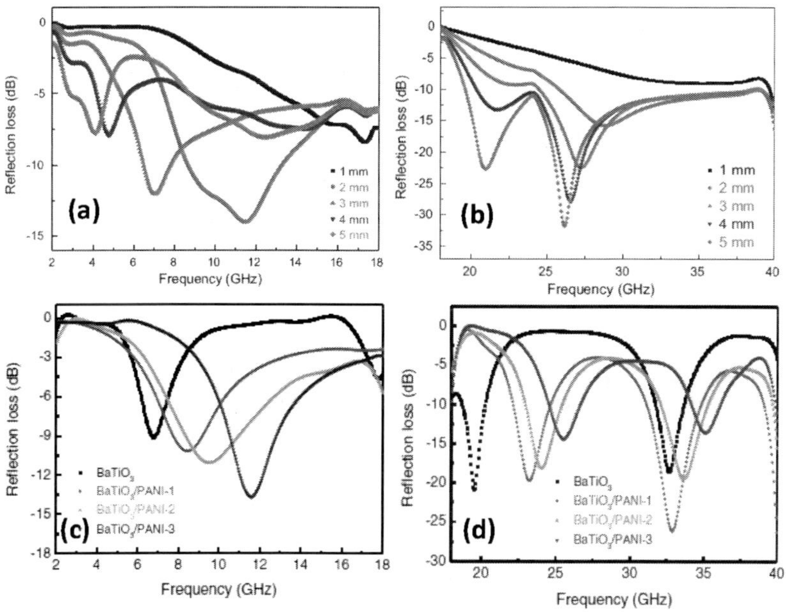

Figure 6. Frequency dependence of the reflection loss for PANI/MWNT-2/epoxy at various sample thickness (d = 1.0, 2.0, 3.0, 4.0 and 5.0 mm) in (a) 2–18 GHz and (b)18-40 GHz range. Reprinted from [42] with permission from RSC. Frequency dependence of the reflection loss for PANI/$BaTiO_3$/epoxy composites in (c) 2–18 GHz and (d)18-40 GHz range. Reprinted from [41] with permission from Springer.

9.4.1.2. EMI Shielding Performance of ICP-Matrix Based Hybrids/Composites

As the SE_R (and related SE_T) is extremely sensitive towards conductivity, therefore, several attempts were also made to further increase the conductivity (and related SE) ICPs by loading of conducting fillers (e.g., metal particles, carbonaceous fillers like graphite, carbon fibers, CNTs, graphene etc.) within microwave non-transparent doped (intrinsically conducting) ICP matrices [4, 7, 9, 11, 29, 32, 37]. Such enhancement (Figure 9.13) of electrical conductivity can be explained on the basis of granular metal/inhomogeneous doping model, which considered that ICPs consists of highly conducting metallic islands dispersed within low conductivity amorphous matrix [7]. It is demonstrated that bridging of these metallic islands by the metallic filler particles, facilitating enhanced inter-particle transport leading to improvement in conductivity. The increase in conductivity is strongly dependent on nature, concentration and aspect ratio of filler particles as well as type and morphology of host ICP matrix.

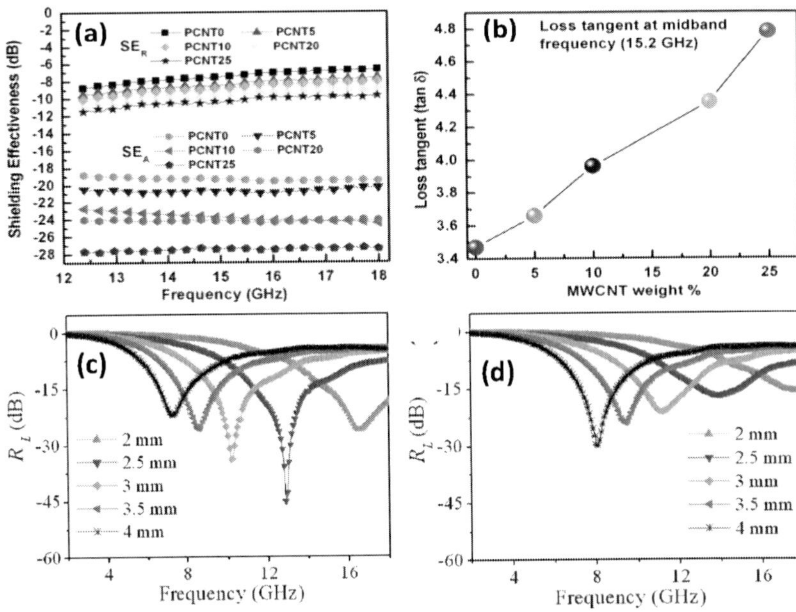

Figure 7. (a) Frequency dependence of losses due to reflection (SE_R) and absorption (SE_A) for MWCNT loaded PANI nanocomposites having different loadings of MWCNT relative to aniline monomer viz. PCNT0 (0.0 wt. %),PCNT5 (5.0 wt. %), PCNT10 (10 wt. %), PCNT20 (20 wt. %) and PCNT25 (25 wt. %) (b) Loss tangent (tan δ) of in-situ synthesized PANI-MWCNT nanocomposites as a function of MWCNT loading. Reprinted from [11] with permission from Elsevier. The reflection loss calculated for the (c) G/PANI nanorod arrays and (d) for the PANI nanorods. Reprinted from [16] with permission from RSC.

It is important to point that the incorporated filler (e.g., metallic fillers, carbonaceous materials like graphite, carbon nanotubes or graphene) can induce polarization and related dielectric losses. Such a polarization occurs due to electrical conductivity differences between ICP and metallic fillers leading to charge localization at interfaces via Maxwell-Wagner-Sillars [7, 44] interfacial polarization phenomenon. Such polarization and related relaxation phenomenon contribute towards energy storage and losses, leading to improvement of SE_A. For example, silver, graphite or carbon black incorporated PANI systems [31, 32] were shown to display SE_T was in the range of 20-50 dB (10 MHz to 1 GHz), due to improvement in conductivity as well as polarization and related losses. Similarly, loading of MWCNTs within PANI is known to display superior shielding performance due to increased conductivity, which resulted in increase in charge transport, polarization, and related relaxation processes [7, 9]. The actual losses can be computed by normalization of these losses with storage terms [i.e., by ratio of dielectric losses/imaginary permittivity (ε'') with dielectric constant/real permittivity (ε')] to quantity loss tangent (tan δ). In case of MWCNT/PANI nanocomposites, improvement of dielectric properties leads to high value of loss tangent which further increases with increase in MWCNT loading. Efforts have also been made to combine ICPs with 2D carbon foams like graphene analogues. For example, perpendicularly grown PANI nanorods on the graphene's [16] surface lead to significantly enhanced EM absorption properties (Figure 7c, RL ~ -45 dB) compared to pure PANI (Figure 7d, RL ~ -33dB), due to combination of electrical conduction and polarization losses.

According to EM theory, the dielectric loss may be contributed by the processes like natural resonance, Debye dipolar relaxation and electron polarization relaxation etc. In the Debye dipolar relaxation regime, the relative complex permittivity can be expressed as [7, 45]:

$$\varepsilon_r^* = (\varepsilon' + j\varepsilon'') = \left[\varepsilon_\infty + \frac{(\varepsilon_s - \varepsilon_\infty)}{(1 + j2\pi f\tau)} \right] \tag{9.78}$$

where f, ε_s, ε_∞ and τ are frequency, static permittivity, relative dielectric permittivity at the high-frequency limit, and polarization relaxation time, respectively, which gives the following relationship between ε' and ε'' [46]:

$$\left[\varepsilon' - \frac{(\varepsilon_s + \varepsilon_\infty)}{2} \right]^2 + [\varepsilon'']^2 = \left[\frac{(\varepsilon_s + \varepsilon_\infty)}{2} \right]^2 \tag{9.81}$$

Thus, the plot of ε' versus ε'' would be a single semicircle, generally denoted as the Cole–Cole semicircle. Each semicircle corresponds to one Debye relaxation process.

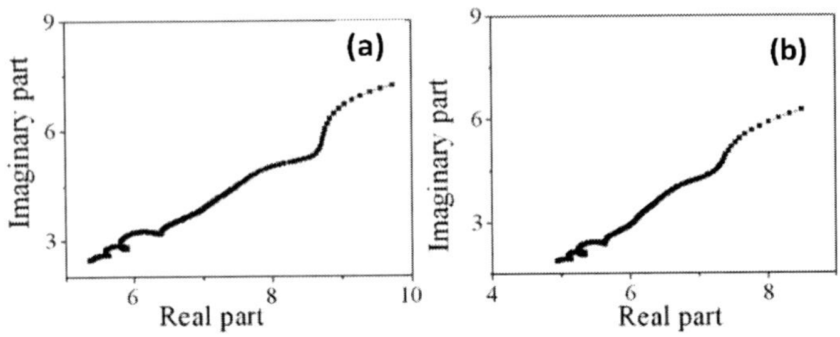

Figure 8. (a) Cole–Cole semicircles for the G/PANI nanorod arrays and (b) PANI nanorod. Reprinted from [16] with permission from RSC.

Further, Cole-Cole plots for PANI/graphene (Figure 8a) displays clear segment of four semicircles, which is correlated with the involvement of multi-dielectric relaxation processes. The similar circles were also found in the PANI (Figure 8b) indicating that the presence of graphene does not increase the number of the relaxation processes. However, the widths of the semicircles of G/PANI nanorod arrays are larger than those of the PANI nanorods, revealing that the presence of graphene improves the intensity of the Debye dipolar relaxation process, that also get reflected in the higher tangent loss of the G/PANI nanorod arrays compared to that of the PANI nanorods. The studies on SWCNT/PANI and graphene/PANI composites[47] have shown that though the conductivity and attenuation of the graphene/PANI (-34.2 dB) composite are better than that of the SWCNT/PANI (-31.5 dB) composite, though absorption proportion of the SWCNT/PANI composite is higher than that of the graphene/PANI composite. This is attributed to presence of Fe-nanoparticles in the SWCNT/PANI composite, which exhibit magnetic losses in addition of dielectric losses.

Studies have shown that in case of in-situ formed PANI/$BaTiO_3$ nanocomposites [28], the SE_R value decreases monotonically with the increase in high dielectric constant $BaTiO_3$ filler, due to decrease in both concentrations as well as mobility of polarons, leading to proportionate deterioration of conductivity and hence SE_R. In contrast, SE_A value first increases upto certain $BaTiO_3$ (-65.6 dB, Figure 9a) and decreases afterwards, suggesting that some other mechanism apart from conductivity also plays an important role in observed SE_A. It was found that for any given composition dielectric constant (ε') decreases with frequency. As the frequency increases, the dipoles present on the chains find it increasingly difficult to maintain in phase orientation with the electric vector of the incident radiation, which is responsible for observed decrease in ε'. Further, for any given composition at any fixed frequency (say 15.2 GHz), the ε' increases (Figure 9b) with loading level of BT upto PBT11 and decreases afterwards. In addition, with the increasing BT content, though the decrement of dipole density (normalized polarons/bipolarons concentration) and a parallel decrement of the carrier mobility were observed, the space charge polarization effect enhances. The net result is observed in terms of enhancement of dielectric constant. The observed reduction of ε' after PBT11 may be ascribed to the phase segregation at higher BT loadings, leading to decrease in number of heterojunctions/grain boundaries and consequent reduction of overall polarization. The dielectric losses (ε'') decreases with frequency (Figure 9c) with well-defined maxima in case of PBT11 and PBT12. The increase in ε'' can be ascribed to the intensification of losses (occurring by molecular friction) due to out of phase motion of dipoles. As the polarization and losses are associated with the rate of absorption and dissipation of energy, the loss factor i.e., loss tangent ($\tan \delta = \varepsilon''/\varepsilon'$) has also been plotted as a function of frequency. The results (Figure 9d) clarify that tan δ registered highest value in case of PBT11 accounting for observed maximum absorption loss (SE_A) of -65.6 dB. Therefore, absorption loss reaches a maximum value only at some optimum filler loading (1:1) after which no further rise in attenuation was observed. Some efforts have also been made in past to synthesize PANI-Ni core shell composites [17] by in-situ polymerization (Figure 9.30) with an aim to match the natural resonance of Ni cores and the dielectric loss of PANi shells. It is observed that permittivity of the Ni/PANi nanocomposite presents dual dielectric relaxations due to cooperative consequence of the core/shell interfaces and the dielectric PANi shells. Additionally, the permeability presents a strong natural resonance around 2–8 GHz leading to microwave magnetic loss. The proper matching of the permittivity and the permeability contributes to enhanced microwave absorption is responsible for return loss of (R_L) of < -10 dB for all samples and -35 dB (Figure 10a) for 5 mm thick nanocomposite sample (S4) containing ~78 wt. Ni nanoparticles. Similarly, PANI-HA based nanocomposites [34] containing TiO_2 and Fe_3O_4 nanoparticles as dielectric filler and magnetic filler, respectively, show good microwave absorption response with attenuation (R_L) of -48.9 dB. Likewise, the Ni/PPY core/shell composites [48] display good attenuation where a Cole-Cole semicircle was applied to explain the permittivity and the reflection loss (R_L) properties of Ni/PPy composites.

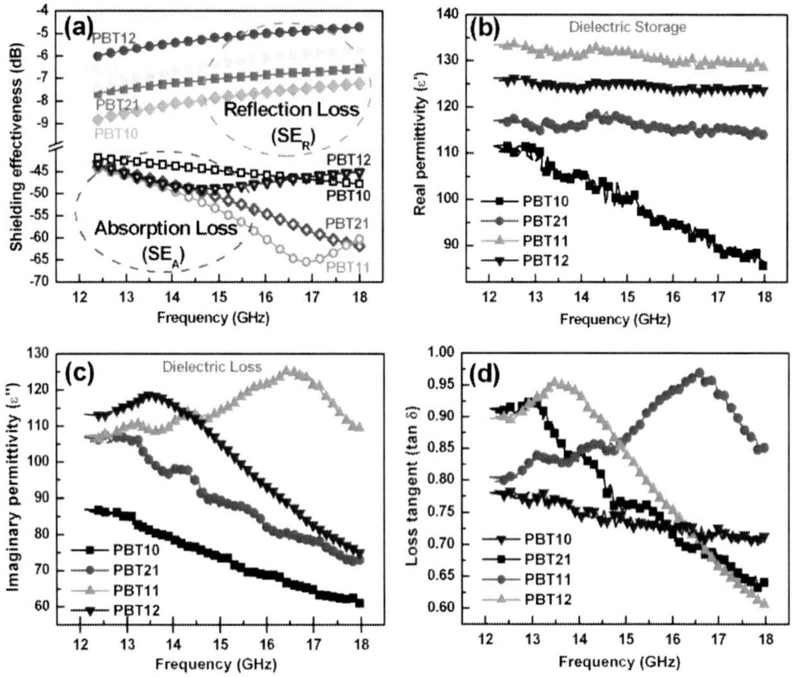

Figure 9. Variation of (a) SE_R and SE_A, (b) real permittivity (ε'), (c) imaginary permittivity (ε"), (d) loss tangent (tan δ) with frequency for PANI and its TBT based NCs Reprinted from [28] with permission from RSC.

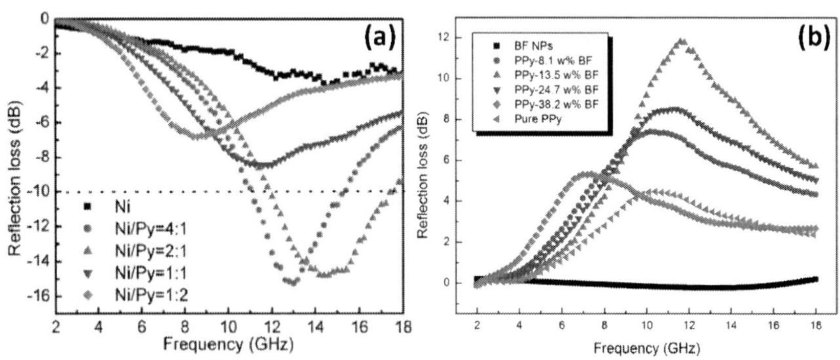

Figure 10. Reflection loss for Ni/PPY composites and Ni powder, with a sample thickness of 2 mm in the frequency range of 2-18 GHz. Reprinted from [48] with permission from RSC. Reflection loss for PPY/BF composites, pure PPY powders, and BF NPs, with a sample thickness of 2 mm in the frequency range of 2-18 GHz. Reprinted from [49] with permission from RSC.

The minimum reflection losses of Ni/PPy composites (Figure 9.31a) from Ni/Py) 4:1, 2:1, 1:1, and 1:2 are -15.2, -14.8, -8.4, and -6.8 dB at 13.0, 14.4, 11.6, and 8.6 GHz respectively, and electromagnetic absorption less than -10 dB is only found for Ni/Py) 4:1 and 2:1, in the 11-15.4 GHz range for the former and 12-17.5 GHz for the latter. The electromagnetic absorption properties of Ni/PPy composites from Ni/Py 4:1 and 2:1 are much better than those of Ni/PANI nanocomposites and BaTiO3/PANI composites at the same thickness in the frequency range of 2-18 GHz.10,12 Similarly, $BaFe_{12}O_{19}$ (BF)/PPY [18] nanocomposites display maximum R_L value of ~12 dB (Figure 10b) for 13.5 wt. % BF loading.

Conclusion

The conducting polymer based systems have emerged as a promising EMI shielding materials due to controllable electrical and dielectric properties. In order to further improve the microwave absorption properties, strategies of combining the conducting polymers with variety of conducting, dielectric and magnetic fillers and or various conventional polymeric matrices, to the form of hybrids and composites, is suggested. It has been found that besides manipulation of electrical properties, such systems often display peculiar dielectric properties due to charge localization and interfacial polarization, ultimately resulting in improved dielectric loses and overall attenuation. It is envisaged that those combination possessing tradeoff between electrical, dielectric, magnetic, electromagnetic and physico-mechanical properties, are suitable for designing technically & commercially viable futuristic shields. In this context, with lot of work already done, still there is enormous scope for identifying and realizing efficient yet scalable and viable shielding materials.

Acknowledgments

The author would like to thank Director CSIR-NPL for necessary support and encouragement for the work.

References

[1] Paul, C. R. *Electromagnetics for engineers: with applications to digital systems and electromagnetic interfe*rence. (John Wiley & Sons, 2004).

[2] Saini, P. & Arora, M. Microwave Absorption and EMI Shielding Behavior of Nanocomposites Based on Intrinsically Conducting Polymers, Graphene and Carbon Nanotubes. in *New Polymers for Special Applications* (ed. De Souza Gomes, A.) 71–112 (InTech, 2012).

[3] Ott, H. W. & Ott, H. W. *Electromagnetic compatibility engineering*. (John Wiley & Sons, 2009).

[4] Saini, P. Electrical Properties and Electromagnetic Interference Shielding Response of Electrically Conducting Thermosetting Nanocomposites. in *Thermoset Nanocomposites* (ed. Mittal, V.) 211–237 (Wiley-VCH Verlag GmbH & Co. KGaA, 2013). doi:10.1002/9783527659647.ch10.

[5] Saini, P., Arora, M., Gupta, G., Gupta, B. K., Singh, V. N., & Choudhary, V. High permittivity polyaniline–barium titanate nanocomposites with excellent electromagnetic interference shielding response. *Nanoscale* 5, 4330 (2013).

[6] Abbas, S. M., Chandra, M., Verma, A., Chatterjee, R. & Goel, T. C. Complex permittivity and microwave absorption properties of a composite dielectric absorber. *Composites Part A* 37, 2148–2154 (2006).

[7] Saini, P. Intrinsically Conducting Polymer-Based Blends and Composites for Electromagnetic Interference Shielding: Theoretical and Experimental Aspects. in *Fundamentals of Conjugated Polymer Blends, Copolymers and Composites* (ed. Saini, P.) 449–518 (John Wiley & Sons, Inc., 2015). doi:10.1002/978111913 7160.ch9.

[8] Bigg, D. M. & Stutz, D. E. Plastic composites for electromagnetic interference shielding applications. *Polym. Compos.* 4, 40–46 (1983).

[9] Saini, P. & Aror, M. Microwave Absorption and EMI Shielding Behavior of Nanocomposites Based on Intrinsically Conducting Polymers, Graphene and Carbon Nanotubes. in *New Polymers for Special Applications* (ed. De Souza Gomes, A.) (InTech, 2012). doi:10.5772/48779.

[10] Colaneri, N. F. & Schacklette, L. W. EMI shielding measurements of conductive polymer blends. *IEEE Trans. Instrum. Meas.* 41, 291–297 (1992).

[11] Saini, P., Choudhary, V., Singh, B. P., Mathur, R. B. & Dhawan, S. K. Enhanced microwave absorption behavior of polyaniline-CNT/polystyrene blend in 12.4–18.0GHz range. *Synthetic Metals* 161, 1522–1526 (2011).

[12] Pud, A., Ogurtsov, N., Korzhenko, A. & Shapoval, G. Some aspects of preparation methods and properties of polyaniline blends and composites with organic polymers. *Progress in Polymer Science* 28, 1701–1753 (2003).

[13] Kamchi, N. E., Belaabed, B., Wojkiewicz, J.-L., Lamouri, S. & Lasri, T. Hybrid polyaniline/nanomagnetic particles composites: High performance materials for EMI shielding. *J. Appl. Polym. Sci.* 127, 4426–4432 (2013).

[14] Liu, C.-D., Lee, S.-N., Ho, C.-H., Han, J.-L. & Hsieh, K.-H. Electrical Properties of Well-Dispersed Nanopolyaniline/Epoxy Hybrids Prepared Using an Absorption-Transferring Process. *J. Phys. Chem. C* 112, 15956–15960 (2008).

[15] Hoang, N. H., Wojkiewicz, J.-L., Miane, J.-L. & Biscarro, R. S. Lightweight electromagnetic shields using optimized polyaniline composites in the microwave band. *Polym. Adv. Technol.* 18, 257–262 (2007).

[16] Yu, H., Wang, T., Wen, B., Lu, M., Xu, Z., Zhu, C., Chen, Y., Xue, X., Sun, C., & Cao, M. Graphene/polyaniline nanorod arrays: synthesis and excellent electromagnetic absorption properties. *J. Mater. Chem.* 22, 21679 (2012).

[17] Dong, X. L., Zhang, X. F., Huang, H. & Zuo, F. Enhanced microwave absorption in Ni/polyaniline nanocomposites by dual dielectric relaxations. *Appl. Phys. Lett.* 92, 013127 (2008).

[18] Xu, P., Han, X., Jiang, J., Wang, X., Li, X., & Wen, A. Synthesis and Characterization of Novel Coralloid Polyaniline/$BaFe_{12}O_{19}$ Nanocomposites. *J. Phys. Chem.* C 111, 12603–12608 (2007).

[19] Hsieh, C. H., Lee, A. H., Liu, C. D., Han, J. L., Hsieh, K. H., & Lee, S. N. Polyaniline nano-composites with large negative dielectric permittivity. *AIP Advances* 2, 012127 (2012).

[20] Hong, Y. K., Lee, C. Y., Jeong, C. K., Lee, D. E., Kim, K., & Joo, J. Method and apparatus to measure electromagnetic interference shielding efficiency and its shielding characteristics in broadband frequency ranges. *Review of Scientific Instruments* 74, 1098–1102 (2003).

[21] Nicolson, A. M. & Ross, G. F. Measurement of the Intrinsic Properties of Materials by Time-Domain Techniques. *IEEE Trans. Instrum. Meas.* 19, 377–382 (1970).

[22] Weir, W. B. Automatic measurement of complex dielectric constant and permeability at microwave frequencies. *Proc. IEEE* 62, 33–36 (1974).

[23] Shacklette, L. W. & Colaneri, N. F. EMI shielding measurements of conductive polymer blends. in [1991] Conference Record. *IEEE Instrumentation and Measurement Technology Conference* 72–78 (IEEE, 1991). doi:10.1109/IMTC.1991.161543.

[24] Taka, T. EMI shielding measurements on poly(3-octyl thiophene) blends. *Synthetic Metals* 41, 1177–1180 (1991).

[25] Lakshmi, K., John, H., Mathew, K. T., Joseph, R. & George, K. E. Microwave absorption, reflection and EMI shielding of PU–PANI composite. *Acta Materialia* 57, 371–375 (2009).

[26] Paligová, M., Vilčákova, J., Sáha, P., Křesálek, V., Stejskal, J., & Quadrat, O. Electromagnetic shielding of epoxy resin composites containing carbon fibers coated with polyaniline base. *Physica A: Statistical Mechanics and its Applications* 335, 421–429 (2004).

[27] Saini, P., Choudhary, V., Vijayan, N. & Kotnala, R. K. Improved Electromagnetic Interference Shielding Response of Poly(aniline)-Coated Fabrics Containing Dielectric and Magnetic Nanoparticles. *J. Phys. Chem.* C 116, 13403–13412 (2012).

[28] Saini, P., Arora, M., Gupta, G., Gupta, B. K., Singh, V. N., & Choudhary, V. High permittivity polyaniline–barium titanate nanocomposites with excellent electromagnetic interference shielding response. *Nanoscale* 5, 4330 (2013).

[29] Chen, K., Xiang, C., Li, L., Qian, H., Xiao, Q., & Xu, F. A novel ternary composite: fabrication, performance and application of expanded graphite/polyaniline/CoFe2O4 ferrite. *J. Mater. Chem.* 22, 6449 (2012).

[30] Tang, X. & Yang, Y. Surface modification of M-Ba-ferrite powders by polyaniline: Towards improving microwave electromagnetic response. *Applied Surface Science* 255, 9381–9385 (2009).

[31] Joo, J. & Lee, C. Y. High frequency electromagnetic interference shielding response of mixtures and multilayer films based on conducting polymers. *Journal of Applied Physics* 88, 513–518 (2000).

[32] Lee, C. Y., Song, H. G., Jang, K. S., Oh, E. J., Epstein, A. J., & Joo, J. Electromagnetic interference shielding efficiency of polyaniline mixtures and multilayer films. *Synthetic Metals* 102, 1346–1349 (1999).

[33] Abbas, S. M., Chandra, M., Verma, A., Chatterjee, R. & Goel, T. C. Complex permittivity and microwave absorption properties of a composite dielectric absorber. *Composites Part A: Applied Science and Manufacturing* 37, 2148–2154 (2006).

[34] Phang, S. W., Tadokoro, M., Watanabe, J. & Kuramoto, N. Effect of Fe_3O_4 and TiO_2 addition on the microwave absorption property of polyaniline micro/nanocomposites. *Polym. Adv. Technol.* 20, 550–557 (2009).

[35] Li, S., Gan, M., Ma, L., Yan, J., Tang, J., Fu, D., Li, Z., & Bai, Y. Preparation and microwave absorbing properties of polyaniline-modified silicon carbide composites. *High Performance Polymers* 25, 901–906 (2013).

[36] Abbas, S. M., Dixit, A. K., Chatterjee, R. & Goel, T. C. Complex permittivity and microwave absorption properties of BaTiO3–polyaniline composite. *Materials Science and Engineering: B* 123, 167–171 (2005).

[37] Saini, P., Choudhary, V., Singh, B. P., Mathur, R. B. & Dhawan, S. K. Polyaniline–MWCNT nanocomposites for microwave absorption and EMI shielding. *Materials Chemistry and Physics* 113, 919–926 (2009).

[38] Naishadham, K. & Kadaba, P. K. Measurement of the microwave conductivity of a polymeric material with potential applications in absorbers and shielding. IEEE Trans. *Microwave Theory Techn.* 39, 1158–1164 (1991).

[39] *Fundamentals of Conjugated Polymer Blends, Copolymers and Composites: Synthesis, Properties and Applications.* (John Wiley & Sons, Inc., 2015). doi:10.1002/9781119137160.

[40] Sudha, J. D., Sivakala, S., Prasanth, R., Reena, V. L. & Radhakrishnan Nair, P. Development of electromagnetic shielding materials from the conductive blends of polyaniline and polyaniline-clay nanocomposite-EVA: Preparation and properties. *Composites Science and Technology* 69, 358–364 (2009).

[41] Ting, T.-H. & Wu, K.-H. Synthesis and electromagnetic wave-absorbing properties of BaTiO3/polyaniline structured composites in 2–40 GHz. *J Polym Res* 20, 127 (2013).

[42] Ting, T. H., Jau, Y. N. & Yu, R. P. Microwave absorbing properties of polyaniline/multi-walled carbon nanotube composites with various polyaniline contents. *Applied Surface Science* 258, 3184–3190 (2012).

[43] Belaabed, B., Wojkiewicz, J. L., Lamouri, S., El Kamchi, N. & Lasri, T. Synthesis and characterization of hybrid conducting composites based on polyaniline/magnetite fillers with improved microwave absorption properties. *Journal of Alloys and Compounds* 527, 137–144 (2012).

[44] Wagner, K. W. Erklärung der dielektrischen Nachwirkungsvorgänge auf Grund Maxwellscher Vorstellungen. *Archiv f. Elektrotechnik* 2, 371–387 (1914).

[45] Cole, K. S. & Cole, R. H. Dispersion and Absorption in Dielectrics I. Alternating Current Characteristics. *The Journal of Chemical Physics* 9, 341–351 (1941).

[46] Cole, K. S. & Cole, R. H. Dispersion and Absorption in Dielectrics II. Direct Current Characteristics. *The Journal of Chemical Physics* 10, 98–105 (1942).

[47] Yuan, B., Yu, L., Sheng, L., An, K. & Zhao, X. Comparison of electromagnetic interference shielding properties between single-wall carbon nanotube and graphene sheet/polyaniline composites. *J. Phys. D: Appl. Phys*. 45, 235108 (2012).

[48] Xu, P., Han, X., Wang, C., Zhou, D., Lv, Z., Wen, A., Wang, X., & Zhang, B. Synthesis of Electromagnetic Functionalized Nickel/Polypyrrole Core/Shell Composites. *J. Phys. Chem. B* 112, 10443–10448 (2008).

[49] Xu, P., Han, X., Wang, C., Zhou, D., Lv, Z., Wen, A., Wang, X., & Zhang, B. Synthesis of Electromagnetic Functionalized Nickel/Polypyrrole Core/Shell Composites. *J. Phys. Chem. B* 112, 10443–10448 (2008).

About the Editor

Dr. Parveen Saini
Conjugated Polymers and Graphene
Technology Group, Solar Hutment,
Polymeric and Soft Materials Section,
Materials Physics and Engineering Division
CSIR-National Physical Laboratory,
New Delhi, India
Email: pksaini@nplindia.org

Index

N

P

R

S

T

V

W